IET ENERGY ENGINEERING SERIES 208

Nanogrids and Picogrids and their Integration with Electric Vehicles

Other volumes in this series:

Volume 1	**Power Circuit Breaker Theory and Design** C.H. Flurscheim (Editor)
Volume 4	**Industrial Microwave Heating** A.C. Metaxas and R.J. Meredith
Volume 7	**Insulators for High Voltages** J.S.T. Looms
Volume 8	**Variable Frequency AC Motor Drive Systems** D. Finney
Volume 10	**SF_6 Switchgear** H.M. Ryan and G.R. Jones
Volume 11	**Conduction and Induction Heating** E.J. Davies
Volume 13	**Statistical Techniques for High Voltage Engineering** W. Hauschild and W. Mosch
Volume 14	**Uninterruptible Power Supplies** J. Platts and J.D. St Aubyn (Editors)
Volume 15	**Digital Protection for Power Systems** A.T. Johns and S.K. Salman
Volume 16	**Electricity Economics and Planning** T.W. Berrie
Volume 18	**Vacuum Switchgear** A. Greenwood
Volume 19	**Electrical Safety: a guide to causes and prevention of hazards** J. Maxwell Adams
Volume 21	**Electricity Distribution Network Design, 2nd Edition** E. Lakervi and E.J. Holmes
Volume 22	**Artificial Intelligence Techniques in Power Systems** K. Warwick, A.O. Ekwue and R. Aggarwal (Editors)
Volume 24	**Power System Commissioning and Maintenance Practice** K. Harker
Volume 25	**Engineers' Handbook of Industrial Microwave Heating** R.J. Meredith
Volume 26	**Small Electric Motors** H. Moczala *et al.*
Volume 27	**AC-DC Power System Analysis** J. Arrillaga and B.C. Smith
Volume 29	**High Voltage Direct Current Transmission, 2nd Edition** J. Arrillaga
Volume 30	**Flexible AC Transmission Systems (FACTS)** Y-H. Song (Editor)
Volume 31	**Embedded generation** N. Jenkins *et al.*
Volume 32	**High Voltage Engineering and Testing, 2nd Edition** H.M. Ryan (Editor)
Volume 33	**Overvoltage Protection of Low-Voltage Systems, Revised Edition** P. Hasse
Volume 36	**Voltage Quality in Electrical Power Systems** J. Schlabbach *et al.*
Volume 37	**Electrical Steels for Rotating Machines** P. Beckley
Volume 38	**The Electric Car: Development and future of battery, hybrid and fuel-cell cars** M. Westbrook
Volume 39	**Power Systems Electromagnetic Transients Simulation** J. Arrillaga and N. Watson
Volume 40	**Advances in High Voltage Engineering** M. Haddad and D. Warne
Volume 41	**Electrical Operation of Electrostatic Precipitators** K. Parker
Volume 43	**Thermal Power Plant Simulation and Control** D. Flynn
Volume 44	**Economic Evaluation of Projects in the Electricity Supply Industry** H. Khatib
Volume 45	**Propulsion Systems for Hybrid Vehicles** J. Miller
Volume 46	**Distribution Switchgear** S. Stewart
Volume 47	**Protection of Electricity Distribution Networks, 2nd Edition** J. Gers and E. Holmes
Volume 48	**Wood Pole Overhead Lines** B. Wareing
Volume 49	**Electric Fuses, 3rd Edition** A. Wright and G. Newbery
Volume 50	**Wind Power Integration: Connection and system operational aspects** B. Fox *et al.*
Volume 51	**Short Circuit Currents** J. Schlabbach
Volume 52	**Nuclear Power** J. Wood
Volume 53	**Condition Assessment of High Voltage Insulation in Power System Equipment** R.E. James and Q. Su
Volume 55	**Local Energy: Distributed generation of heat and power** J. Wood
Volume 56	**Condition Monitoring of Rotating Electrical Machines** P. Tavner, L. Ran, J. Penman and H. Sedding
Volume 57	**The Control Techniques Drives and Controls Handbook, 2nd Edition** B. Drury
Volume 58	**Lightning Protection** V. Cooray (Editor)
Volume 59	**Ultracapacitor Applications** J.M. Miller

Volume 62 **Lightning Electromagnetics** V. Cooray
Volume 63 **Energy Storage for Power Systems, 2nd Edition** A. Ter-Gazarian
Volume 65 **Protection of Electricity Distribution Networks, 3rd Edition** J. Gers
Volume 66 **High Voltage Engineering Testing, 3rd Edition** H. Ryan (Editor)
Volume 67 **Multicore Simulation of Power System Transients** F.M. Uriate
Volume 68 **Distribution System Analysis and Automation** J. Gers
Volume 69 **The Lightening Flash, 2nd Edition** V. Cooray (Editor)
Volume 70 **Economic Evaluation of Projects in the Electricity Supply Industry, 3rd Edition** H. Khatib
Volume 72 **Control Circuits in Power Electronics: Practical issues in design and implementation** M. Castilla (Editor)
Volume 73 **Wide Area Monitoring, Protection and Control Systems: The enabler for Smarter Grids** A. Vaccaro and A. Zobaa (Editors)
Volume 74 **Power Electronic Converters and Systems: Frontiers and applications** A. M. Trzynadlowski (Editor)
Volume 75 **Power Distribution Automation** B. Das (Editor)
Volume 76 **Power System Stability: Modelling, analysis and control** A.A. Sallam and B. Om P. Malik
Volume 78 **Numerical Analysis of Power System Transients and Dynamics** A. Ametani (Editor)
Volume 79 **Vehicle-to-Grid: Linking electric vehicles to the smart grid** J. Lu and J. Hossain (Editors)
Volume 81 **Cyber-Physical-Social Systems and Constructs in Electric Power Engineering** S. Suryanarayanan, R. Roche and T.M. Hansen (Editors)
Volume 82 **Periodic Control of Power Electronic Converters** F. Blaabjerg, K.Zhou, D. Wang and Y. Yang
Volume 86 **Advances in Power System Modelling, Control and Stability Analysis** F. Milano (Editor)
Volume 87 **Cogeneration: Technologies, Optimisation and Implentation** C. A. Frangopoulos (Editor)
Volume 88 **Smarter Energy: from Smart Metering to the Smart Grid** H. Sun, N. Hatziargyriou, H. V. Poor, L. Carpanini and M. A. Sánchez Fornié (Editors)
Volume 89 **Hydrogen Production, Separation and Purification for Energy** A. Basile, F. Dalena, J. Tong and T.N.Veziroğlu (Editors)
Volume 90 **Clean Energy Microgrids** S. Obara and J. Morel (Editors)
Volume 91 **Fuzzy Logic Control in Energy Systems with Design Applications in Matlab/Simulink®** İ. H. Altaş
Volume 92 **Power Quality in Future Electrical Power Systems** A. F. Zobaa and S. H. E. A. Aleem (Editors)
Volume 93 **Cogeneration and District Energy Systems: Modelling, Analysis and Optimization** M. A. Rosen and S. Koohi-Fayegh
Volume 94 **Introduction to the Smart Grid: Concepts, technologies and evolution** S.K. Salman
Volume 95 **Communication, Control and Security Challenges for the Smart Grid** S.M. Muyeen and S. Rahman (Editors)
Volume 96 **Industrial Power Systems with Distributed and Embedded Generation** R Belu
Volume 97 **Synchronized Phasor Measurements for Smart Grids** M.J.B. Reddy and D.K. Mohanta (Editors)
Volume 98 **Large Scale Grid Integration of Renewable Energy Sources** A. Moreno-Munoz (Editor)
Volume 100 **Modeling and Dynamic Behaviour of Hydropower Plants** N. Kishor and J. Fraile-Ardanuy (Editors)
Volume 101 **Methane and Hydrogen for Energy Storage** R. Carriveau and D. S-K. Ting
Volume 104 **Power Transformer Condition Monitoring and Diagnosis** A. Abu-Siada (Editor)
Volume 106 **Surface Passivation of Industrial Crystalline Silicon Solar Cells** J. John (Editor)

Volume 107 **Bifacial Photovoltaics: Technology, applications and economics** J. Libal and R. Kopecek (Editors)
Volume 108 **Fault Diagnosis of Induction Motors** J. Faiz, V. Ghorbanian and G. Joksimović
Volume 110 **High Voltage Power Network Construction** K. Harker
Volume 111 **Energy Storage at Different Voltage Levels: Technology, integration, and market aspects** A.F. Zobaa, P.F. Ribeiro, S.H.A. Aleem and S.N. Afifi (Editors)
Volume 112 **Wireless Power Transfer: Theory, Technology and Application** N. Shinohara
Volume 114 **Lightning-Induced Effects in Electrical and Telecommunication Systems** Y. Baba and V. A. Rakov
Volume 115 **DC Distribution Systems and Microgrids** T. Dragičević, F. Blaabjerg and P. Wheeler
Volume 116 **Modelling and Simulation of HVDC Transmission** M. Han (Editor)
Volume 117 **Structural Control and Fault Detection of Wind Turbine Systems** H.R. Karimi
Volume 119 **Thermal Power Plant Control and Instrumentation: The control of boilers and HRSGs, 2nd Edition** D. Lindsley, J. Grist and D. Parker
Volume 120 **Fault Diagnosis for Robust Inverter Power Drives** A. Ginart (Editor)
Volume 121 **Monitoring and Control using Synchrophasors in Power Systems with Renewables** I. Kamwa and C. Lu (Editors)
Volume 123 **Power Systems Electromagnetic Transients Simulation, 2nd Edition** N. Watson and J. Arrillaga
Volume 124 **Power Market Transformation** B. Murray
Volume 125 **Wind Energy Modeling and Simulation Volume 1: Atmosphere and plant** P.Veers (Editor)
Volume 126 **Diagnosis and Fault Tolerance of Electrical Machines, Power Electronics and Drives** A.J.M. Cardoso
Volume 128 **Characterization of Wide Bandgap Power Semiconductor Devices** F. Wang, Z. Zhang and E.A. Jones
Volume 129 **Renewable Energy from the Oceans: From wave, tidal and gradient systems to offshore wind and solar** D. Coiro and T. Sant (Editors)
Volume 130 **Wind and Solar Based Energy Systems for Communities** R. Carriveau and D. S-K. Ting (Editors)
Volume 131 **Metaheuristic Optimization in Power Engineering** J. Radosavljević
Volume 132 **Power Line Communication Systems for Smart Grids** I.R.S. Casella and A. Anpalagan
Volume 134 **Hydrogen Passivation and Laser Doping for Silicon Solar Cells** B. Hallam and C. Chan (Editors)
Volume 139 **Variability, Scalability and Stability of Microgrids** S. M. Muyeen, S. M. Islam and F. Blaabjerg (Editors)
Volume 143 **Medium Voltage DC System Architectures** Brandon Grainger and Rik De Doncker (Editors)
Volume 145 **Condition Monitoring of Rotating Electrical Machines** P. Tavner, L. Ran and C. Crabtree
Volume 146 **Energy Storage for Power Systems, 3rd Edition** A.G. Ter-Gazarian
Volume 147 **Distribution Systems Analysis and Automation 2nd Edition** J. Gers
Volume 151 **SiC Power Module Design: Performance, robustness and reliability** A. Castellazzi and A. Irace (Editors)
Volume 152 **Power Electronic Devices: Applications, failure mechanisms and reliability** F. Iannuzzo (Editor)
Volume 153 **Signal Processing for Fault Detection and Diagnosis in Electric Machines and Systems** M. Benbouzid (Editor)
Volume 155 **Energy Generation and Efficiency Technologies for Green Residential Buildings** D. Ting and R. Carriveau (Editors)
Volume 156 **Lithium-Ion Batteries Enabled by Silicon Anodes** C. Ban and K. Xu (Editors)
Volume 157 **Electrical Steels, 2 Volumes** A. Moses, K. Jenkins, Philip Anderson and H. Stanbury
Volume 158 **Advanced Dielectric Materials for Electrostatic Capacitors** Q Li (Editor)

Volume 159 **Transforming the Grid Towards Fully Renewable Energy** O. Probst, S. Castellanos and R. Palacios (Editors)
Volume 160 **Microgrids for Rural Areas: Research and case studies** R.K. Chauhan, K. Chauhan and S.N. Singh (Editors)
Volume 161 **Artificial Intelligence for Smarter Power Systems: Fuzzy Logic and Neural Networks** M.G. Simoes
Volume 166 **Advanced Characterization of Thin Film Solar Cells** N Haegel and M Al-Jassim (Editors)
Volume 167 **Power Grids with Renewable Energy** Storage, integration and digitalization A.A. Sallam and B. Om P. Malik
Volume 169 **Small Wind and Hydrokinetic Turbines** Philip Clausen, Jonathan Whale and David Wood (Editors)
Volume 170 **Reliability of Power Electronics Converters for Solar Photovoltaic Applications** F. Blaabjerg, A.l. Haque, H. Wang, Z. Abdin Jaffery and Y. Yang (Editors)
Volume 171 **Utility-scale Wind Turbines and Wind Farms** A. Vasel-Be-Hagh and D. S.-K. Ting
Volume 172 **Lighting interaction with Power Systems, 2 volumes** A. Piantini (Editor)
Volume 174 **Silicon Solar Cell Metallization and Module Technology** T. Dullweber (Editor)
Volume 180 **Protection of Electricity Distribution Networks, 4th Edition** Juan Gers and Edward Holmes
Volume 182 **Surge Protection for Low Voltage Systems** A. Rousseau (Editor)
Volume 184 **Compressed Air Energy Storage: Types, systems and applications** D. Ting and J. Stagner
Volume 193 **Overhead Electric Power Lines: Theory and practice** S. Chattopadhyay and A. Das
Volume 194 **Offshore Wind Power Reliability, availability and maintenance, 2nd edition** P. Tavner
Volume 198 **Battery Management Systems and Inductive Balancing** A. Van den Bossche and A. Farzan Moghaddam
Volume 199 **Model Predictive Control for Microgrids: From power electronic converters to energy management** J. Hu, J. M. Guerrero and S. Islam
Volume 204 **Electromagnetic Transients in Large HV Cable Networks: Modeling and calculations** Ametani, Xue, Ohno and Khalilnezhad
Volume 212 **Battery State Estimation: Methods and Models** S. Wang
Volume 905 **Power system protection, 4 volumes**

Nanogrids and Picogrids and their Integration with Electric Vehicles

Surajit Chattopadhyay

The Institution of Engineering and Technology

Published by The Institution of Engineering and Technology, London, United Kingdom

The Institution of Engineering and Technology is registered as a Charity in England & Wales (no. 211014) and Scotland (no. SC038698).

First published 2022

The Institution of Engineering and Technology
Michael Faraday House
Six Hills Way, Stevenage
Herts, SG1 2AY, United Kingdom

www.theiet.org

British Library Cataloguing in Publication Data
A catalogue record for this product is available from the British Library

ISBN 978-1-83953-482-9 (hardback)
ISBN 978-1-83953-483-6 (PDF)

Typeset in India by MPS

With respectful obeisances to the Almighty -
full of Absolute, Consciousness, and Bliss,
and sincere gratitude to all Well-wishers,
this work is dedicated to my Parents:

Mrs Mira Chattopadhyay
Mr Krishna Mohan Chattopadhyay
Mrs Puspa Bhattacharyya
Mr Sisir Bhattacharyya

Contents

Preface

To engineer the world in a better way from threats of pollution, the whole world is looking for alternate solutions of energy resources and greener transportation to achieve carbon minimization and sustainable development. In this scenario, grids and electric power networks are undergoing decentralized advancement. On the other hand, the mode of mobility in both urban and rural areas is changing from conventional oil-based vehicles to hybrid and electrical electric vehicles. Electric mobility is the future and small grids have become common at present power delivery systems. The future of mobility is demanding to incorporate the change of grid technology that will offer more flexibility to the transport networks. Thus nano and pico grid concepts of power delivery are going to be more effective for The Future of Mobility and Transport (FoMT) application.

Given this, the book will focus on the emerging topic "Nanogrids and Picogrids and their Integration with Electric Vehicles." It covers its' various aspects including modeling, case studies, performance analysis that will help advanced students, researchers, and practitioners.

The book is planned into ten chapters. The first chapter will focus on the paradigm shifts in the field of environment, energy, mobility, and the basic need for Nanogrids and Picogrids. Chapter 2 presents the resources of nanogrid and picogrid. The resources are discussed from the perspective of nanogrid and picogrids. In Chapter 3, circuit modeling is described for AC and DC models of nanogrids and picogrids. It is then followed by a description of the performance analysis of nanogrids and picogrids in Chapter 4. Transients and steady behaviors are covered for different load conditions. Chapter 5 deals with power quality (PQ) issues where different possible PQ disturbances, their types, effects, and possible solutions are covered. Then safety, faults, and protection schemes are discussed in Chapter 6 concerning nano and pico models of grids. The utilization of nanogrids and picogrids is discussed in Chapter 7 that includes electric vehicles. It also includes the discussion on utilities considering application in cities, suburban and rural areas.

Chapter 8 is dedicated to discussing the transition from oil to electric vehicles and the present scenario of electric mobility. Integration of nanogrid and picogrid with an e-vehicle network is discussed in Chapter 9. Grid integration with electric vehicles depends largely on the design of the charging system. In Chapter 10, grid integration with charging stations includes the effective use of the Internet of Things (IoT), artificial intelligence, and cloud computing and will give some directions for future research.

In the book, Nanogrids and Picogrids have been presented lucidly with many case studies and examples. Integration with electric vehicles has been presented along with applications of IoT and cloud computation. Useful information and standards have been presented for ready reference.

Hope that the book will benefit advanced students, researchers, and practitioners in the field. Any constructive suggestion is always welcome.

Dr Surajit Chattopadhyay, CEng(UK), FIE(I), MIET
GKCIET, WB, India
Member, IET Communities Committee—South Asia
Past Hon. Secretary, IET Kolkata Local Network, India

Acknowledgment

In presenting the book *Nanogrids and Picogrids and their Integration with Electric Vehicles*, I have received supports from different people and societies and sincerely acknowledge all the support obtained for this work.

I express my sincere thanks to Dr Christoph von Friedeburg, Sr Commissioning Editor for his continuous engagement in giving shape to the book in its present form. I also sincerely thank Ms Olivia Wilkins and other supporting staff of IET book publication for their constant engagement in the work.

I express my sincere thanks to Mr Shekhar Sanyal, Country Head, IET India, his staff members, and Mr R. N. Rajpoot, Chairman, and all other members of IET Communities Committee—South Asia for providing good networking opportunities through IET to carry out and present the work. I am also thankful to Professor Samarjit Sengupta (Past Chairman) and all other expert members of the executive committee of the IET Kolkata Network for sharing their experiences on different occasions.

I thank Dr Santanu Chattopadhyay of NSOU and Mr Badri Prasad Majumder, Senior Consultant, Infor Global Solutions Inc., for sharing their thoughts and experience on many techno-environmental issues.

Last but not the least; I am grateful to (Late) Mrs Renuka Bandyopadhyay, Mrs Anila Halder, Mrs Karuna Bandyopadhyay, (Late) Mr. Panchu Gopal Chattopadhyay, (Late) Mr. Chanchu Gopal Chattopadhyay, my parents, spouse, and all other family members for inspiring me and thankful to Aditi, Aditya, Aritra, Arundhati, Adrita, and Atreya for holding their patience during writing the book.

Surajit Chattopadhyay
Bagnan Chaitanya Bati, Hooghly,
West Bengal, India

Author's Biography

Surajit Chattopadhyay (Ph.D., CEng(UK), FIE(I), MIET) is an associate professor in the Department of Electrical Engineering at Ghani Khan Choudhury Institute of Engineering and Technology, India. His field of interest includes smart grid topology, electric mobility, overhead electric power lines, electric power quality, fault diagnosis, signal analysis, robotics, sensors, and IoT application. He has authored/co-authored one patent, four books and around 140 papers published in International and National Journals and conferences and edited one book. He is a member of the IET Communities Committee, South Asia, a member of Solar Energy and Future of Mobility and Transport (FoMT) working groups, and an executive committee member of the IET Kolkata Network, and former hon. secretary (2013–2016) and YP-chair (2012–2013), IET Kolkata Network. He served as lead Guest Editor for IET Nanodielectrics and IET Smart Grid. He received many awards and best papers' recognition. He was recognized as an outstanding reviewer of IEEE Transactions on Instrumentation and Measurements in 2020. He received Railway Board's Prize (India) in 2021.

Chapter 1

Introduction

This chapter provides a general introduction of "Nanogrids and Picogrids and their Integration with Electric Vehicles." It starts with the present scenario of the world environment. After that, advancement in computation technology has been discussed. Then, one by one, a paradigm shift in the energy market, grid topology, and mobility have been discussed. At last, chapters of the book at a glance have been presented.

1.1 World environment

The world environment is facing a big challenge of climate change that concerns variations of weather events persisting for long period. Anthropogenic change of climate results in an increase of surface temperature known as global warming. The study has shown that climate variations are largely influenced by human activity causing pollution, an increase of greenhouse gases, etc. Extreme variations and random variations (also known as noise) are observed in the regular and periodic environment characteristics in the time–space–temperature scale. It is affecting the normal ecosystem, vegetation, wildlife, and civilization in many ways. Two major contributors to the increase of pollution and greenhouse gas are the thermal industry and the transport sector. World leaders have now joined their hands with researchers to resolve this issue to save the world by providing a green-sustainable and a better environment.

1.2 Paradigm shift in the energy market

The energy market deals with the supply and trade of energy. It has gone through reformation in many countries. For a long period, it depended on fossil fuels. Then nuclear energy has taken a major role in bulk generation. Renewable resources entered the energy market. In the last few years, the focus has shifted toward renewable energy sources mainly due to the issues of market profit, price rise, less or non-availability, and international dependency on fossil fuels. Major renewable contributors are hydroelectric, wind, and solar energy. Among these three resources, solar energy is available almost everywhere and, therefore, becomes the first choice of an alternate energy resource.

The energy market plays one of the leading roles in the economy. For example, it is the third-largest industry in the United States. Among various energy resources, the growth of the renewable energy market, especially, solar and wind, has been observed very fast in recent years [1,2]. Support of regulatory bodies, banks, and the advancement of technology has become the enabler toward the growth of this sustainable form of energy.

Different statistical data on energy are published by International Energy Agency [2]. Approximately 80% of energy is found as a fossil. China, United States, and Arab states generate around half of the total fossil fuels. However, the growth rate of solar and wind energy is high, approximately 20% per year. North America shows high-energy consumption per person, whereas the index is low in developing countries. However, developing countries show good and rapid growth in renewable energy consumption

Fossil fuel is a major contributor to carbon dioxide emissions. In 2019, fossil fuel contributed around 38 Giga-tons carbon emissions. In 2020, shares of different energy resources in world primary energy consumption are as follows:

- Oil: 31.2%.
- Coal: 27.2%.
- Natural gas: 24.7%.
- Nuclear: 4.3%.
- Hydro: 6.9%.
- Other renewables: 5.7%.

In 2018, shares of different resources in world electricity generation (26,700 TWh) are as follows:

- Coal: 38%.
- Gas: 23%.
- Nuclear: 10%.
- Oil: 3%.
- Hydro and other: 19%.
- Solar photovoltaic (PV) and wind: 7%

Electricity generation in 2020 for 15 countries is shown in Figure 1.1.

"New" renewables refer mainly to solar wind and geothermal energy. It is expected that they will contribute more than 80% to electric power generation by 2040. Solar PV-based generation in 2020 for 15 countries is shown in Figure 1.2.

1.3 Paradigm shift in grids topology

A cluster of electric energy sources through power networks with enhanced capacity is known as a grid. With the rapid growth of solar energy-based electricity generation in the global energy market, grid networks are undergoing paradigm shifts. The early, generation did not take place in consumer sites as well as by consumers. But, the use of renewables, mainly solar energy, has changed the

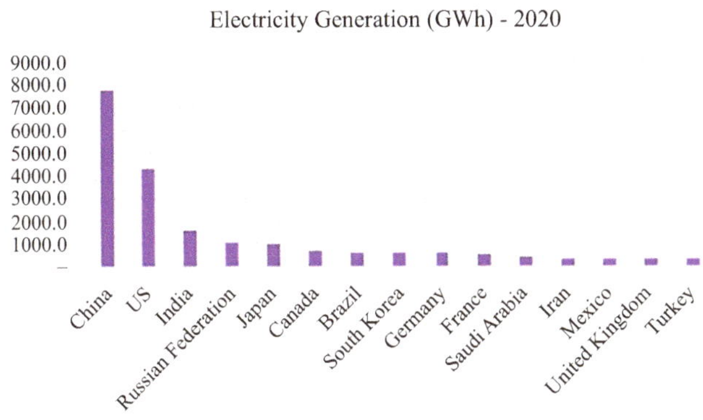

Figure 1.1 Electricity generation in 2020

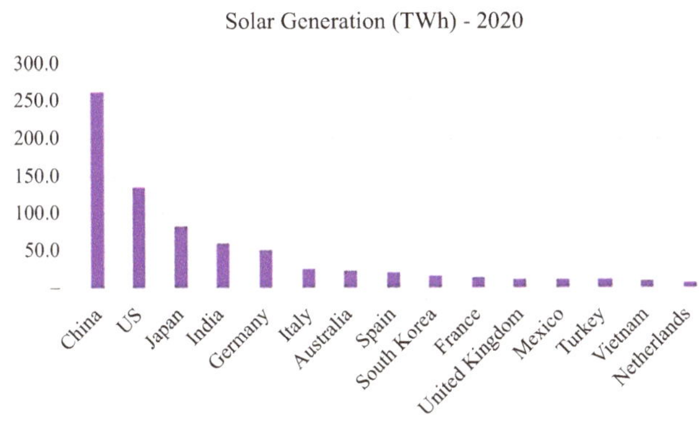

Figure 1.2 Solar generation in 2020

concept. The generation takes place also in the load sides and by the consumers. Thus, the generation has become truly distributed in nature.

On the other hand, the size of generating stations and their capacity are becoming smaller. As a result, the small grid concept has been introduced. Micrograds are already playing important role in electric networks. Based on utility, feasibility, and economy, renewable energy, mainly, solar-based grids are becoming smaller than microgrids. Nanogrids and picogrids are smaller versions of microgrids taking an important role to meet consumer demand. As nanogrids and picogrids are very small in size, the installation cost is less and, therefore, their numbers are increasing rapidly.

1.4 Paradigm shift in mobility

The transport sector is a big contributor to greenhouse gasses causing global warming, climate change, and pollution. Therefore, the transport sector is also undergoing a paradigm shift. Oil-based vehicles are shifted toward hybrid and electric vehicles (EVs). The change has taken place globally concerning both urban and rural areas. While the rate of increase of oil-based vehicle production tends to decline, EV production is increasing at a high rate. The effect is observed in other fields also. Planning for smart cities and smart villages includes the use of EVs. The EV has become the solution for the future of mobility. This demands pollution-free, cost-effective, and time-optimized management of EV charging.

1.5 Advancement in computation and communication techniques

Modern supervisory systems and controllers involve lots of data. For handling large data, many advanced mathematical tools have been introduced. Machine learning tools are enabling extraction of main featured from large data sets very easily. Controllers are added with features of artificial intelligence. On the other hand, communication has become faster. Internet of Things (IoT) is being effectively used in various applications. Advanced computation and communications are together offering cloud computing that has enabled remote monitoring, large data access, handling, and control of many systems. Therefore, it is the right time to utilize those advanced communication networks and computation techniques in managing distributed grids like nanogrids and picogrids remotely.

1.6 Nanogrids and picogrids and their integration with EVs

While electric mobility is a green transport solution, and small grids have become common at present power delivery systems. However, the future of mobility is demanding the change of grid technology that will offer more flexibility for the diverse nature of the EV and electric mobile network. Thus, nanogrid and picogrid systems are going to be more effective for the Future of Mobility and Transport (FoMT) applications.

1.7 Focus of the book

This book will focus on modeling, case studies, and performance analysis on the areas of the emerging topic "Nanogrids and Picogrids and their Integration with Electric Vehicles" that will help advanced students, researchers, and practitioners. The book is planned into 10 chapters. As we are living in the era of microgrid, the discussion starts from the microgrid. Then focus is set on the nanogrids and picogrids, discussion on their definition, why they are needed, and a few highlights of recent advancement. Nanogrid refers to a single unit of reliable electric power that must have at least one load (sink of power, that may be a storage unit) and at least one gateway to the outside; electricity may or may not be present. Pico grids have a much smaller capacity but have the flexibility to be mobile.

Then, Chapter 2 describes the resources of nanogrid and picogrid. The resources are discussed from the perspective of nanogrid and picogrids.

In Chapter 3, circuit modeling is discussed for AC and DC models of nanogrids and picogrids.

Modeling is followed by a description of the performance analysis of nanogrids and picogrids in Chapter 4. Transients and steady behaviors are covered for different load conditions.

Then, in Chapter 5, power quality (PQ) issues are covered. Different possible PQ disturbances, their types, effects, and possible solutions are covered followed by some recent advancements for diagnosis and quality monitoring.

Then safety, faults, and protective measures are discussed in Chapter 6 concerning nano and pico models of grids.

Then utilization of nanogrids and picogrids is discussed in Chapter 7 that includes EVs. It also includes the discussion on utilities considering application in cities, suburban, and rural areas.

Then Chapter 8 is dedicated to discussing the transition from oil to EV and the present scenario of electric mobility.

Integration of nanogrid and picogrid with an e-vehicle network is discussed in Chapter 9. A case study is made considering the mobility characteristics of the e-vehicle network.

Grid integration with EVs depends largely on the design, operation, and supervision of the charging system. Therefore, the development of the charging system is covered with a case study in Chapter 10 that includes the effective use of IoT and cloud computing followed by some trends for research.

References

[1] bp Statistical Review of World Energy 2021, Statistical Review of World Energy, BP p.l.c., London, UK.

[2] https://www.iea.org/fuels-and-technologies/electricity.

Chapter 2
Energy resources for nanogrids and picogrids

This chapter describes energy resources that are suitable for electric energy generation. Different types of energy resources used for electric power generation have been briefed. Based on the type of energy resource, plants may be of different types like a steam power plant, nuclear power plant, diesel–electric power plant, gas turbine power plant, hydro-electric power plant, magneto-hydro-dynamic (MHD) power plant, thermoelectric power plant, wind power plant, tidal power plant, geothermal power plant, etc. All these types have been described. Then, different aspects of solar thermal photovoltaic (PV)-based power plants have been described. Battery-based storage system is discussed. Different features of solar PV systems used as energy resources for nanogrids and picogrids have been presented.

2.1 Introduction

Energy resources for electric power generation are broadly divided into two categories: conventional and non-conventional. Conventional energy resources include fossil fuels. Non-conventional energy resources include solar tidal, geothermal, magneto-hydro-dynamic (MHD), etc.

In the early days, the conventional grid depends largely on conventional energy resources and partly on non-conventional energy resources. But, in the last few decades, dependency on non-conventional energy resources has increased and is still increasing day by day.

While discussing nanogrids and picogrids, various combinations are being observed and are possible. Broadly we can divide the integration of nanogrids and picogrids with energy resources into two types of combinations:

(a) Nanogrid and picogrid integration with two or multiple conventional and non-conventional energy resources.
(b) Nanogrid and picogrid integration with two or multiple numbers of only non-conventional energy resources.

Different types of energy resources may be listed based on utilities as follows:

- energy resources for steam power plant;
- energy resources for nuclear power plant;
- energy resources for diesel–electric power plant;
- energy resources for gas turbine power plant;

- energy resources for hydro-electric power plant;
- energy resources for MHD power plant;
- energy resources for thermoelectric power plant;
- energy resources for thermionic power plant;
- energy resources for wind power plant;
- energy resources for tidal power plant;
- energy resources for geothermal power plant;
- energy resources for solar thermal power plant;
- energy resources for solar photovoltaic (PV) power plant;
- energy resources for battery-based storage or chemical power plant.

2.2 Energy resources for steam power plant

Steam power plant contributes to major electric power need of the world. However, the percentage share of the steam power plant is decreasing rapidly. This type of conventional power plant offers overall efficiency of about 28–30% or less. Concerning hydroelectric power plants, it requires less space and cost of initial installation. But running cost for electric power generation in the steam power plant is higher than that of a hydroelectric power plant. Steam power plant faces challenges of air pollution, adequate water requirement, and availability of skilled manpower. Site selection for this type of plant considers the following major issues:

- fuel supply,
- distance from load center,
- water availability,
- land availability,
- disposal of ash, and
- skilled manpower.

The operation of a steam power plant depends on the type of fuel fed to the boiler for steam generation. Mainly, three types of fuels are used as follows:

- coal,
- oil, and
- gas.

2.2.1 Coal

This type of fossil fuel is found in nature accumulated under surface as a result of transformation for millions of years. Natural coal consists of carbon along with hydrogen, sulfur, nitrogen, moisture, etc. Different types of coals with enthalpy range at combustion are presented in Table 2.1.

Other than the enthalpy of combustion, the following features are considered for the selection of coal:

- Calorific value.
- Environment exposure capability.

- Sulfur content.
- Impurity or ash.
- Grinding ability.

Table 2.1 Different types of coal and range of their enthalpy of combustion

Coal type	Enthalpy of combustion (kJ/kg)
Peat	3 000
Lignite	13 000–18 000
Sub-bituminous coal or black lignite	18 000–23 000
Bituminous coal	23 000–34 000
Semi-bituminous coal	27 000–35 000
Semi-anthracite coal	33 000–35 000
Super-anthracite coal	>35 000

2.2.2 Oil

Naturally available oils are also used in steam power plants. For example, petroleum is widely used. The use of oil-based fuel in steam power plants reduces ash handling units and pollution as compared to coal. But, expenditure for heat production is more than that of coal. Also, it contains much hydrogen which increases the moisture absorption ability of flue gases reducing the overall efficiency. The selection of oil depends on the availability, cost of transportation, and overall efficiency of the plant.

2.2.3 Gas

Gas can be used in a steam power plant. Both natural and manufactured gases may be used for feeding energy to the boiler unit. However, natural gas is preferred to manufactured gas. The use of gas has the difficulty of transportation needed, especially for a bulk generation. The selection of gas depends on availability, calorific value, cost of transportation, etc.

2.3 Energy resources for nuclear power plant

Nuclear fuel is treated as an energy resource for the generation of a bulk amount of power. Nuclear materials offer huge energy through the chain nuclear reaction. The energy generation capacity of 1 kg uranium is equivalent to 3 000 Ton of good quality coal materials. As fossil fuel reserves are decreasing rapidly, nuclear power becomes a good alternative to meet the demand for bulk power. Also, it does not have hazards like coal handling, huge ash, flue gases, air pollution, etc. The initial build-up cost of the nuclear power plant is much higher than that of a coal-based steam power plant. However, its operating and running costs are low. Moderator is used for controlling nuclear reaction and power generation. The use of nuclear power for electric power generation faces challenges of exposure of radio reaction and disposal of radio products. The shielding of equipment and operating personals from radio-active rays is another big challenge. Lead, concrete, thick steel, cadmium, etc. are normally used for safety and shielding.

2.4 Energy resources for diesel–electric power plant

The diesel–electric power plant uses a diesel engine as a prime mover where diesel is used as fuel. The capacity of a diesel-based power plant may vary from very small to large scale. Selection of this type of energy resource involves other factors like nature of demand, availability of diesel, cost of transportation, water availability, distance from load center, population of the area, etc. Fuel efficiency is higher than that of the steam power plant. It offers less air pollution, fire hazards, and the cost of operating persons. But the cost of diesel is a little high and fluctuating in nature in the global market.

2.5 Energy resources for gas turbine power plant

Fossil fuels are used for feeding gas turbine-driven power plants. All three categories of fossil fuels like coal, oil, and natural gas may be used for gas turbines. But very often coal is not chosen to avoid coal and ash handling units. Oil, gas, and gas-oil are preferred to coal for gas turbines. Methane-based gases are commonly used in gas turbines. This type of power plant needs less space per kW power output. It faces the challenge of corrosion, dust deposition, higher air rate, etc.

2.6 Energy resources for hydro-electric power plant

Energy stored in water is utilized in hydroelectric power plants. It offers clean and pollution-free energy resources. Power generation (P) depends on water head (H), water flow rate (Q), and a specific weight (W). Mathematically,

$$P \propto HQW$$

or,

$$P = HQW \times \eta \times 9.81 \times 10^{-3}\ \text{kW}$$

where η refers to plant efficiency.

Hydroelectric power generation suffers from the high cost of initial installation but offers the advantage of low running costs. Availability of water is the most important consideration for setting up the hydroelectric power plant. Often water is not available at or near the load center; in that case cost of transportation of electrical energy increases. These types of power plants use mainly the following types of turbines: Pelton turbine, Francis turbine, Kaplan turbine, Propeller turbine, etc. Based on water-head, hydro-electric power plants are classified as the low head (up to 30 m), medium head (30–300 m), and high head (greater than 300 m). The capacity of a hydroelectric power plant ranges from a small size like 1 MW to a very large size of about several hundred MW capacities.

2.7 Energy resources for MHD power plant

In a MHD power plant, conducting gases are passed through a high magnetic field at high temperature and high pressure. High temperature is produced by the combustion of fossil fuels and temperature is increased in the range of 2 000–6 000 K. The temperature should be controlled such that gas gets ionized. From the moving ionized gas, electric energy is collected and the direct current (DC) is found. Then DC is converted to alternating current (AC) at desired voltage level by the converter. A special feature of this generation is the absence of a turbine-based mechanical energy conversion unit.

Generated power (P) is directly proportional to the specific electrical conductivity (σ) and is proportional to the square of magnetic field intensity (B) and the velocity (v) of gas.

Mathematically,

$$P \propto \sigma$$
$$P \propto B^2$$
$$P \propto v^2$$

Therefore,

$$P \propto \sigma B^2 v^2$$
$$P = K_1 \sigma B^2 v^2$$

where K_1 is the constant of proportionality. It depends on the load voltage and open-circuit voltage of the generator.

MHD power plant offers elimination of moving parts and high efficiency. Control of generation is fast which is suitable for meeting the power demand which may vary rapidly. Also, it offers a competitive price of electricity concerning coal-based power generation. However, it faces technical challenges like generation of high temperature, corrosions, overall commercial viability, etc.

2.8 Energy resources for thermoelectric power plant

Utilizing the "Seebeck effect," electric energy is produced from a combination of hot and cold junctions. Solar radian is used for making hot junctions and cooling water is used for the cold junction. Semiconductor materials are used for building high-temperature gradients by their thermal insulation property. Some commonly used semiconductor materials are germanium silicide, bismuth telluride, zinc antimonide, telluside, neelium, etc. This type of power generation offers an efficiency of about 10%. But it generates power without creating any pollution. This type of plant can be built with a low capacity of about 1 kW at a low cost. Initially, DC is produced and then converted into AC as desired.

2.9 Energy resources for thermionic power plant

Gas or vapor is used for the generation of electric energy through the thermo-ionic process. Gas is placed in between anode and cathode. The cathode is heated to emit electrons. Ionized gases move toward the anode. Energy generation is free from pollution. The output power is DC in nature and then converted into AC as required. It offers about 15% of overall efficiency.

2.10 Energy resources for wind power plant

The wind is an endless natural resource and it is used as a resource material for generating electric power through wind power plants. Energy stored in the wind is converted into electric energy. Wind resource is plenty in nature and electric power generation is non-polluting. The use of wind energy has become very popular and, as a result, the percentage of wind power generation is increasing rapidly. Based on the electrical power-generating capacity, wind turbines are classified into the following main categories:

- small (range: 10–25 kW),
- medium (range: 50–150 kW),
- large (range: 250–1 000 kW), and
- very large (range: 2 000 kW or above).

Two types of turbines are used to collect energy from wind as follows:

- Horizontal axis wind turbine (HAWT): The rotation axis is in parallel with the air stream.
- Vertical axis wind turbine (VAWT): The rotation axis is perpendicular to the air stream.

The main challenge of using wind energy for the generation of electrical power is that wind speed and its direction are unpredictable. Proper conversion and control mechanisms are required to generate power at constant frequency and to synchronize with the grid.

2.11 Energy resources for tidal power plant

Energy stored in the tides is utilized for the generation of electric energy in tidal power plants. Tides are endless in number and this type of electric energy generation is non-polluting. It requires proper selection of sites based on coastal areas or seasides having easy access to tidal features for power generation.

2.12 Energy resources for geothermal power plant

Thermal energy stored under the earth's surface level is used for electric power generation. Site selection is very important for setting up such a power plant. Electric power generation through this process is normally non-polluting.

2.13 Energy resources for solar thermal power plant

Chains of nuclear reactions in the sun are the source of enormous energy radiation from the sun. These radiations move to space. Only a small part of the rays goes to the earth. Then, it goes through reflection and refraction through the earth's atmosphere and then again gets reflected by the earth's surface. At every stage, absorption and loss of energy take place. Around 10% of the total energy that falls on the earth becomes available in the air. Solar energy available on the earth's surface can be utilized as an alternate source of energy for replacing fossil fuels whose resources are limited.

Solar thermal energy is utilized for heating purposes, desalination, etc. Various types of solar heat collectors, solar pumps, etc. have been introduced. Similarly, the thermal energy of solar radiation is used for electric energy generation in solar thermal power plants.

2.13.1 Advantages and limitations

Major advantages of solar thermal energy are as follows:

- freely available in nature;
- limitless resource and there is no threat of ending;
- pollution-free;
- available everywhere.

Some limitations of solar thermal energy are as follows:

- Intensity depends on the climatic and geographic conditions.
- It is not available in the night-time.
- It is less or non-availability in the presence of cloud.
- It is distributed in nature. Available solar thermal energy per unit area is small.

2.13.2 Working topology of solar thermal electric power plant

Working topology is shown in the block diagram in Figure 2.1. The working topology of the solar thermal plant follows the following steps:

- collection of solar thermal energy;
- heat exchange;
- conversion of thermal energy to dynamic energy;
- transfer of dynamic energy to the rotor of the generator;
- generation of electrical energy.

Collection of solar thermal energy: Different types of collectors are designed for collecting solar thermal radiation. In collectors, thermal energy is collected by different types of fluid circulated naturally or with the help of an additional pump system.

Conversion of thermal energy to dynamic energy: Heat collected by fluids from solar radiations is transmitted to the boiler section. The boiler generates the necessary pressure difference for the rotation of the turbine.

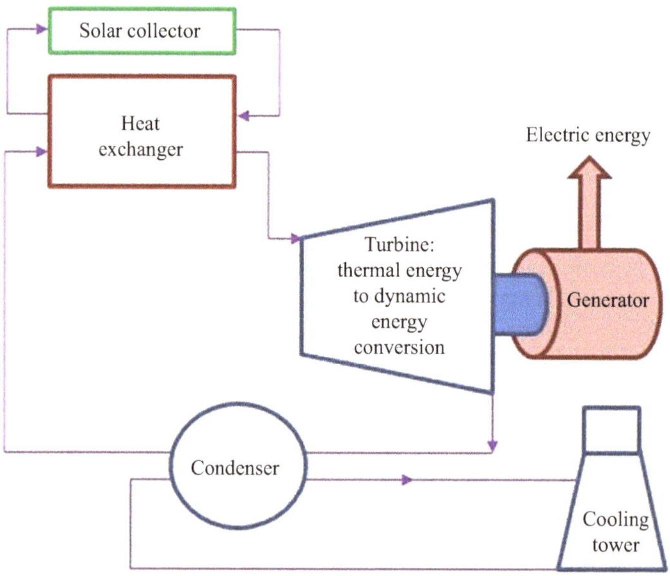

Figure 2.1 Working topology of solar thermal electric power plant

Transfer of dynamic energy to the rotor of the generator: The turbine is coupled with the rotor of the generator. Thus, it transfers mechanical energy to the rotor through the shaft for rotation.

Generation of electrical energy: Generator unit converts mechanical energy to electrical energy. The generated power is alternating in nature and transmitted to the load section as desired.

2.13.3 Different types of collectors of solar thermal energy

Different types of collectors used for collecting solar thermal energy may be categorized into two major groups as follows:

(a) flat collector;
(b) concentric collector:
  (i) parabolic cylindrical collector and
  (ii) paraboloid mirrors.

2.13.4 Flat collector

In a flat type collector, solar rays fall on a flat plate. At the bottom surfaces, tubes carrying fluid are placed. Fluid materials collect thermal energy from the rays. Natural circulation or forced circulation by the pump is used for fluid to collect and transfer the energy. Mainly two types of flat collectors are used as follows:

- horizontally placed flat collected;
- rotating and inclined flat collector.

Horizontally placed flat collector: In this type of collector, the collecting plate is placed horizontally with the earth's surface. The main advantages of this type of collection are its simple construction and low cost. The main limitation of this scheme is that it cannot capture the maximum amount of thermal energy from the inclined solar rays. As the sun moves from east to west, the angle of incidence varies from time to time. When rays fall in a perpendicular direction at noontime, maximum heat energy can be collected. At other times, when rays fall on the horizontally placed plate with different angles of incidence, some parts of rays reflect resulting in a lesser collection of heat energy.

Rotating and inclined flat collector: In this type of collector, the collection plate is rotated from morning to evening in such a way that solar rays fall on the plate in a perpendicular direction. This reduces reflection and increases the amount of heat extraction. It is costlier than a simple horizontally plate flat type collector.

2.13.5 Parabolic cylindrical collector

The parabolic cylindrical collector is a concentrating type collector. The collecting plates/tubes are placed at the focus area of the cylindrical and parabolic reflector. Sunrays falling with different angles of incidences get reflected and pass through the focused area. From the focused area, heat is collected.

2.13.6 Paraboloid mirrors

Paraboloid mirror collector is also concentrating type collector. The collecting plates/tubes are placed at the focus area of the paraboloid reflector. Sometimes arrays of mirrors are used. Sunrays falling with different inclinations get reflected and pass through the focus area. Heat is collected from the focused area.

2.14 Energy resources for solar PV power plant

Solar PV-based power plants have become most popular and are increasing at a faster rate throughout the world. It is plenty in nature and there is no threat of ending solar energy. It is non-polluting. DC electric power is produced through PV conversion, this is then converted into AC and transformed into desired voltage levels as required.

2.14.1 PV cell

PV cells received solar energy and produce electric energy. It works on the principle of the PV effect. It is often referred to as a transductor. It is made of semiconductor materials with PN junction structures. By different types of impurities, P-type and N-type material blocks are formed. For example, silicon doped with boron can produce P-type and silicon doped with phosphorous or arsenic can produce N-type materials. P-type and N-type blocks are characterized by sufficient numbers of positively charged holes and negatively charged electrons. By these two blocks, a PN junction is formed. Based on the diffusion or movement of holes and free electrons, a potential barrier (E_g) is formed in the junction.

When this PN junction diode is exposed to sunlight, both P-type and N-type blocks receive sunlight. Sunrays consist of packets of energy, known as a photon. Each photo has the energy of amount E_P given as follows:

$$E_P = hv$$

where h is called Plank's constant; v is the frequency of the light.

If the photon falls on the P side, the valance electron will absorb photons and move toward the N side. On the other hand, if the photon falls on the N side, holes will absorb photon energy and will move toward the P side. These movements of holes and free electrons develop an electric field in a direction opposite to the field developed by diffusion.

Thus, when sunrays are received by the P-type and N-type blocks, a potential difference is developed across the P and N sides. This potential difference increases with the increase of solar electromagnetic radiation. Each cell can develop up to 1 V.

Thus, a PV cell can be represented as the symbol of voltage source as shown in Figure 2.2.

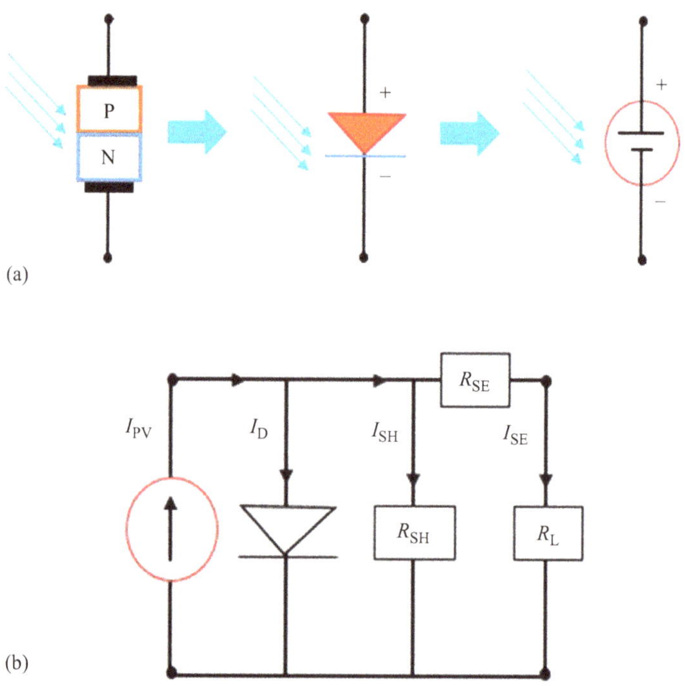

Figure 2.2 (a) Symbol of PV cell and (b) equivalent circuit of PV cell

2.14.2 Characteristics

2.14.2.1 PN junction characteristics without illumination

Characteristics of a PN junction are shown in Figure 2.3. At first, “no illumination” by sunrays has been considered here. It looks like *V–I* characteristics of a PN junction diode. If voltage is applied in forward-biased conditions, it will soon start conduction. However, in reverse-biased conditions, very small leakage current flows. When the voltage across the diode is zero, the current becomes zero and may be considered as an ideal short circuit condition.

2.14.2.2 Shorted PN junction with illumination

Now if the PN junction is illuminated at short-circuited conditions, it will absorb a photon. As a result, short circuit current will flow (in the direction same as reverse biased direction) as shown in Figure 2.4. With the increase of illumination level, the magnitude of short circuit current will also increase.

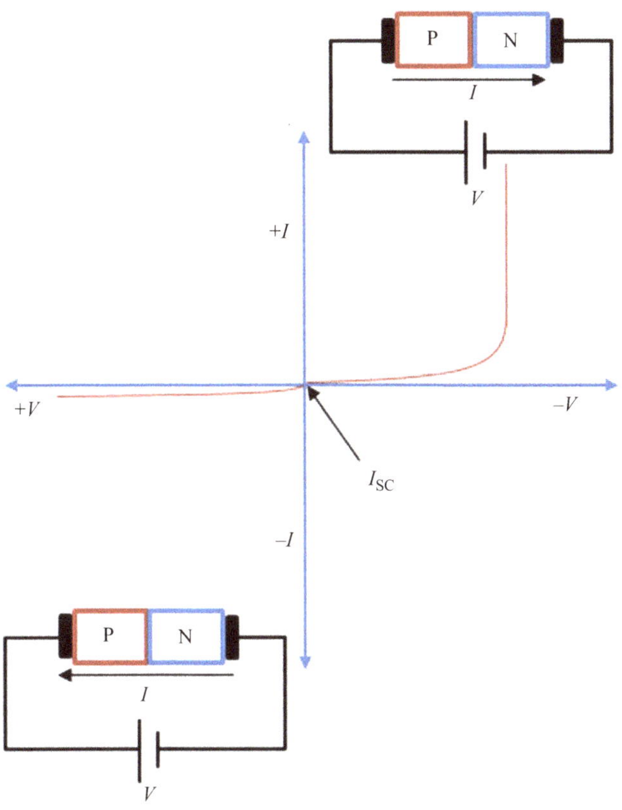

Figure 2.3 Characteristics of PN junction without illumination (i.e., in dark)

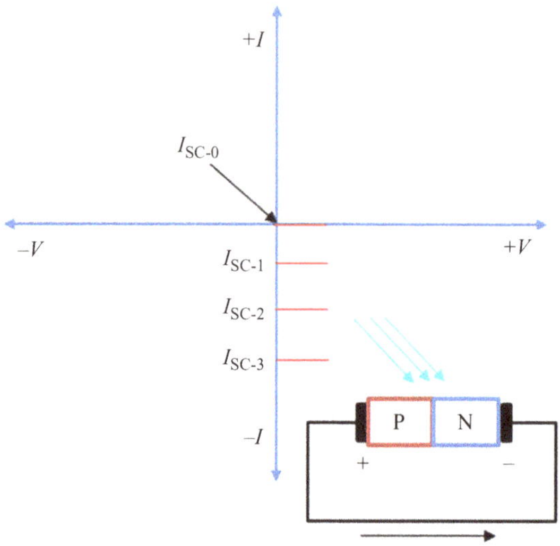

Figure 2.4 Short circuit current of PN junction under dark and solar illumination

2.14.2.3 Shorted PN junction with illumination with small voltage in the forward direction

If a small positive voltage is applied across the PN junction, the short circuit current will reduce as shown in Figure 2.5. The voltage corresponding to zero current is known as open-circuit voltage.

2.14.2.4 PV PN junction at illumination

When the PV PN junction is illuminated, the P side will become positive and open circuit voltage is observed across the junction as shown in Figure 2.6. If the junction is shorted, a short circuit current will flow. With the decrease of current by inserting load resistance, the voltage will increase and reach the maximum open-circuit voltage.

2.14.2.5 Power

Power can be determined from the V–I characteristic. Consider "A" is the operating point as shown in Figure 2.7. Then, power at operating point A will be

$$P_A = V_A I_A$$

2.14.2.6 Maximum power point

The operating point in V–I characteristics of a PV cell when it delivers maximum power to load is known as the maximum power point (MPP). Voltage versus power along with the V–I characteristics is shown in Figure 2.8. The MPP has been shown by the point "P_m" as in Figure 2.8.

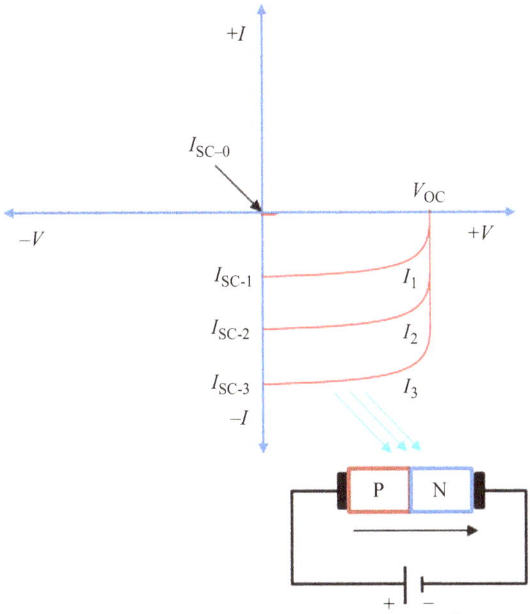

Figure 2.5 *V–I characteristics of PN junction under solar illumination with forward-biased voltage*

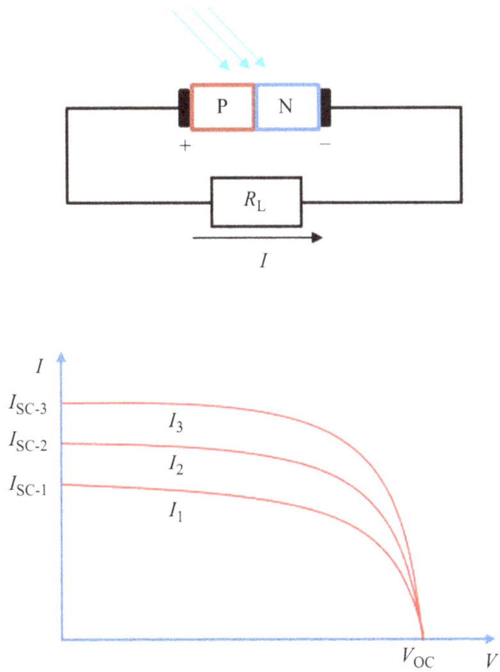

Figure 2.6 *Solar PV cell with resistive load and its V–I characteristics under different solar illumination*

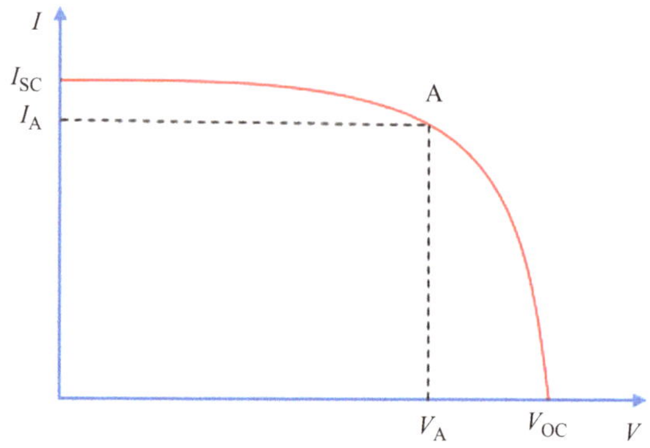

Figure 2.7 V–I characteristics of solar PV cell

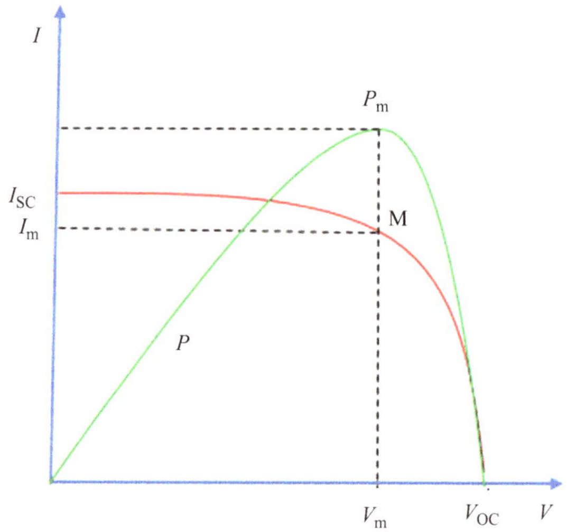

Figure 2.8 V–I and V–P characteristics of solar PV cell

Maximum power may be calculated as follows:

$$P_m = V_m I_m$$

where V_m is the voltage at the point of maximum power; I_m is the current at the point of maximum power.

2.14.2.7 Fill factor

Fill factor is defined as the ratio of maximum power (P_m) to the product of short circuit current I_{SC} and open-circuit voltage V_{OC}. Mathematically,

$$Fill\ factor = \frac{P_m}{V_{OC}I_{SC}} = \frac{V_m I_m}{V_{OC}I_{SC}}$$

Thus, the fill factor determines the outcome-based quality of a PV cell. Its value ranges from 0.5 to 0.85.

2.14.2.8 Efficiency

The efficiency of a PV cell may be defined as the ratio of output power to the absorbed solar power. Let, P_{SP} is the absorbed solar power. Therefore, efficiency may be written as follows:

$$\eta = \frac{\text{output power}}{\text{solar power}} = \frac{P_{out}}{P_{SP}} = \frac{VI}{P_{SP}}$$

If the PV cell operates at the maximum power point, the efficiency will be maximum and can be written as follows:

$$\eta_{\max} = \frac{\text{maximum output power}}{\text{solar power}} = \frac{P_m}{P_{SP}} = \frac{V_m I_m}{P_{SP}} = \text{fill factor} \times \frac{V_{OC}I_{SC}}{P_{SP}}$$

2.14.2.9 Voltage versus power

Voltage versus maximum power is shown in Figure 2.8. It shows that with the increase of voltage, power increases. After reaching maximum power, it starts falling sharply.

2.14.2.10 Illumination versus maximum power

Voltage versus current and maximum power for different illumination (W/m^2) levels is shown in Figure 2.9. With the increase of illumination, maximum power

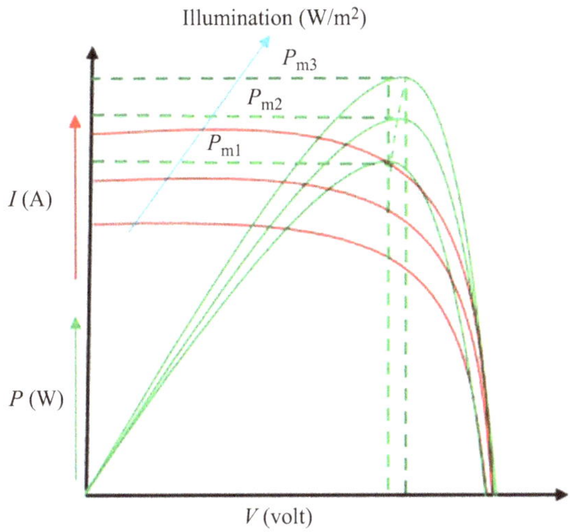

Figure 2.9 Illumination versus maximum power characteristics of solar PV cell

and voltage at maximum power point are increasing. This indicates that sufficient illumination level plays an important role in the generation of electric energy using solar PV cells.

2.14.2.11 Temperature versus maximum power

Voltage versus current and maximum power for different temperature (°C) levels is shown in Figure 2.10. Maximum power and voltage at maximum power point decrease with the increase of temperature. This indicates that temperature level plays an important factor to be considered in the generation of electric energy using solar PV cells. Maximum power point tracking (MPPT)-based controller considers both illumination and temperature for obtaining maximum power point region in operation.

2.14.3 Module, panel, string, arrays

2.14.3.1 Module

A single PV cell has a very small voltage developing and current delivering capacity, a module of solar PV cells is formed to improve the power delivering ability. The output power of a single cell is too small and also there are lots of difficulties in operating a single cell in an open atmosphere where it may be encountered by dust, moisture, mechanical damage, etc. Module refers to the combination of several cells in compact formed with transparent cover. For example, 32 cells can be combined to form one module of a particular output voltage, say, 12 V. Generally, two types of module structures are found as follows:

- Square-shaped module.
- Off square-shaped module.

Ideally, the output power capacity of the module should be equal to the algebraic sum of the output power of individual cells. However, the output power

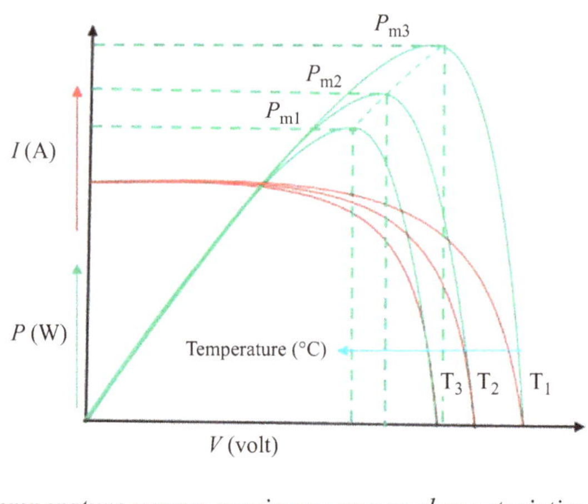

Figure 2.10 Temperature versus maximum power characteristics of solar PV cell

capacity of the module is less than the algebraic sum of the output power of individual cells due to different internal losses. Let

P_i is the output power of ith cell;
P_m is the output power of module;
N is the total number of cells in one module.
Then, the output power of a single module will be

$$P_m \leq \sum_{i=1}^{N} P_m$$

2.14.3.2 Panel

PV panel refers to multiple numbers of PV modules connected in series and parallel combinations to deliver desired output current at a desired voltage level. The increase of the number of PV modules in series combination increases the output voltage and the increase of several parallel paths of PV modules connected in series increases the current delivering capacity. The sample panel of PV modules is shown in Figure 2.11. The small panel may consist of a single module of cells.

2.14.3.3 String

PV string normally refers to multiple numbers of PV panels connected in a series combination to deliver output current at desired voltage level. Panels may be placed at different locations. An increase in the number of PV panels in series combination increases the output voltage, sample string of PV panels is shown in Figure 2.11.

2.14.3.4 Array

PV array refers to multiple numbers of PV strings connected in parallel or both series and parallel combination to deliver desired output current at desired voltage level. An increase in the number of PV strings in parallel combinations increases the output current delivering capacity. A sample array of PV strings is shown in Figure 2.11.

2.14.4 Cell mismatching

Cell mismatching refers to the mismatch of properties of cells of a single module. It occurs when cells of different characteristics are combined to form a module. It results in the following main problems:

- *Undesired output voltage and unequal voltage distribution*: For the same load, voltage distribution will not be uniform. The output voltage will not be the same as desired.
- *Less output power*: Due to unequal cell voltage distribution some may deliver peak output power and others will give less output power. Mathematically, the output power of a single module will be

$$P_m < \sum_{i=1}^{N} P_m$$

- *Different short circuit nature*: At ideal conditions, during a short circuit, there will be no output power when all cell voltages become zero. But for the

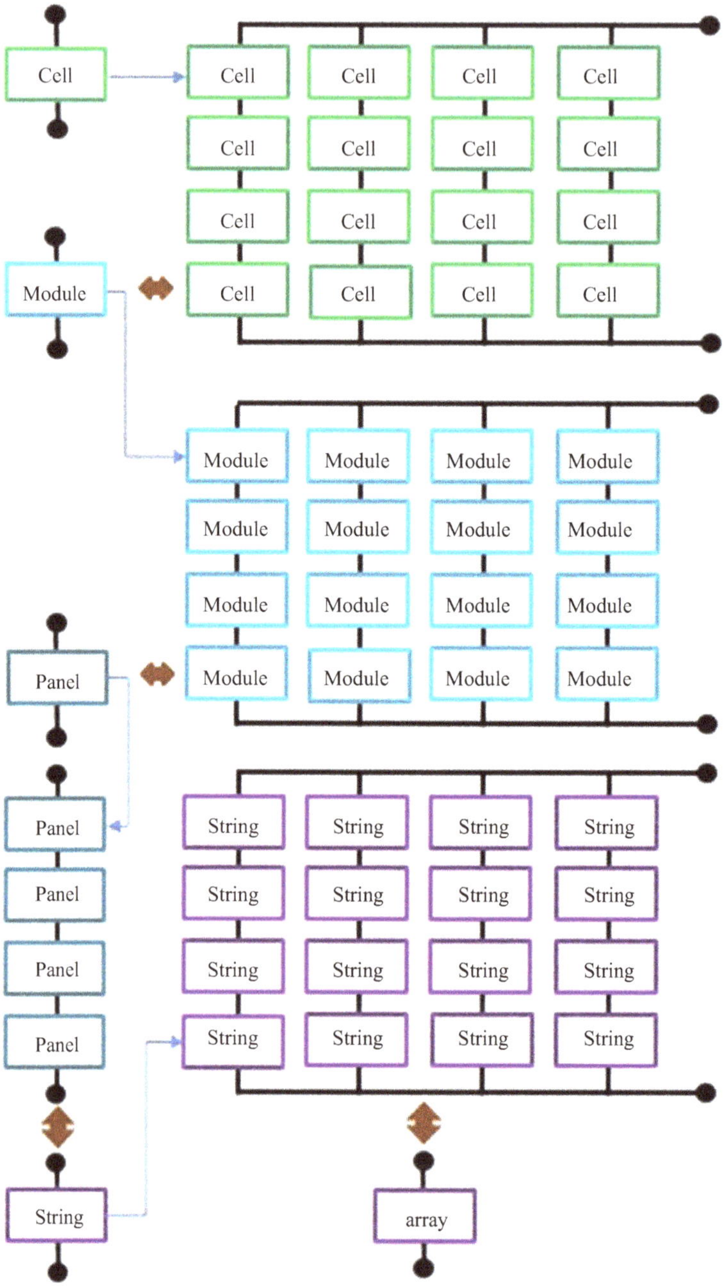

Figure 2.11 Logical block diagram of solar PV module, panel, string, and array

module with cells of different characteristics, when the module is shorted some cells will show small positive voltage and some will show negative voltage. Cells of the shorted module showing positive voltage generate power and cells of the shorted module showing positive voltage will dissipate power.

2.14.5 Advantages and disadvantages

The major advantages of solar PV-based power generation are as follows:

- Power is generated from solar energy that is plenty in nature.
- Energy resource if freely available.
- Energy generation is non-polluting.
- Running cost is almost absent.
- It requires very little maintenance and PV cell last for long.

However, solar PV-based energy generation has some major disadvantages or limitations as follows:

- The efficiency of conversion is low.
- As sunlight is distributed, large open or rooftop areas are required for bulk generation.
- The initial cost is high.
- As it is dependent on sunlight, storage units are required for operation when sunlight is not available.
- As power produced is of DC in nature, grid integration requires proper technical care.

2.14.6 Classification of PV cell

The classification of PV cells based on various aspects is presented in Table 2.2.

Table 2.2 Classification of PV cell

Serial no.	Classification based on	Classifications
1	Active material type	• Single-crystalline silicon PV cell • Multi-crystalline silicon PV cell • Gallium arsenide PV cell • Amorphous silicon PV cell • Copper indium PV cell • Organic PV cell
2	Active material thickness	• Thin material PV cell • Bulk material PV cell
3	Structure of junction	• PN homojunction PV cell • PN heterojunction PV cell • PN multi-junction PV cell • Schottky (metal-semiconductor) junction PV cell • P-intrinsic-N (PIN) junction PV cell

Table 2.3 Classification of PV resource system

Serial no.	Type	Subcategory
1	Central solar power system	• Large capacity (>2 MW) • Medium capacity (<2 MW)
2	Distributed solar power system	• Stand alone: for a separate run • Distributed grid-connected: for micro-, nano-, and pico-grid, ranges from several Watts to 25/50 kW • Small localized grid-connected: ranges up to 1/5 kW

2.14.7 Specification properties of PV cell

The main properties of PV cells that may be used for specifications of a cell are as follows:

- short-circuit current,
- open-circuit voltage,
- maximum power,
- fill factor,
- efficiency,
- doping type, and
- cell structure.

2.14.8 Classification of solar PV-based resource system

PV cell-based solar power resources may broadly be classified into two categories: central solar power system and distributed solar power system. Central solar power system has comparatively greater capacity and can be divided into two subcategories as large and medium. Distributed solar power systems may be subdivided into three categories: stand-alone, distributed grid-connected, and small grid-connected. Classifications of PV-based solar power stations are presented in Table 2.3.

2.15 Biomass

Biomass is treated as a big potential for energy resources in various forms. The main advantage of biomass is that as solar energy, it is available plenty in nature throughout the world. It can be found in the following major areas:

- charcoals,
- crop residues,
- fuel wood,
- cow dice,
- energy crop,
- agricultural residue,

- bio-diesel,
- urban waste, and
- aquatic plants.

Energy extraction from biomass can be done by thermochemical, biochemical, and direct combustion or incineration process. Some physical processes like pelletization or pulverization, briquetting, etc. may also be done for energy extraction.

The major advantages of biomass energy are as follows:

- The biomass resource is plenty in nature.
- It is renewable.
- It pollutes less.
- It offers natural storage.
- The cost is low.

However, energy extraction from biomass has some major limitations like as follows:

- Energy density is low.
- It involves high labor costs.
- The slow rate of energy conversion does not make it suitable for rapidly varying loads.

Biomass can be utilized in producing chemicals, fertilizers, etc. It can be utilized to produce fuel that can be stored and can also be used to obtain thermal energy. Thermal energy produced by biomass can be used to generate electrical energy.

2.16 Fuel cell

A fuel cell converts chemical energy into electric energy directly. It works on the principle of the flow of free electrons produced by chemical reactions. Cathode and anode act as terminals of the fuel cells. From electrodes, power is collected.

2.16.1 Major advantages of fuel cells

The major advantages of fuel cells are as follows:

- *High efficiency*: In a steam power plant, heat is produced using chemical energy, the thermal energy is converted into mechanical energy and then electrical energy. At each stage, loss occurs lowering the overall efficiency. But, in fuel cells, electric energy is produced directly from chemical energy which reduces the loss and increases efficiency.
- *Energy per kg*: Electrical energy production per kg is higher than that of the steam power plant.
- *Portability*: As fuel cells are available in lighter weight and small in size, they can be easily transported from one place to another place.
- *Pollution*: Fuel cells produce very little pollution.

2.16.2 Major disadvantages of fuel cells

Major disadvantages are as follows:

- *Cost*: Cost per unit of energy is high in fuel cell generated power.
- *Short life*: The life span of the fuel cell is less than other power generation systems. Therefore, it requires replacement at regular intervals.
- *DC generation:* Output power in a fuel cell is DC in nature. Therefore, the inverter is required where AC is required.
- *Low voltage*: The output voltage of the fuel cell is low as compared to conventional power generation. Therefore, after inversion, a stepping-up arrangement is required.

2.16.3 Cell bank

Bank of fuel cells is formed to increase the power delivering capacity. Cell bank refers to multiple numbers of cells connected in series and parallel combinations to deliver desired output current at a desired voltage level. Increase of the number of cells in series combination increases the output voltage, increase of parallel path of cells connected in series increases the current delivering capacity.

2.16.4 Fuels used in fuel cells

Mainly three types of fuels are used in fuel cells as follows:

(a) *Hydrogen and oxygen*: Hydrogen and oxygen are common types of fuels widely used in fuels cell. They can be made easily available and also offer efficient power generation.
(b) *Hydrocarbon*: Hydro-carbon materials are also used as fuel. However, they are found less reactive than hydrogen-based cells. Commonly used hydrocarbons are methane, ethane, benzene, etc.
(c) *Compromised fuel:* Chemical reactivity of these materials is less than hydrogen and oxygen but higher than hydro-carbons. Commonly used compromise fuels are methanol, hydrazine, ammonia, etc.

2.16.5 Electrodes

Electrodes are used to collect power. Nickel, platinum, etc. are used as electrodes. As electrodes, nickel is widely used in "Hydrox" fuel cells. Carbon catalyzed platinum is used as electrodes in phosphoric acid fuel cells.

2.17 Energy storage systems

Various attempts have been taken to store energy in different forms and to use them at the time of requirement. Different methods have been introduced. However, their effectiveness depends on many parameters including cost and space. Major ways for storing energy may be classified as follows:

- mechanical process for energy storing;
- chemical process for energy storing;
- thermal process for energy storing;
- electrostatic process for energy storing;
- electromagnetic process for energy storing.

2.17.1 Mechanical process for energy storing

Commonly used mechanical approaches are

- energy storage by pumped water;
- energy storage by compressed air;
- energy storage by a flywheel.

Energy storing in terms of potential energy is an easy approach that can be converted into kinetic energy to rotate the turbine and generate electrical energy as required. For big applications, pumped water storage can be used.

2.17.2 Chemical process for energy storing

Chemical process for energy storing requires chemicals and involves chemical reactions. Commonly used techniques are using

- battery storage and
- hydrogen storage.

The energy-storing chemical process needs regular charging and replacement or maintenance of chemicals used for the purpose after regular time intervals. Chemical reaction of reversible manner is often used for energy storing.

2.17.3 Thermal process for energy storing

In a thermal process for energy storing, liquids or gases are used to be heated up and store the energy. Stored heat energy may be used directly or through the heat exchanger. By the flow of fluids, thermal energy is converted to mechanical energy to rotate turbines and generate electric power when it is required. For storing, water, ammonia, etc. are used.

2.17.4 Electrostatic process for energy storing

In this process, energy is stored in the form of static electrical energy. The field and the potential gradient produced by this are utilized to generate current and electric power to feed loads. Super-capacitor is used to store a large amount of energy. However, large-scale storage is not cost-effective.

2.17.5 Electromagnetic process for energy storing

In this type, energy is stored in the form of electromagnetic energy. Large-sized reactors may be designed for this purpose. The main challenge is the short duration of energy storage. Cost is also high in this way of storing energy.

2.17.6 Difference between fuel cell and battery

Battery and fuel cells are both used for energy resources in nanogrid and picogrid applications. The main difference between them is in the batteries that aim to store electric energy. Regular charging is required and therefore additional charging circuit is required. A fuel cell converts chemical energy directly to electrical energy and it needs a regular supply of fuel. Therefore, an additional fuel supply is required.

2.18 Summary

This chapter has covered different aspects of energy resources used for electric power generation. Features, advantages, and limitations of conventional, and non-conventional types of energy resources have been discussed. The discussion has revealed that solar PV, battery, and fuel cell-based energy resources can be effectively utilized for nanogrid and picogrid applications. Different features of these resources have been presented in detail. Characteristics of PV cells have been presented. Solar PV module, panel, string, and array have been introduced. Fuel cells and different energy storage systems have been discussed conserving their importance concerning the grid applications. Though there are many types of energy resources, and energy storage systems, solar PV-based alternate sources of energy and battery systems are found suitable for nanogrids and picogrids applications.

Further reading

[1] R. A. Dunlap, *Renewable Energy: Volume 1: Requirements and Sources*, Morgan & Claypool, 2020.

[2] M. Vilathgamuwa, D. Nayanasiri, and S. Gamini, *Power Electronics for Photovoltaic Power Systems*, Morgan & Claypool, 2015.

[3] S. Beeby and N. White, *Energy Harvesting for Autonomous Systems*, Artech, 2010.

[4] A. Phil Weicker, *Systems Approach to Lithium-Ion Battery Management*, Artech, 2013.

[5] Y. Chen, "Energy Sources," in *Energy Harvesting Communications: Principles and Theories*, Wiley, 2019, pp. 5–17, DOI: 10.1002/9781119383062.ch2.

[6] M. G. Simes and F. A. Farret, "Modeling Alternative Sources of Energy," in *Modeling Power Electronics and Interfacing Energy Conversion Systems*, IEEE, Wiley-IEEE Press, 2017, pp. 203–226, DOI: 10.1002/9781119058458.ch9.

[7] B. K. Bose, "Power Semiconductor Devices for Smart Grid and Renewable Energy Systems," in *Power Electronics in Renewable Energy Systems and Smart Grid: Technology and Applications*, IEEE, Wiley-IEEE Press, 2019, pp. 85–152, DOI: 10.1002/9781119515661.ch2.

[8] B. K. Bose, "Energy, Environment, Power Electronics, Renewable Energy Systems, and Smart Grid," in *Power Electronics in Renewable Energy Systems and Smart Grid: Technology and Applications*, IEEE, 2019, pp. 1–83, DOI: 10.1002/9781119515661.ch1.

[9] Q-C. Zhong and T. Hornik, "Current Proportional - Integral Control," in *Control of Power Inverters in Renewable Energy and Smart Grid Integration*, IEEE, 2012, pp. 250–258, DOI: 10.1002/9781118481806.ch15.

[10] S. Rao, S. Katoch, V. Narayanaswamy, G. Muniraju, C. Tepedelenlioglu, A. Spanias, and P. Turaga, *Machine Learning for Solar Array Monitoring, Optimization, and Control*, Morgan & Claypool, 2020.

[11] H. Braun and M. Banavar, *Andreas Spanias, Signal Processing for Solar Array Monitoring, Fault Detection, and Optimization*, Morgan & Claypool, 2012.

[12] P. R. Wolfe, "Economics of Solar Generation," in *The Solar Generation: Childhood and Adolescence of Terrestrial Photovoltaics*, IEEE, Wiley-IEEE Press, 2018, pp. 95–104, DOI: 10.1002/9781119425618.ch6.

Chapter 3
Modeling of nanogrids and picogrids

This chapter starts with a description of gradual changes in the grid structure. Then nanogrid (NG) and picogrid (PG) are focused. The general structure, components of nano and pico grid (NPG) systems have been presented. Different types of grid integration with NG and PG have been discussed. Different aspects of NPG components have been discussed and their models have been provided. Converters have been classified and various types of converters have been described with the configuration and properties. NPG has been classified and different grid types have been discussed. Useful topologies of different grids have been presented. In addition to that, general load flow for the common AC and DC networks has been provided.

3.1 Introduction

The journey of electric power started with conventional energy sources. For long, wide progress of the different types of conventional energy resources and power system networking have been observed. The gird concept was introduced for integrating conventional sources for setting up large power banks. It has increased the reliability and sustainability of the power system. With the introduction of different non-conventional energy resources, the grid concept has been upgraded to the smart grid. Recent advancement keeps all sources i.e., conventional and non-conventional sources in parallel. The existence of a conventional grid is still present. Along with conventional grids, other types of grids are coming to play an important role in the power industry. The concept of microgrids (MGs) has been accepted rapidly and widely. In todays' world, MG has confirmed its position along with conventional grids in the power system network. With the development of new resources, innovative converters, metering systems, and change of demand nature, nanogrids (NGs) and picogrids (PGs) have started their journey and are growing very fast. It is expected that combining all, the aim of the super grid can be reached very soon. The progress of grid technology with time is shown in Figure 3.1.

3.2 Nanogrid

NG refers to a reliable and small grid model that must have at least one load (sink of power, which could be storage) and at least one gateway to the outside.

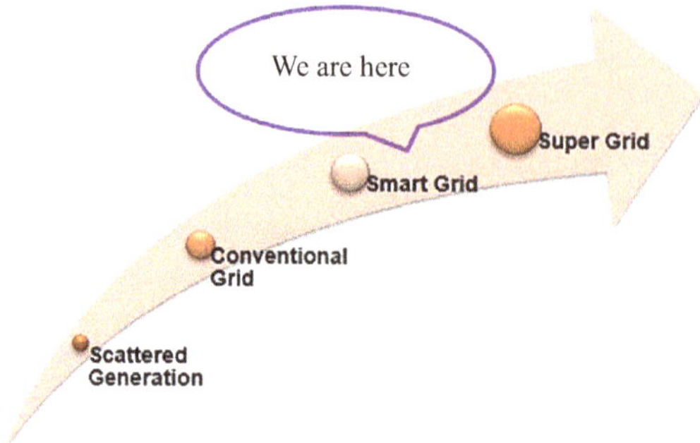

Figure 3.1 Progress of grid technology

Sometimes the term is also used for localized generation and distribution of loads less than 20–25 kW; however, no such rigid demarcation has been found in this regard. NG can provide both direct current (DC) and alternate current (AC) as an output and can be integrated with a cost-effective battery model and external grid utilities.

The main properties of NGs may be listed as follows:

- It includes battery and solar photovoltaic (PV) or renewable energy resources.
- It offers localized power distribution. It covers load(s) distributed over a small area.
- It has at least single control or administration unit.
- It has at least one gateway to be connected with other grids like one or more other NG, MG, and/or conventional grids. However, it can be operated in isolation.
- It has a small capacity, say up to 20–25 kW. However, no rigid and common demarcation is accepted by all.

NG definition: NG may be defined as a grid that is capable of delivering power to a small area and load having at least one gateway to be connected with other grids and may have at least one control unit and the ability to run in isolation.

3.3 PGs

PG is a very small grid model having a capacity of less than NG. Concerning NG and MG, it has a small capacity but is very effective and smart in small area operation. The main properties of PGs may be listed as follows:

- It includes batteries and mainly solar PV as local energy resources.
- It covers load(s) distributed over a very small area that may be treated as lumped. There may be a single load unit to take power from PG.
- It has single control or administration unit.

- It has at least one gateway to be connected with other grids like one or more other PGs, NGs, and/or MGs. However, it can be operated in isolation.
- It must have the capacity for mobility. It can be moved to other places if required.
- It has a very small capacity, say up to 5 kW. However, no rigid and common demarcation is accepted by all.

PG definition: PG may be defined as a grid that is capable of delivering power to a very small area and load having single-control unit, at least one gateway to be connected with other grids, and has the ability of easy mobility or portability (that may or may not be required in a particular application) from one place to another place and run in isolation.

3.4 Comparison among NG, PG, and MG

The MG, NG, and PG have many common as well as different features. A comparison among them is presented in Table 3.1.

Nanogrids are small energy grids, powered by various generators often including photovoltaics. For example, a nanogrid might supply a village in a rural area and allow that village to trade its surplus energy. A picogrid is still a smaller energy grid. IRENA defines NG as systems handling up to 5 kW of power while

Table 3.1 Important difference between NG, PG, and MG

Property	NG	PG	MG
Definition	A grid that is capable of delivering power to a small area and load having at least one gateway to be connected with other grids and may have at least one control unit and the ability to and run in isolation. It may be considered as one type of small MG with centralized management and control	A grid that is capable of delivering power to a very small area and load having single control unit, at least one gateway to be connected with other grids, and has the ability of mobility or portability from one place to another place and runs in isolation	Small grid structure having all possible small capacity resources with centralized or decentralized control and management
Local resource	Battery, solar PV cells, and renewable energy resources	Battery and solar PV cells	Battery, all possible renewable energy may be considered
Distribution	It offers localized power distribution. It covers load(s) distributed over a small area	It covers load(s) distributed over a very small area that may be treated as lumped. There may be a single load unit to take power from PG	It offers a wide area as well as localized power distribution

(Continues)

Table 3.1 (*Continued*)

Property	NG	PG	MG
Control unit	It has at least single-control or administration unit	It has single-control or administration unit	It may have multiple-control units
Gateway	It has at least one gateway to be connected with other grids like one or more other NG, MG, and/or conventional grids	It has at least one gateway to be connected with other grids like one or more other PGs, NG, and/or MG	It may have one or multiple gateways at different locations to be connected with other grids
Operation in isolation	It can be operated in isolation	It can be operated in isolation	It can be operated in isolation
Capacity	It has a small capacity, up to 20–25 kW. However, no rigid and common demarcation is accepted by all	It has a very small capacity, up to 5 kW. However, no rigid and common demarcation is accepted by all	It has a higher capacity, up to 250/500 kW. However, no rigid and common demarcation is accepted by all
Mobility or portability	It does not have the capacity for mobility or portability. It cannot be moved from one to another place if required	It must have the capacity for mobility. It can be moved to other places if required	It does not have the capacity for mobility or portability. It cannot be moved from one to another place if required
EV charging	Normally, 2,3,4-wheeler, or above ranging from the small segment to large segment	Normally, 2,3,4-wheeler ranging from the small segment to the large segment	Any EV segment can be charged

PG handles up to 1 kW. Other practices follow NG up to 20/25 kW and PG up to 1/5 kW. However, it should be noted that kW demarcation is not rigid or fixed criteria to deal with NG and PG as compared to their other features.

3.5 Main parts for NG and PG model

Both nano and pico grids (NPGs) are found to use conventional technologies as well as use modern and advanced approaches to make them smart. The simplified structure of typical NGs or PG is shown in Figure 3.2. Major parts of these grids can be listed as follows:

- resources,
- converter,
- load,
- gateway,
- control and load management (CLM) unit,
- metering and supervision unit, and
- protection unit.

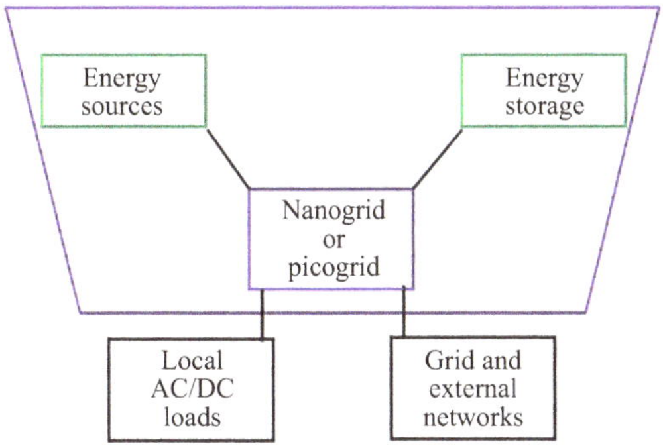

Figure 3.2 General common structure of NG or PG

Modeling of NPG involves modeling all member units and their effective integration. All parts of NPGs have been discussed in the following sections.

3.6 Resources

For NGs and PGs, various combinations of resources are being observed and are possible. Broadly we can divide two types of resource combinations:

(a) NG and PG integrated with two or multiple conventional and non-conventional energy resources.
(b) NG and PG integrated with two or multiple numbers of only non-conventional energy resources.

Solar PV and battery storage act as DC-type local energy resources. However, small AC resources are also used that may be single-phase or three-phase and synchronous or asynchronous. Among different energy resources, wind turbine (AC-type) and solar PV (DC-type)-based energy resources are commonly used. Both are non-polluting to the environment. All over the world, solar energy is given utmost importance as an important resource unit for NPG systems. In some countries, government and regulatory bodies are providing additional support for using solar energy resources. It is observed that each house has a rooftop area, at least that can be utilized for collecting solar energy using PV cells. An increase of solar energy is through the formation of NPG that will certainly reduce the pressure of increasing power demand to the conventional grid and it is hoped that in near future, non-industrial loads may be taken care of by NPG and MG at large.

3.7 Power electronics components

Power electronics devices form basic components of converters and controllers of NPG systems. Different types of power electronics components commonly used for switching purposes in NPG systems have been presented in the following subsections.

3.7.1 Power diode

The power diode is made of P-type and N-type semiconductor layers. Basic features of power diodes may be listed as follows:

- It allows unidirectional flow of power.
- It can allow the flow of high current.
- It can withstand a high reverse voltage.
- In special type diodes, a light-activated operation is possible.
- In the Schottky diode, a faster operation is possible.
- It finds wide application in converters, freewheels, switching circuits, etc.

3.7.2 Schottky diode

Schottky diode is a special type diode with the following features concerning normal PN junction diode:

- It needs less amount of recovered charges.
- It does not depend on the reverse rate of change of current (di/dt).
- Rectification by Schottky diode depends mainly on majority carries and self-junction-capacitance decides recovery effect.
- It offers a low forward voltage drop.
- It has a large leakage current.

3.7.3 Silicon-controlled rectifier

Silicon-controlled rectifiers are widely used in converters. The trade name of silicon-controlled rectifier is thyristor; the symbol of thyristor is shown in Figure 3.3.

- It has four layers and three terminals (anode, cathode, and gate).
- It can operate in ON and OFF modes.

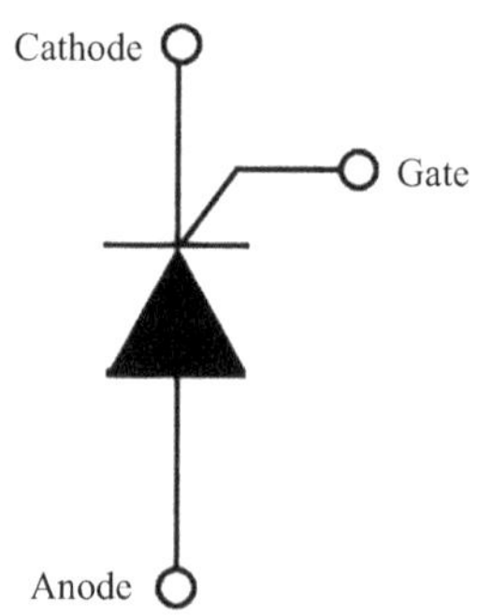

Figure 3.3 Symbol of thyristor

- The power flow is unidirectional.
- Control on ON and OFF is possible.
- It can also be modeled by two transistors.
- Switching operation is possible in the following ways:
    - Thermal—by controlling temperature.
    - Light—by allowing light to fall on junctions.
    - Voltage—by controlling voltage.
    - *dv*/*dt*—by controlling the rate of change of voltage.
    - Gate current—by controlling gate current.
    - Natural commutation: turn off by natural commutation at line frequency (phase-controlled thyristor).

3.7.4 Gate turn off thyristor (GTO)

Turn on and turn off are done by a positive gate signal and negative gate signal, respectively. Thus, it avoids the complexity of forced commutation. It offers faster turn off, lighter weight, high anode current to gate current ratio i.e., ON state gain. However, the OFF state gain is low. The pulsed signal can be applied to the gate for switching operation. It finds the application in the high-frequency switching operation.

3.7.5 Fast switching thyristor (FST)

These types of thyristor offer high-speed switching. They are very suitable in converter applications special in inverters and choppers. They are also known as inverter thyristor.

3.7.6 Reverse conducting thyristor (RCT)

RCT is made of a combination of a thyristor and a diode. SCR is connected with a diode in an anti-parallel direction as shown in Figure 3.4. It allows the reverse flow of current during thyristor turn-off. It reduces steady-state reverse blocking voltage (1–2 V).

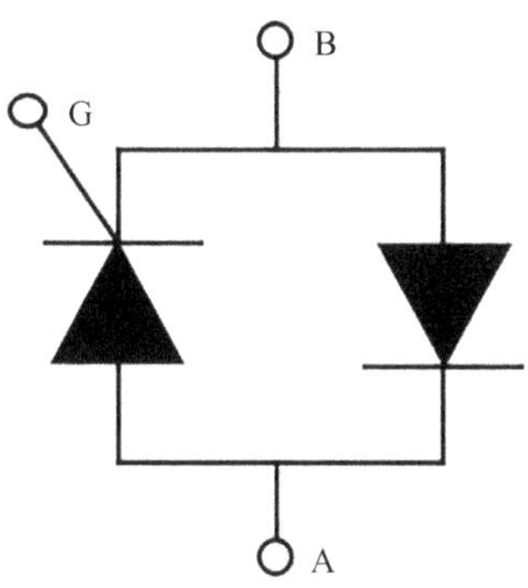

Figure 3.4 Reverse conducting thyristor

3.7.7 Static induction thyristor

Operation static induction thyristor is mainly influenced by the flow of minority carriers. It offers low ON-state resistance. This reduces the voltage drops. Turn on and turn off are done by the positive and the negative gate voltage, respectively. It offers fast switching and can withstand a high rate of change of current and voltage (di/dt or dv/dt).

3.7.8 Light-activated silicon-controlled rectifier (LASCRs)

Triggering is done by lights. It finds applications in high-voltage applications like in high-voltage DC (HVDC) converters, VAR control, etc. Also, this can be utilized in automation and home appliances.

3.7.9 FET-controlled thyristor

In a FET-controlled thyristor, turn-off cannot be controlled by a gate signal. It is similar to thyristor in parallel with metal oxide semiconductor field effect transistor (MOSFET). It finds applications where isolation is required.

3.7.10 Triac

Triac (symbol shown in Figure 3.5) is used for controlling the directional flow of current. Typical output for sinusoidal input is shown in Figure 3.6. It has the following main features:

- It has three terminals. It may be considered as bidirectional triode thyristor (BTD).
- Power flow is bidirectional.
- It can be compared with two anti-parallel thyristors.
- It is suitable for low-power applications like control of the motor fan, light, etc.
- Switching operations in the triac are often unsymmetrical. This switching operation can cause unwanted electromagnetic interference with other circuits.
- Not suitable for very high-power applications.

3.7.11 Bipolar junction transistor (BJT)

- Its operation is governed by the flow of holes and electrons through P-type and N-type semiconductor blocks.

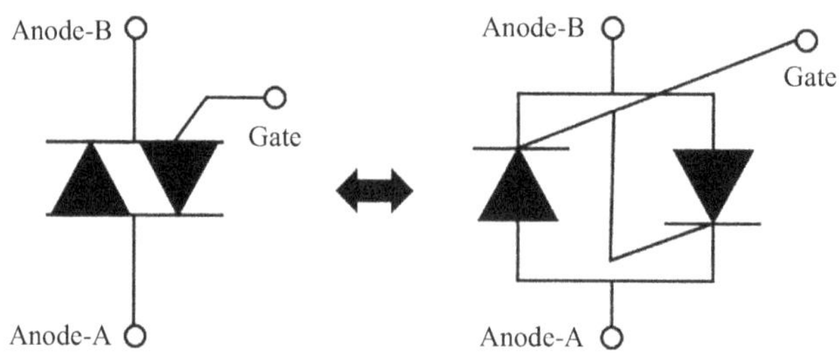

Figure 3.5 Symbol of triac

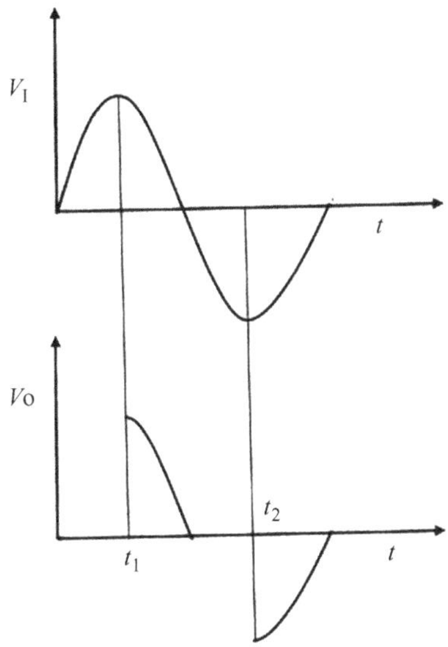

Figure 3.6 Typical output of triac for sinusoidal input with resistive load

- It has three terminals: base, emitter and collector, the current relation is

$$I_E = I_B + I_C \tag{3.1}$$

- It can operate in different modes like common emitter, common base, or common collector.
- It is characterized by two factors as follows:
  Transport factor—a ratio of the collector current and the emitter current:

$$\alpha = I_C I_E \tag{3.2}$$

  Current gain—a ratio of the collector current to the base current:

$$\beta = \frac{I_C}{I_B} = \frac{\alpha}{1 - \alpha} \tag{3.3}$$

- The switching speed is low.
- It can be operated with a high current and high voltage (HV).

3.7.12 Insulated gate bipolar transistor (IGBT)

IGBT is mainly used in fast switching applications. It has the following main features:

- It has three terminals.
- Very useful as an electronic switch.
- It offers fast switching.

- It offers high efficiency.
- Suitable for applications in fast switching operations like lamp ballasts, electric cars, and variable frequency drives *VFDs*.
- It is suitable for high power and HV application (e.g. in 6–10 kV, etc.).
- It offers the features of both MOSFET and BJT.

3.7.13 MOSFET

MOSFET has the following main features:

- It has three outer terminals: source, drain, and gate (Figure 3.7).
- It delivers a high current.
- It offers a high switching speed.
- It is comparatively expensive.

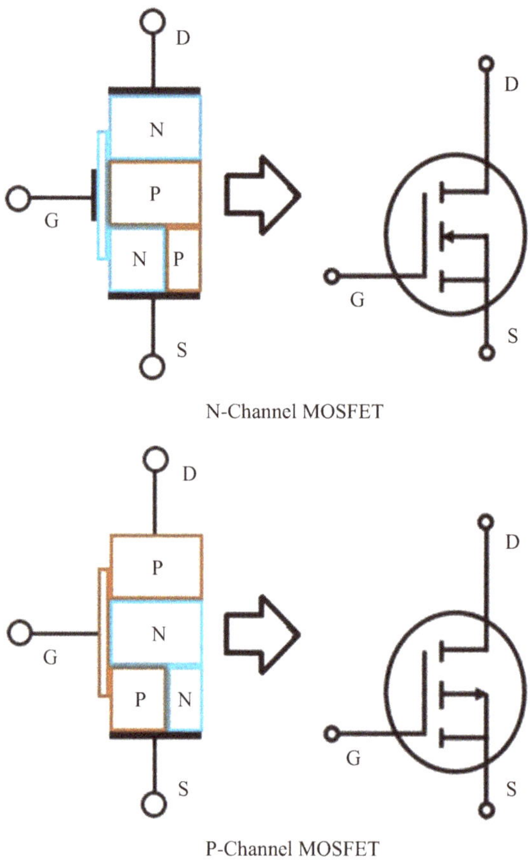

Figure 3.7 Layout and symbols of N-channel MOSFET and P-channel MOSFET

3.8 Converter

Converter plays an important role in the design and operation of NPG. The inclusion of resources and grids is governed by converter configuration. Coordination must be there among resource units for selecting types, design parameters, and specifications of converters. Mainly following types of converters are used in NPG applications:

- Inverter.
- Rectifier.
- Transformer.
- Chopper.
- Resonant converter.

3.8.1 Inverter

An inverter is a power electronic device that converts DC to AC at desired frequency and voltage. Inverter outcome is judged by the following major parameters:

- output voltage and current waveforms;
- RMS of output voltage and current signals;
- harmonic distortion of output voltage and current waveforms.

3.8.1.1 Classification of inverters

Many types of inverters are made based on the requirements. Inverters are classified based on different aspects as presented in Table 3.2.

3.8.1.2 VSI

VSIs refer to the inverters that are fed by DC sources having a constant voltage magnitude (Figure 3.8(a)). In ideal modeling internal impedance of the input source is assumed zero (corresponding voltage is often known as stiff voltage), however, in practical cases, input sources have very small impedance properties. The output voltage is controlled and fully dependent on controls of the switching elements used in the inverter unit.

3.8.1.3 CSI

CSIs refer to the inverters that are fed by DC sources having constant current magnitude (Figure 3.8(b)). In ideal modeling internal impedance of the input source is assumed "infinity" (the corresponding current is often known as stiff current), however, in most practical cases, input sources have very high impedance properties. The output current is fully dependent on source capacity and can be controlled by the switching operation in the inverter unit.

The difference between the VSI and the current inverter is presented in Table 3.3.

Table 3.2 Classification of inverters

S. no.	Classification based on	Categories
1	Input source	• Voltage source inverter (VSI) • Current source inverter (CSI)
2	Number of output phase	• Single-phase inverter • Three-phase inverter
3	Commutation method	• Line commuted inverter • Force commuted inverter
4	Connection of thyristor and commutating elements	• Series inverter • Parallel inverter • Bridge inverter
5	Output waveform	• Square wave inverter • Quasi/semi sine wave inverter • Pure sine wave inverter
6	Operational mode	• Standalone inverter • Grid inverter
7	Grid connection	• Central inverter • String inverter • Micro-inverter
8	Number of intermediate voltage levels	• Two levels inverter • Multiple levels inverter
9	PWM features	• Single PWM inverter • Multiple PWM inverter • Sinusoidal PWM inverter • Modified sinusoidal inverter

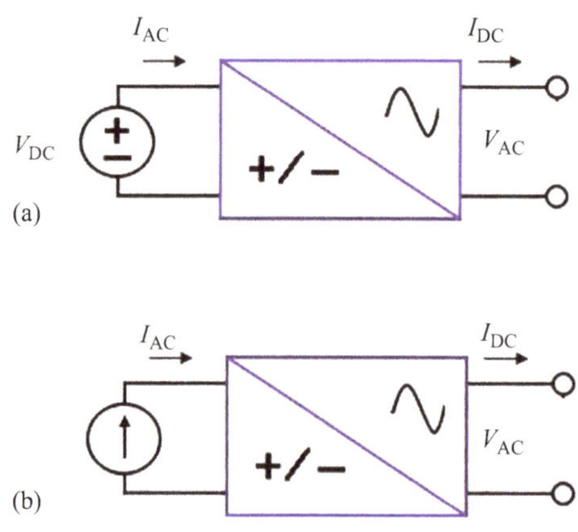

Figure 3.8 (a) VSI and (b) CSI

Table 3.3 Difference between VSI and CSI

VSI	CSI
Input is a voltage source	Input is a current source
The impedance of the input source is ideally zero	The impedance of the input source is ideally infinity
The impedance of the input source is small in reality	The impedance of the input source is very high in reality
Output voltage does not depend on the load	Output current does not depend on the load

3.8.1.4 Single-phase inverter

A single-phase inverter offers single-phase output. This type of inverters may be made in two ways as follows:

- Half-bridge inverter.
- Full-bridge inverter.

The conduction state comes after 180° intervals. The frequency is designed as either 50 Hz or 60 Hz based on the other systems. If output voltage levels do not match with gateway bus for grid connection or load bus, transformers are used with suitable turns ratio.

3.8.1.5 Three-phase inverter

A three-phase inverter offers three-phase output. This type of inverter may be logically thought of as a combination of three single-phase inverters.

Considering three numbers' phases, total time (T) can be equally divided into six parts i.e., $T/6$. The conduction state of one phase comes after 120 modes of intervals of the next phase and so on. The frequency is designed as either 50 Hz or 60 Hz based on the other systems. If output voltage levels do not match with gateway bus for grid connection or load bus, transformers are used with suitable turns ratio.

3.8.1.6 Line commuted inverter

The switching thyristor in the inverter unit needs to be commuted. An inline commuted inverter is done by the negative part of the alternating voltage. Current goes to natural zero. Line voltage is accessed for this purpose. Line commutation does not create any additional loss of power.

3.8.1.7 Forced commuted inverter

In the line commuted inverter, commutation is done by applying an external source. The current does not go to natural zero. Forced commuted inverters use forced commutation that creates an additional loss of power.

3.8.1.8 Series inverter

In a series inverter, for each unit, two semiconductor-based switches (normally, thyristor) are used. One switch (say, T_1) is connected in series with resistance–

inductance–capacitance (RLC) combination, the other (say, T_2) is connected in parallel to them as shown in Figure 3.9. On a time scale, it operates in two steps of operation by making conduction through alternate switches:

- Step 1: T_1 is switched on, T_2 is switched off.
- Step 2: T_1 is switched off, T_2 is switched on.

If both switches are made ON at the same time, it will make the source short and damage it. Therefore, in series inverter, precaution should be taken so that both switches are not ON at the same time. This type of inverter can be effectively used where high current (like induction heating) and high frequency may be required.

3.8.1.9 Parallel inverter

In a parallel inverter, two switches are connected with two terminals of a center-tapped transformer as shown in Figure 3.10. One capacitor is connected across the terminals of the primary side of the transformer. The source is connected through an inductor in between switches and the center tap of the transformer. This inductor helps in limiting the current. The inverter operates in two steps as follows:

- Step 1: T_1 is switched on, T_2 is switched off.
- Step 2: T_1 is switched off, T_2 is switched on.

Parallel inverters offer simple commutation and a stable voltage across the load.

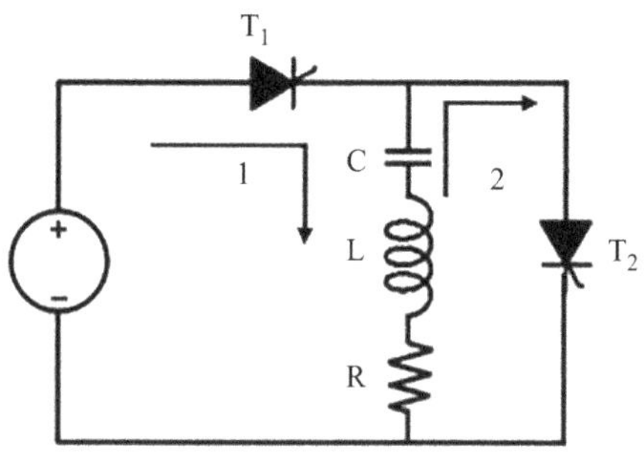

Figure 3.9 Series inverter

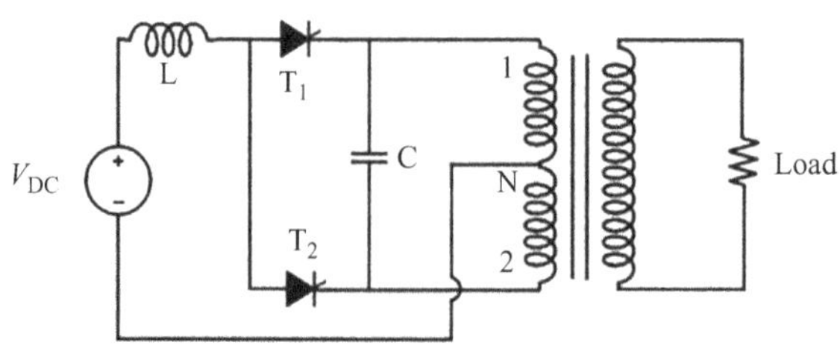

Figure 3.10 Parallel inverter

3.8.1.10 Bridge inverter

Bridge inverter may be single-phase or three-phase and half bridge or full bridge. A half-bridge single-phase connection is shown in Figure 3.11. Two switches are used in two loops through a common load. The source voltage is divided into two equal parts. Half of the supply voltage feeds power to each loop. Two switches are made ON in an alternate way making an alternate direction of flow of current through the loads. The inverter operates in two steps as follows:

- Step 1: T_1 is switched on, T_2 is switched off.
- Step 2: T_1 is switched off, T_2 is switched on.

Full-bridge connections for single-phase and three-phase are shown in Figures 3.12 and 3.13, respectively. Here the whole supply voltage has been applied across the bridge and no division of voltage is required like half-bridge connection. Switches are made ON in an alternate way for inversion of voltage across the load. As a switch, any semiconductor-based switch can be used like MOSFET, thyristor, BJT, etc. Diodes are connected for commutation and circulation of free energy.

3.8.1.11 Standalone inverter

In the standalone operation of an inverter, separate storage is present. Energy resources are utilized to charge the storage unit. The storage unit feeds power to the inverter. Inverter delivers power to the load at a desired frequency and voltage. A

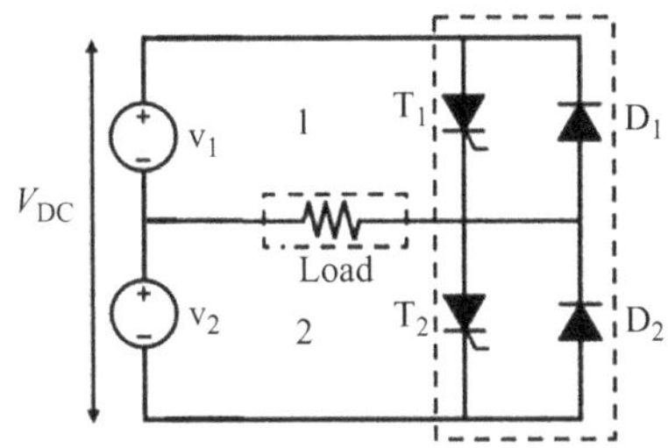

Figure 3.11 Half-bridge inverter

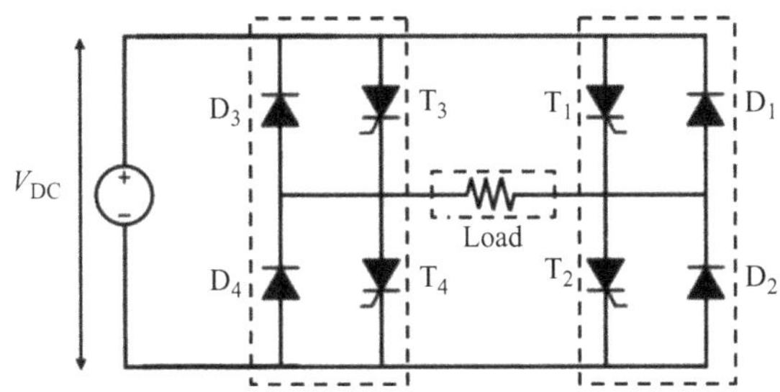

Figure 3.12 Full-bridge single-phase inverter

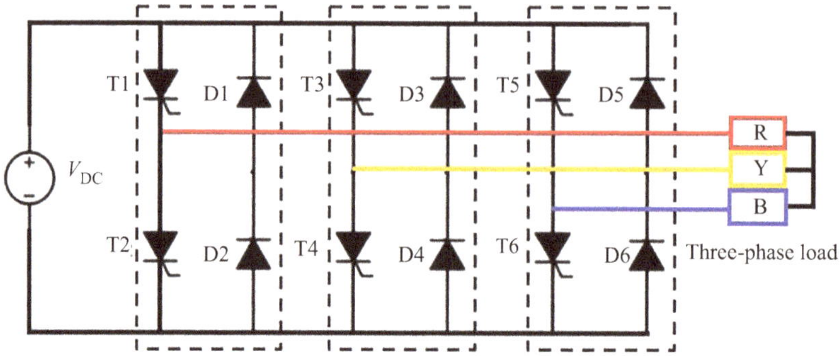

Figure 3.13 Full-bridge three-phase inverter

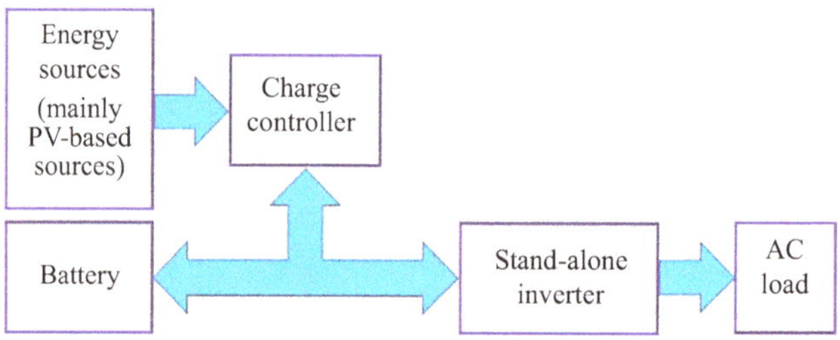

Figure 3.14 Schematic diagram for a standalone inverter

schematic diagram of a standalone inverter is shown in Figure 3.14. It can feed power when the source is not available. For this reason, this type of inverter is also known as a stand-alone inverter. The inverter is not connected with the grid and, therefore, it is also known as an off-grid inverter. As stand-alone inverters are not connected to the utility, grid inverters do not require to have synchronizing property. This type of inverter is very simple in configuration and widely used in small applications.

3.8.1.12 Grid inverter

In the grid inverter, energy resources of the grid are utilized to feed power to the inverter. Inverter delivers power to the grid as well as to the load at a desired frequency and voltage. The voltage level is adjusted by a transformer. Sometimes, additional and separate storage is present. Energy responses are utilized to charge the storage unit that delivers power to the input of the inverters if there is an insufficient state of energy resource. A schematic diagram of the grid inverter is shown in Figure 3.15.

3.8.1.13 Central inverter

A schematic diagram of the central inverter is shown in Figure 3.16. Here one big inverter is used and placed in a central position logically. All energy resources are

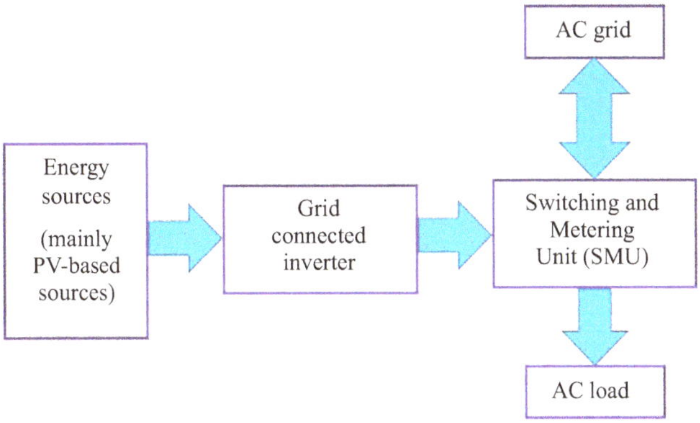

Figure 3.15 Schematic diagram for grid inverter

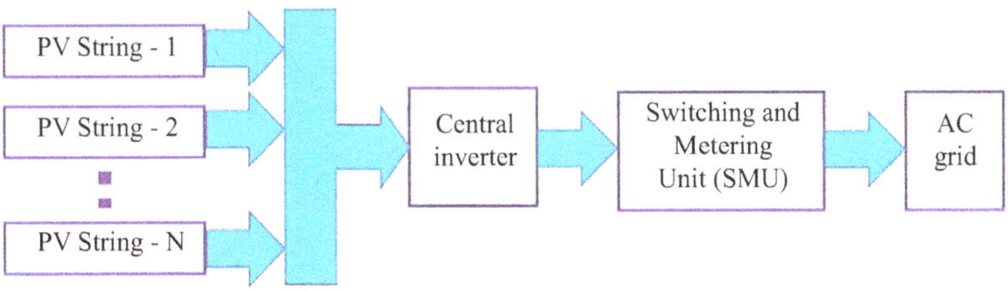

Figure 3.16 Schematic diagram for central inverter

utilized to feed power to the central inverter. The inverter is connected with loads and/or a grid through a transformer. The use of a central inverter makes the local grid configuration simple and compact.

3.8.1.14 String inverter

A schematic diagram of the string inverter is shown in Figure 3.17. Here multiple inverters are used. Each inverter takes power from a separate energy resource unit of the same or different categories. The use of a string of inverters increases the overall grid capacity. The maintenance of the temporary absence of one or few inverters does not hamper much the operation. Thus, the string inverter improves the reliability of the grid system.

3.8.1.15 Microinverter

A schematic diagram of the very small inverters used for MGs, NGs, and PGs is shown in Figure 3.18 (a). Here inverter(s) of small capacity is used. Inverters are attached at the bottom of the PV panels as shown in Figure 3.18(b). It takes power directly from PV panels made of a string of single or few panels and delivers power to the load. Based on the number of panels and capacity, different types of micro,

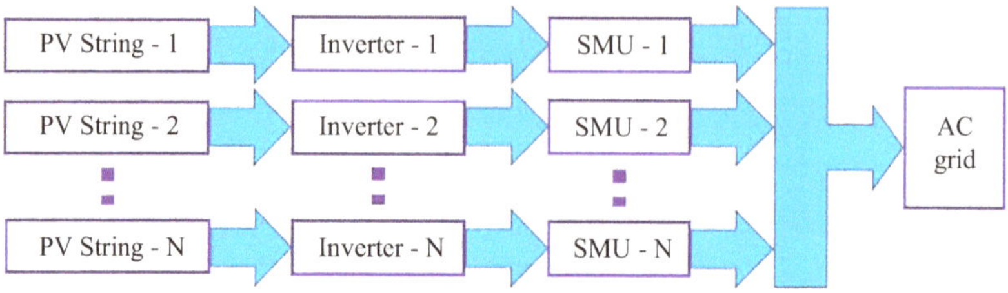

Figure 3.17 Schematic diagram for string inverter

(a)

(b)

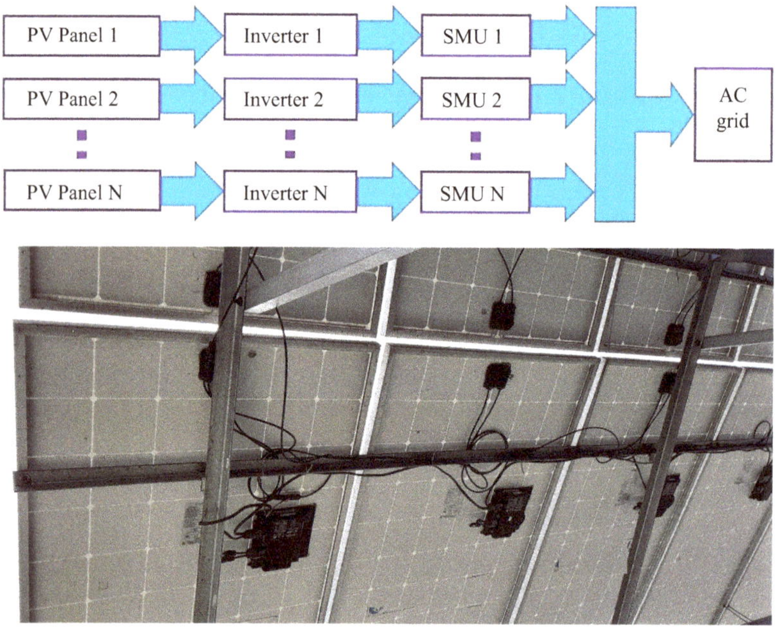

Figure 3.18 Microinverters in AC grid: (a) schematic diagram and (b) connections at the bottom of PV panels

nano, and pico AC grids can be formed. It may be connected with other low-voltage (LV) grids. The major advantages of these inverters are as follows:

- It is very simple to configure.
- Very often microinverters are single phase.
- The cost is comparatively low.
- Affordable for small and home applications.

Specifications of a microinverter include both DC and AC features of the inverter. Specifications are mentioned at standard test conditions (STCs). STC for solar PV panels refers to the condition of irradiance of 1000 W/m^2, cell temperature of 25 °C, and air mass coefficient ($AM = 1.5$).

Air mass coefficient is defined as follows:

$$AM = \frac{L_i}{L}$$

where L_i is the length of the path of incident radiation inclined at a certain angle with earth's surface; L is the length of the path of incident radiation perpendicular to the earth's surface.

For solar PV generation, $AM = 1.5$ is almost universal. An example of microinverter specifications at STC is given below:

DC specifications:

- maximum power—375 Wp
- maximum power voltage—39.00 V
- maximum power current—9.62 A
- short circuit current—9.90 A
- open circuit voltage—44.5 V
- maximum system voltage—1 000 V

AC specifications:

- maximum AC output power (continuous)—290 W
- AC output voltage—185–275 V
- AC output current (max) —1.26 A
- AC output frequency—45–55 Hz
- Operating temperature— −40 °C to +65 °C

3.8.1.16 Bimodal inverter

A bimodal inverter can operate either as stand-alone or as grid-connected mode. Its operation changes based on the object for power feeding. It may be AC load or may be grid. This type of inverter is more flexible. It can run on off-grid mode as well as it can run in synchronism with grids.

3.8.1.17 Square wave inverter

This type of inverter is designed to generate square waveforms as an output. Sample output waveforms of square wave inverters are shown in Figure 3.19.

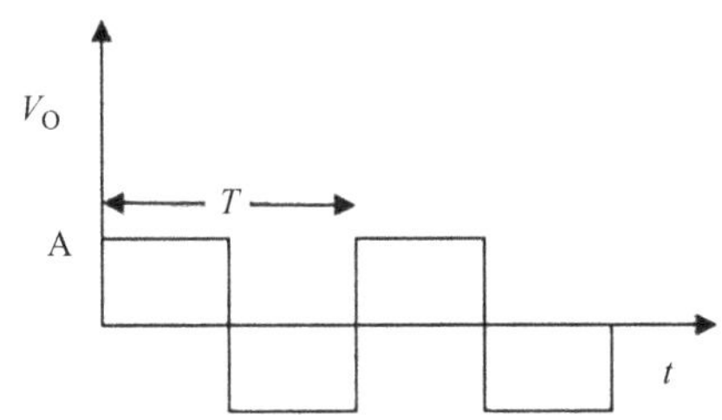

Figure 3.19 Output waveform of square wave inverters

3.8.1.18 Quasi/semi-sine wave inverter

This type of inverter is designed to generate quasi sinusoidal waveforms as an output. Sample output waveforms of quasi sine wave inverters are shown in Figure 3.20.

3.8.1.19 Pure sine wave inverter

This type of inverter is designed to generate pure sinusoidal waveforms as an output. Sample output waveforms of pure sine wave inverters are shown in Figure 3.21.

3.8.1.20 Inverter classification based on number of levels of a staircase output waveform

Based on the number of levels of the staircase output waveform, inverters are divided into two categories:

- Two levels inverter.
- Multilevel inverter.

Two levels inverter

If the staircase output waveforms have one positive step and one negative step, then the inverters are called two-step inverters. Considering zero levels, sometimes these are called three-steps or three-levels inverters.

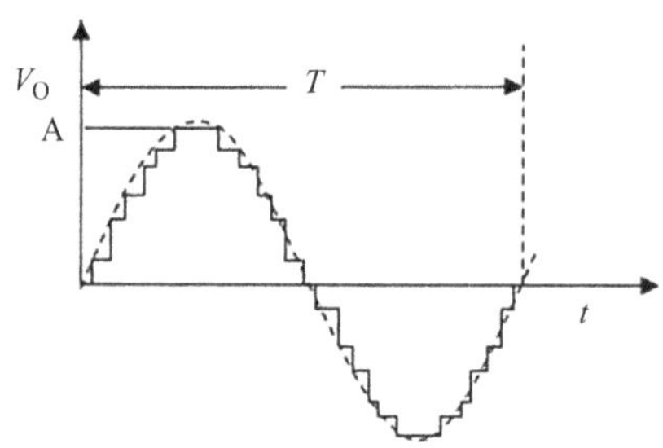

Figure 3.20 Output waveform of quasi sine wave inverters

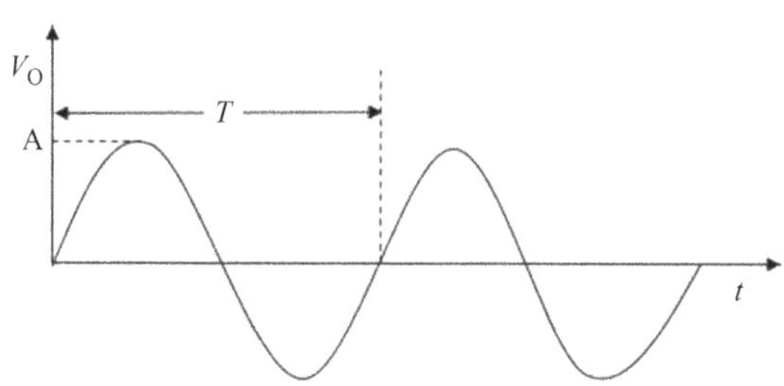

Figure 3.21 Output waveform of pure sine wave inverters

Table 3.4 Differences between two-step inverter and multilevel inverters

Two-step inverters	Multilevel inverters
The number of steps is 2 or 3 (considering zero levels)	The number of steps is 2 or 3 (considering zero levels)
Simple in logic	Complex
The output waveform is far away from its sinusoidal nature	The output waveform is more closed to the sinusoidal nature
Harmonic distortion is high	Harmonic distortion is low
The output voltage is low and normally non-adjustable	The output voltage is high and adjustable
Voltage regulation facility is not normally incorporated	Voltage regulation can be done
Overall efficiency and power quality are poor	Overall efficiency and power quality are better than two-step inverters

Multilevel inverter

If the staircase output waveforms have more than zero-step levels (considering zero-step level), then the inverters are called multilevel inverters. Multilevel inverters give better waveform quality two-step inverters that are more closed to a sine wave and provide comparatively less harmonic distortion.

Differences between two-step inverter and multilevel inverters are presented in Table 3.4.

3.8.1.21 Pulse width modulation (PWM) inverter

PWM type of inverters uses a modulation of pulse width for the generation of output waveform. It controls the ON and the OFF logic of the switching circuits of the inverter. Based on the output waveform, PWM inverters are divided into three categories as follows:

- Single PWM inverter.
- Multiple PWM inverter.
- Sinusoidal PWM inverter.

Modified sinusoidal inverter refers to a modified version of sinusoidal PWM. This type is capable of controlling power delivery and the power factor of the delivered power.

3.8.1.22 Series resonant inverter

Resonant condition is created in the inverter that influences the operation. Elements creating resonance and elements for switching operation are connected in series. It creates a natural "zero current" through switching paths. It helps in delivering a high sinusoidal current at high frequencies (20–100 kHz). Broadly resonant type inverters may be classified into two categories as follows:

- Inverters with unidirectional switches.
- Inverters with bidirectional switches.

This type of inverter finds applications in fluorescent lighting, induction heating, etc.

3.8.1.23 Design considerations for inverter

Design of inverter is the most important job in designing or selecting inverter for grid applications. Before performing this job, we must have adequate information about the resource available, grid specification to be linked with load specifications. Then following features should be considered for the selection or design of a particular inverter unit:

- specification of input resource;
- number of output phases;
- output voltage;
- power and current rating;
- switching element;
- commutation type and commutation circuits;
- allowed distortion level;
- harmonic distortion;
- harmonics distortion reducing methods;
- filters;
- snubber and other protective measures.

3.8.2 Rectifier

Rectifier units are needed for the connection of AC resources, DC loads as well as for connecting other AC grids through gateways. These rectifiers can be of two categories:

- multi-phase or three-phase to DC rectifier;
- single-phase to DC rectifier.

Based on power electronics elements, rectifiers can be made into uncontrolled rectifiers and controlled rectifiers. uncontrolled rectifiers are made of diodes. It may be a half-wave or full-wave rectifier. Single-phase full-wave rectifiers can be made using two diodes with a center-tapped transformer or by four diodes with a transformer without tapping.

3.8.2.1 Half-wave uncontrolled rectifier

Let, the applied input voltage is

$$v(t) = V \sin \omega t \tag{3.4}$$

where $\omega = 2\pi f$, f is the frequency of the input voltage.

Let T is the period. If the load is resistive, the output DC voltage for a half-wave uncontrolled rectifier can be written as follows:

$$V_{DC} = \frac{1}{T}\int_0^{T/2} v(t)dt = \frac{-V}{T}\left(\cos\frac{\omega T}{2} - 1\right) \tag{3.5}$$

$$T = \frac{1}{f} = \frac{2\pi}{w} \tag{3.6}$$

$$V_{DC} = 0.318\ V \tag{3.7}$$

Current can be written as follows:

$$I_{DC} = \frac{V_{DC}}{R} = \frac{0.318\ V}{R} \tag{3.8}$$

RMS voltage can be written as follows:

$$V_{RMS} = \sqrt{\frac{1}{T}\left\{\int_0^{T/2} v(t)^2 dt\right\}} = \frac{V}{2} \tag{3.9}$$

RMS current can be written as follows:

$$I_{RMS} = \frac{V_{RMS}}{R} \tag{3.10}$$

[for half-wave rectifier, the load voltage is zero between $T/2 < t < T$].

Let, the load is inductive, then output DC voltage for half-wave uncontrolled rectifier can be written as follows:

$$V_{DC} = \frac{\omega}{2\pi}\int_0^{\pi+\varphi} v(t)dt = \frac{V}{2\pi}\{1 - \cos(\pi + \varphi)\} \tag{3.11}$$

where,

$$T = \frac{1}{f} = \frac{2\pi}{w}$$

Current can be written as follows:

$$I_{DC} = \frac{V_{DC}}{R} = \frac{V}{2\pi R}\{1 - \cos(\pi + \varphi)\} \tag{3.12}$$

3.8.2.2 Full-wave-uncontrolled rectifier

Let, the applied input voltage is

$$v(t) = V \sin \omega t$$

where $\omega = 2\pi f$, f is the frequency of the input voltage.

Let, T is the period. If the load is resistive, the output DC voltage for a full-wave uncontrolled rectifier can be written as follows:

$$V_{DC} = \frac{2}{T}\int_0^{T/2} v(t)dt = \frac{2V}{\pi} = 0.636\ \mathrm{V} \tag{3.13}$$

The corresponding current can be written as follows:

$$I_{DC} = \frac{V_{DC}}{R} = \frac{0.636\ \text{V}}{R} \tag{3.14}$$

RMS voltage can be written as follows:

$$V_{RMS} = \sqrt{\frac{2}{T}\left\{\int_0^{T/2} v(t)^2 dt\right\}} = \frac{V}{\sqrt{2}} = 0.707\ \text{V} \tag{3.15}$$

Corresponding RMS current can be written as follows:

$$I_{RMS} = \frac{V_{RMS}}{R} = \frac{0.707\ \text{V}}{R} \tag{3.16}$$

3.8.2.3 Controlled rectifiers

Controlled rectifiers offer the following features:

- The output voltage can be controlled by switching operations.
- It may operate in one, two, or four quadrants.
- Three types of controlled rectifiers are available as follows: semi-converter, full converter, and dual converter.
- Semi-converter gives output voltage and current in single polarity. Thus, it operates in one quadrant.
- Full-converter gives output voltage with two polarities (either positive or negative), but current in single polarity. Thus, it can operate in two quadrants.
- Full-converter gives output voltage with two polarities (either positive or negative) and also gives current in two polarities (either positive or negative). Thus, it can operate in four quadrants.

Types, design parameters, and specifications of rectifiers should be chosen to keep coordination with supply, load, and control systems. Rectifier units are also the source of harmonics which are mainly dominated by sub- and fractional harmonics.

3.8.3 Transformer

A transformer is needed after inversion or before rectification to make the interconnection of two sections of different voltage levels. This helps to serve the following purposes:

- For making the connection between AC voltage levels of NPG and AC voltage levels of other NPG, MG and conventional grid which that are going to be integrated.
- For connection of AC load of different voltages.
- For connection of AC resource with particular voltage levels. This is rarely used in comparison with the above purposes.

3.8.4 Chopper

Chopper is used to convert one DC voltage level to another DC voltage level. This may be required for connecting DC resources as well as DC loads. Control mechanism and quality of chopper circuit are important for NPG applications. For PG applications, the chopper unit plays a very important role in load management. A chopper is characterized by a high speed of switching operation. Choppers may be classified into three main categories as follows:

- Boost chopper.
- Buck chopper.
- Buck-boost chopper.

Boost chopper steps up the voltage level while buck chopper steps down the voltage level. The buck-boost chopper can perform both stepping up and stepping down the voltage levels as required.

Waveforms of different types of choppers are shown in Figure 3.22.

Cuk chopper regulator: Cuk chopper regulator has similarities with a buck-boost chopper. Both of them can step up or step down voltage. Cuk chopper regulator gives output voltage with polarity opposite to the polarity of input voltage.

Chopper conversion may be implemented in two broad ways as follows:

- *Single-stage conversion*: it gives small power delivering capacity.
- *Multi-stage conversion*: it improves power delivering capacity.

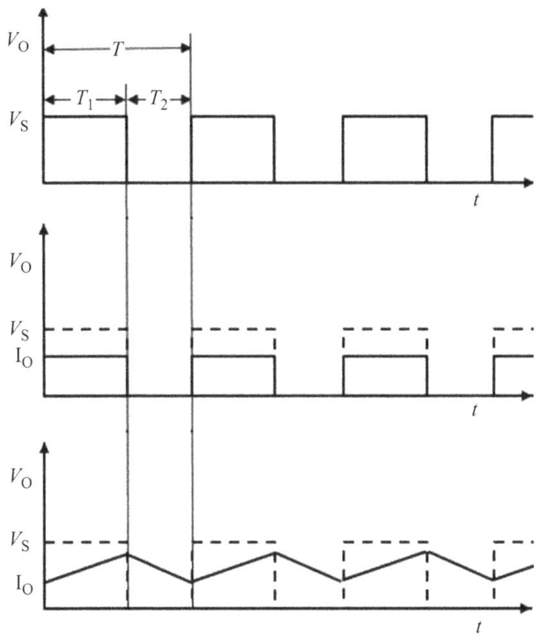

Figure 3.22 Sample voltage waveforms of different types of choppers

Multi-stage conversion can easily be implemented by thyristor-based switching circuits.

In closed-loop control of chopper circuits, the output voltage is monitored. Based on the feedback signal, the controller performs switching operations. Impulse and resonant commutation are often used for switching operations. External input power source and number of stages decide the power delivering capacity of the chopper.

3.8.5 Resonant converters

A resonant converter refers to a converter unit with inductance and capacitance-based resonant circuitry. It helps in providing sinusoidal oscillation of voltage and currents during resonant switching operations. Resonant converters may be classified as follows:

- Resonant AC to DC converter.
- Resonant DC to DC converter.
- Resonant DC to AC converter.

3.8.6 Switched-mode power supply (SMPS)

A SMPS is designed to deliver desired voltage that is required by a specific load unit. Though SMPS having DC output is widely used, outputs of the following types may occur:

- DC output.
- AC output.
- Both DC and AC outputs.

Similarly, the input to SMPS may be DC or AC. Proper conversions are to be made. To obtain desired output waveforms, any one of the following methods is normally used:

- PWM.
- Frequency modulation.
- Matrix conversion.
- Space vector modulation.

Matrix conversion and space vector modulation techniques are found suitable for AC output and are becoming more popular. "Venturi analytical method" is also seen to be used in SMPS control algorithms.

3.9 Load

Load of NPG may be AC load or DC load. For NG, AC loads are normally single phase; however, they may be three-phase. But for PG, if the loads are AC, most of the practical cases are single phase.

3.10 Gateway

Gateways create the path of power flow between a particular NG and PG with other grid(s). The connecting grid may be the same type of grid or MG or conventional grid. Power flow through the gateway may be bidirectional or unidirectional. Smart metering and monitoring at the gateway are essential for smart NG and smart PG which will relate power trading, load management, protection, and reliability of the overall system. In most cases, the gateway exists adjacent (just after/before) to the converter unit. The operation of the converter unit and the control unit is closely related to the power flow through the gateway.

3.11 Control and load management (CLM) unit

Separate CLM units are not mandatory for setting up NG or PG. But the grid must aim to be smart. In coordination with the metering and the supervisory unit, this CLM unit controls power flow through the gateway with the outer grid and manages load distribution.

In conventional smart NPG, CLM units are designed numerically based on microcontrollers, microprocessors, or computer-based circuit networks. But in advanced NPG applications, CLM is compatible with remote control through the Internet of Things (IoT) features. The control unit decides the following main operations:

- commutations;
- control of time and current; and
- control of other power parameters.

3.11.1 Commutation

The control unit decides the methods for commutation in different converter switching circuits. Different types of commutation techniques are used in converter applications. Commonly used approaches are as follows:

- line commutation,
- forced commutation,
- gate-controlled commutation,
- thermal commutation, and
- light-activated commutation.

Forced commutation can be following categories:

- self-commutation,
- impulse commutation,
- external pulse commutation,
- resonant commutation,
- complementary commutation,
- line-side commutation, and
- load-side commutation.

3.11.2 Control of time and current

In the control unit, two control methods are performed for the converter section

- control of time ration and
- control of current.

Control of time ratio is done in two ways as follows:

- by control of ON and OFF time known as PWM;
- by control of frequency (1/time period (T)) known as frequency modulation.

Control of current limit is done by control of ON and OFF switch, freewheels diodes, and by use inductors. It ensures the safety of the devices.

3.11.3 Control of other power parameters

The control unit of a grid system also controls the voltage, power factor, phase angles, and power. Other than converter control circuits, sometimes separate units are used for phase power factor and different types of powers in the paths are connected with other grids and local loads.

3.12 Distribution network

NG and PG have a distribution network covering a small area. This distribution network should be carefully planned and designed. The distribution network should be equipped with metering and monitoring units and be connected to the CLM unit for proper load distribution. Care should be taken so that each section of the distribution network is covered by zones of protection.

3.13 Metering and supervision unit (MSU)

NPG normally has a MSU but it is not mandatory. For smart NPG, smart MSU (SMSU) is essential. In advanced NGP applications, SMSU may be connected with IoT. MSU or SMSU should be well equipped with sensor units and data processing units. Sometimes, metering and supervision are done at the site of the grid from remote ends.

3.14 Protection unit

Like conventional grid, protections are needed for each member unit of the NPG. However, the design and the art of protection depend on the requirement, importance, and budget of the NPG. Source unit, converter unit, load side, and gateways should be protected. Sometimes, the control unit, itself needs to be protected. The protection system is designed based on the possibility of the occurrence of different types of faults. Faults may be internal to the grid or external to the grid. Identification of the type and location of faults is essential.

3.15 Smart NG and PG and their essential additional components

Smart NG and smart PG refer to NG and PG equipped advanced sensor and metering arrangement all feasible features of smart grid offering smart supervisory control load management. Smart nano and pico grids (smart NPG) have all common features of NPG. In addition to that, it must have the following essential components to become smart:

- smart monitoring and control;
- ICT and IoT;
- smart energy meter;
- virtual storage system.

3.16 AC grid

AC grid is normally used where loads are mainly AC in nature. AC grid is shown in Figure 3.23. Converter output is connected with the AC bus. AC bus is connected to load directly or through a transformer. In addition, the converter output AC bus is connected to the gateway bus. Gateway bus is connected directly or through the transformer to outer grids. Sources feeding power to the converter, in most cases, is

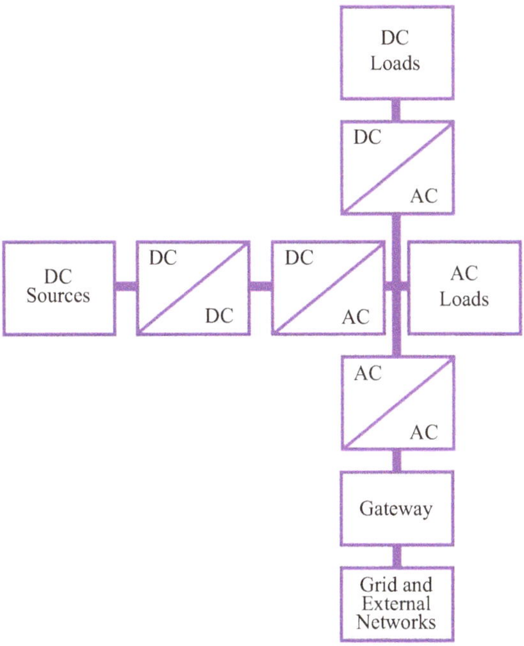

Figure 3.23 AC grid model

DC and then the converter is chosen as an inverter. If any source is AC of different voltage levels, then a transformer with proper transformation ratio will be required to fed power to the AC bus. In few cases, both DC source and AC source can feed power to the NG, then both inverter and transformer may be required. However, for PG applications, the use of AC grid resources is very rare.

3.17 DC grid

DC grid is normally used where loads are mainly DC in nature. DC grid is shown in Figure 3.24. Converter output is connected with a DC bus. DC bus is connected to the load directly or through the chopper unit. In addition, the converter output DC bus is connected to the gateway bus through the inverter unit. Gateway bus is connected directly or through a transformer to outer grids. Sources feeding power to the converter, in most cases, is DC and then power can be fed directly or through the chopper unit. If any source is AC of a different voltage level, then a proper rectifier unit will be required to feed power to the DC bus. In few cases, both DC source and AC source can feed power to the NG, then both chopper unit and rectifier unit may be required. However, for PG applications, the use of AC grid is comparatively less.

3.18 Composite AC–DC grid

Composite AC–DC NPG may be used where both AC load and DC load are present. It may be considered that two NPGs together are making on single NPG and act keeping full coordination to each other.

Resources of the grid are normally DC types; however, AC sources can be connected to feed power to the grid if required. The composite grid is shown in Figure 3.25. This type of composite grid uses both rectifier and inverter units.

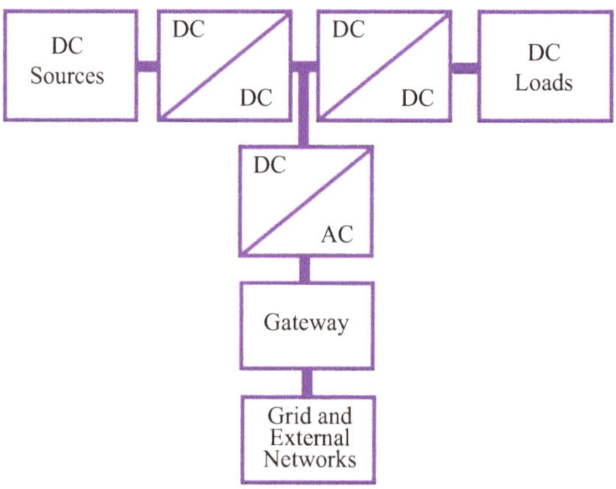

Figure 3.24 DC grid model

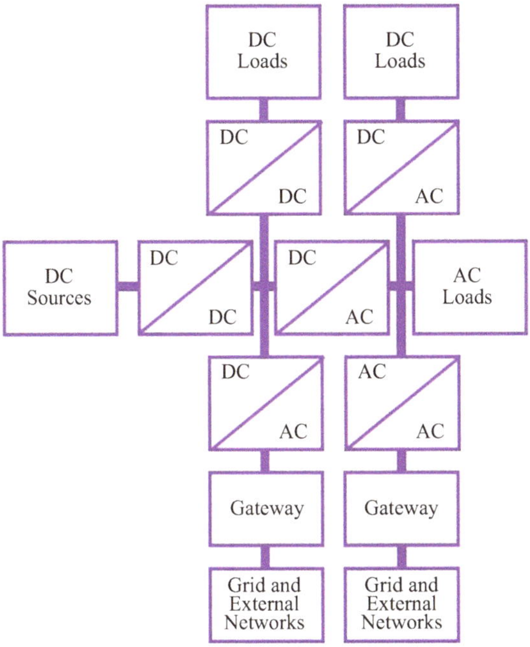

Figure 3.25 Composite AC–DC grid model

Chopper is used to adjust DC voltage level and transformer is used to connect to the outer grid through the gateway.

3.19 Comparison between DC and AC NGs

Comparison between AC and DC NG is not universal. It depends on the resource availability, load characteristics, and application area. A comparison should cover four aspects: efficiency, acceptability, safety, and quality.

Efficiency: The efficiency of DC NG is normally higher than AC NG for a similar type of application. The main reason behind this is that DC–AC conversion produces more conversion losses (around 14–20%) than DC–DC conversion. This makes AC NG comparatively less efficient.

Acceptability: This aspect depends on customers' needs. In the present scenario, most of the loads are compatible with the AC supply. Therefore, if DC NGs are applied, either additional conversions will be required, or load will have to be made compatible with DC supply. This will increase an additional burden for the application of DC NG that is normally absent for AC NG.

Safety: DC NG are more close to different types of ground and other short circuit faults that are comparatively difficult to detect and hence protection becomes comparatively easier in AC NG.

Quality: Providing power of high quality is a big issue. Harmonics-free AC power of perfect sinusoidal nature at a rated frequency (single frequency with no harmonic components) is a bigger challenge of AC NG than providing chopped step response of DC power.

3.20 Grid topology

The topology of grids of different voltage levels is an important matter of concern while interconnections are made. The topology of a particular grid decides its own operation as well as has considerable influence on the operation of other connected grids. Resource and load distribution, dealing with fault situations, and information sharing are planned and executed on topologies of the connected grid. So, the topologies of different types of grids have been briefly presented in the following sections.

3.20.1 General considerations

Grid topology refers to the general structure of sub-members in a grid deciding the flow of operations of all grid components ensuring safe, stable, and reliable grid outcomes. The topology of grids should consider the following aspects in general:

- voltage levels;
- number of incomers;
- nature of resources;
- intermediate substations;
- out-going loads;
- feeders;
- connection of distributed generations (DGs);
- sensor and actuators;
- switchgear and protections;
- telecommunications and IoT applications.

3.20.2 HV and extra-high voltage (EHV) grid topology

HV and EHV grids receive energy from generators. They can also other grids to make a super or more powerful grid. An example of an HV or EHV grid network is shown in Figure 3.26. It receives power from two generating stations and one other grid. This type of grid sends power to long transmission networks and is not connected with the distribution system directly. Multiple grids may be connected to form a grid of higher capacity.

3.20.3 Medium-voltage (MV) grid topology

MV grids involve intermediate power transmission at MV. An example of an MV grid is shown in Figure 3.27. MV grids are normally found in the distribution networks or in between transmission and distribution networks. It normally receives power through transmission networks by substation arrangements and does

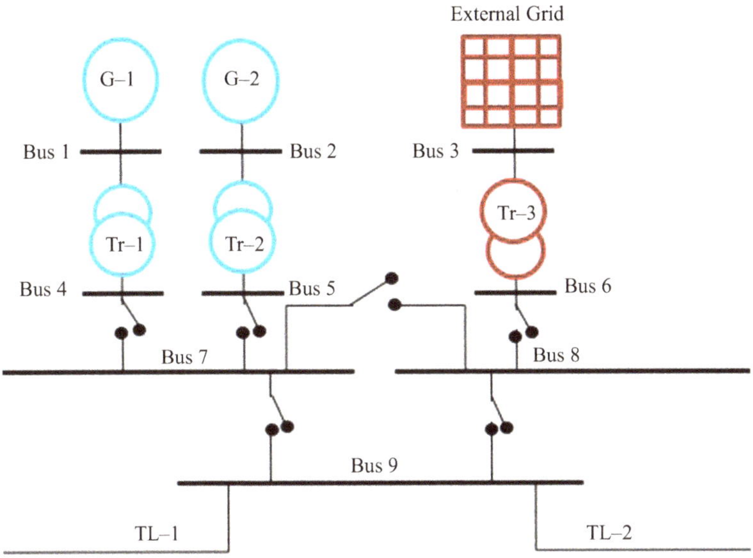

Figure 3.26 Sample HV or EHV grid structure

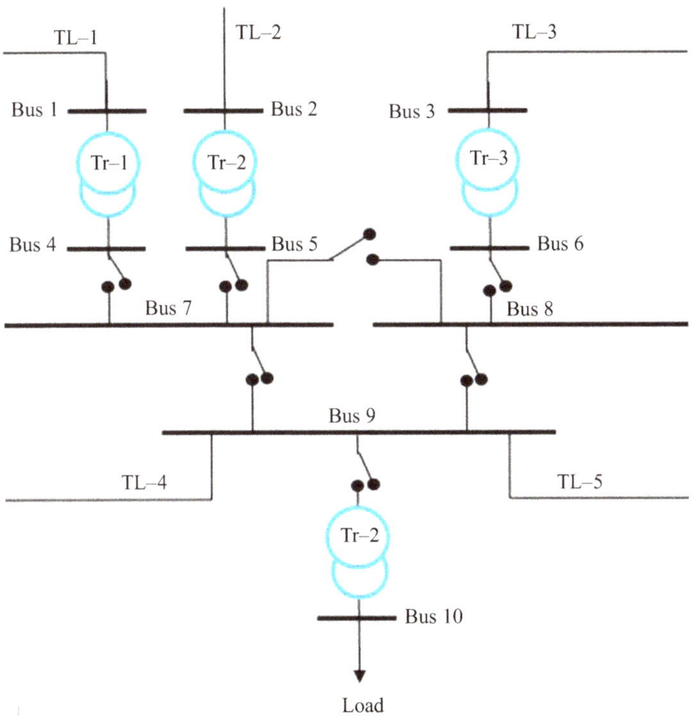

Figure 3.27 Sample MV grid structure

not power directly from generating stations. Normally, this type of grid does not involve is power distribution. However, it can deliver bulk power to the industry.

3.20.4 LV grid topology

LV grid topology deals with short transmission and power distribution. An example of an LV grid is shown in Figure 3.28. LV grids take power from different medium voltage networks. They do not receive power directly from generating stations. Then they feed powers to different industrial or domestic loads mainly through short-type transmission lines. Power distribution to consumers is the main job of the LV grid. The conventional LV grid is of AC-type. However, with the advancement of NGs and PGs, LV grids of both AC-type and DC-type are used to meet different load demands.

3.20.5 HVDC grid topology

HVDC grids are found in the long transmission networks. An example of an HVDC grid is shown in Figure 3.29. They receive power from HV or EHV transmission systems through the substation and converter arrangements and feed DC power to HVDC transmission networks. HVDC grids are not found in medium or low-voltage

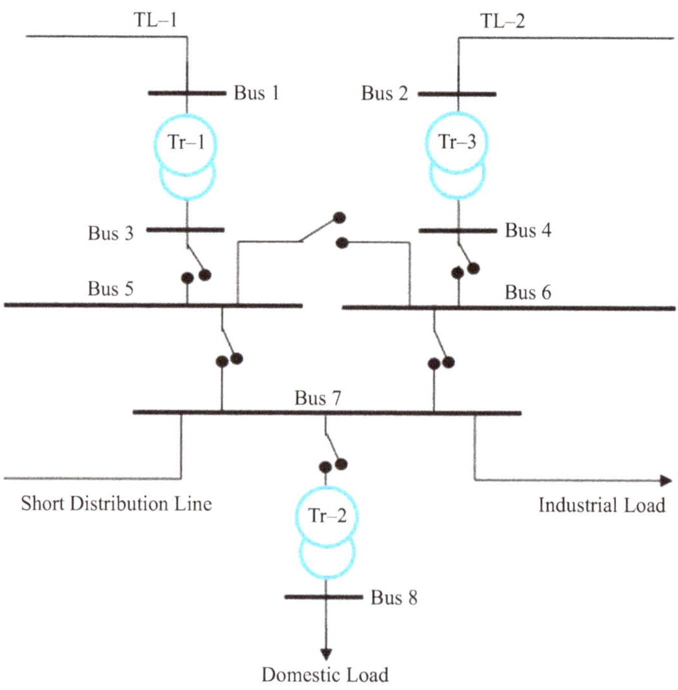

Figure 3.28 Sample LV grid structure

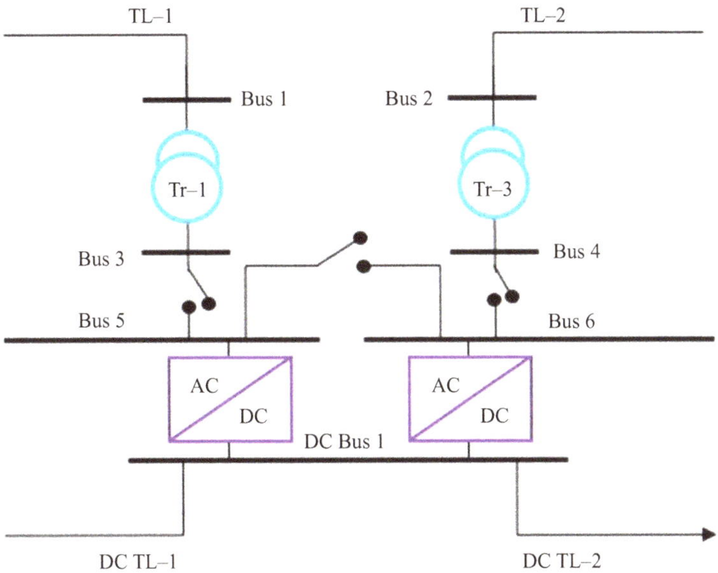

Figure 3.29 Sample HVDC grid structure

distribution systems. Dealing with converter control is an important issue in HVDC grids.

3.20.6 Hybrid grid topology

A hybrid grid involves feeders of different voltage levels. It collects energy from different sources at different voltage levels. Also, it delivers energy to different feeders, loads, and distribution sections at multiple voltage levels. In some cases, it may connect with DC links.

3.20.7 Smart grid topology

Reliability and continuity are the two most important requirements of a power supply system. With this aim, grids are being converted or upgraded into smart grids. It should handle variable, increasing, and extreme conditions of the demand efficiently. Along with this, it aims at minimizing carbon emission, integrating renewable sources, and reduction of the tariff. It allows DG and opens the door of both way communications between suppliers and customers. The main advantageous features of the smart grid are as follows:

- use of more digital devices,
- use of smart meters,
- information sharing with customer,
- better demand response,
- better load management,

- more power safety and security,
- better self-restoration capacity, and
- remote monitoring and remote control.

Additional main components of the smart grid are as follows:

- smart metering;
- monitoring and digital control;
- interconnected transmission facility;
- advanced communication, IoT with remote monitoring and control opportunity;
- virtual storage facility;
- flexible distributed load management.

3.20.8 Topology for NG and PG

3.20.8.1 General structure

The topological structure of NG and PG has been described with the help of different subunits of the system. The general structure of a NG station (NGS) may be built up as shown in Figure 3.30. Renewable sources, non-renewable sources, and battery storage may act as energy input. It may be noted that in practice, one or many types of sources may be present. NGS will feed power to local loads as well as can feed power to a grid that is connected to external networks.

The general structure of a PG station (PGS) may be built up as shown in Figure 3.31. Like previous renewable sources, non-renewable sources and battery storage may act as an energy input. It may be noted that in practice, one or many types of sources may be present. PGS will feed power to local loads as well as can feed power to the grid that is connected to external networks.

In practice, the presence of non-renewable sources is rare for PGSs, and therefore, the structure may be modified as shown in Figure 3.32.

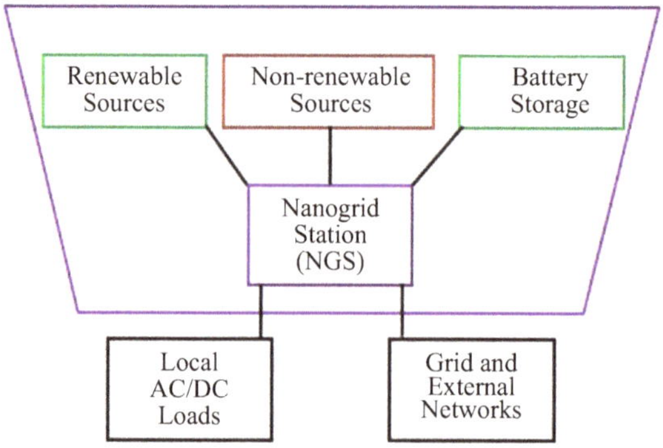

Figure 3.30 General structure of nanogrid station

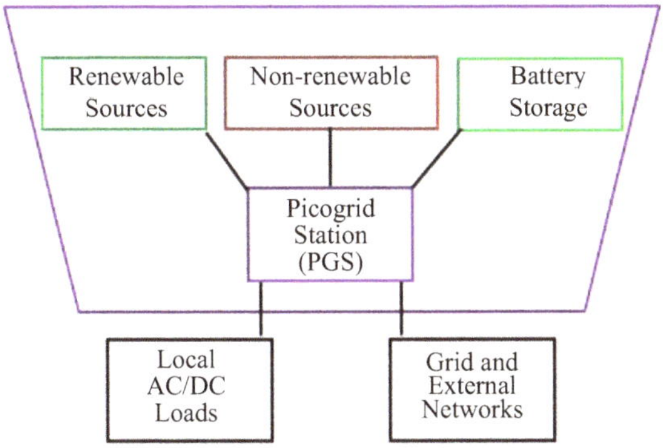

Figure 3.31 General structure of picogrid station

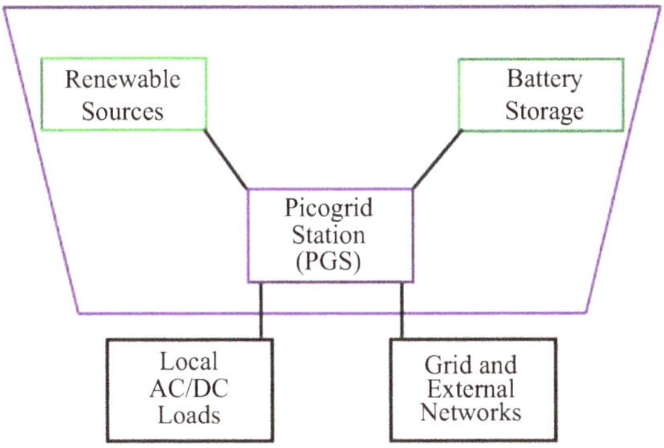

Figure 3.32 General structure of picogrid station considering only renewable sources and battery storage unit

3.20.8.2 Grid integration

Multiple PGSs may be integrated to form the NG as shown in Figure 3.33. It provides the advantages of DG, distributed monitoring, and control. It also improves the overall load handling capacity. Let there is N number of pico grids:

P_{PG1} = power capacity of picogrid – 1
P_{PG2} = power capacity of picogrid – 2
P_{PG3} = power capacity of picogrid – 3
P_{PGN} = power capacity of picogrid – N

Then, the capacity of the overall NG system may be written as follows:

$$P_{NG} = P_{PG1} + P_{PG2} + P_{PG3} + \ldots + P_{PGN} = \sum_{i=1}^{i=N} P_{PGi} \tag{3.17}$$

Similarly, multiple NG stations (NGS) may be integrated to form the MG as shown in Figure 3.34. It provides all the advantages of DG, monitoring, and

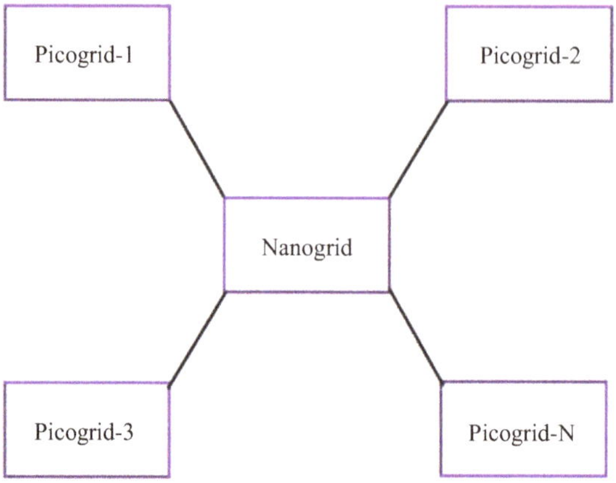

Figure 3.33 Picogrids integration for making up nanogrid

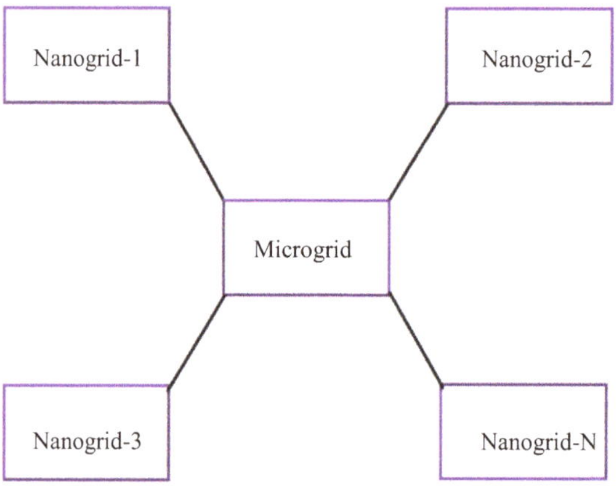

Figure 3.34 NGs integration for making up a microgrid

control. Also, the integrated system improves the overall load handling capacity. Let, there is N number of nano grids.

P_{NG1} = power capacity of picogrid – 1
P_{NG2} = power capacity of picogrid – 2
P_{NG3} = power capacity of picogrid – 3
P_{NGN} = power capacity of picogrid – N

Then, the capacity of the overall MG system may be written as follows:

$$P_{MG} = P_{NG_1} + P_{NG_2} + P_{NG_3} + \ldots + P_{NG_N} = \sum_{i=1}^{i=N} P_{NG_i} \tag{3.18}$$

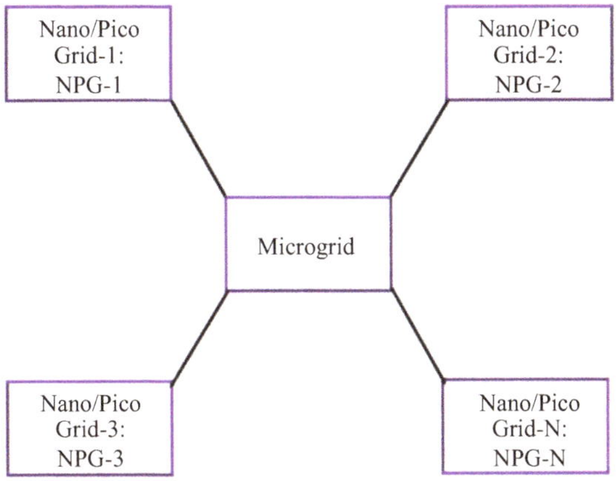

Figure 3.35 NGs or PGs integration for making up a microgrid

Similarly, multiple either NGs or PGs may be integrated to form a MG as shown in Figure 3.35. It provides all the previously mentioned advantages. Also, the integrated system improves the overall load handling capacity. Let there is N number of NGs or PGs:

P_{NPG1} = power capacity of picogrid – 1
P_{NPG2} = power capacity of picogrid – 2
P_{NPG3} = power capacity of picogrid – 3
P_{NPGN} = power capacity of picogrid – N

Then, the capacity of the overall MG system may be written as follows:

$$P_{MG} = P_{NPG1} + P_{NPG2} + P_{NPG3} + \ldots + P_{NPGN} = \sum_{i=1}^{i=N} P_{NPGi} \tag{3.19}$$

3.20.8.3 Grid's internal structure

The generalized internal structure of NG or PG is shown in Figure 3.36. It has eight sub-members as follows:

(a) sources,
(b) converter unit,
(c) controller unit,
(d) link bus,
(e) local load distribution center,
(f) gateway,
(g) measurement and metering unit, and
(h) interfacing unit for local and remote monitoring.

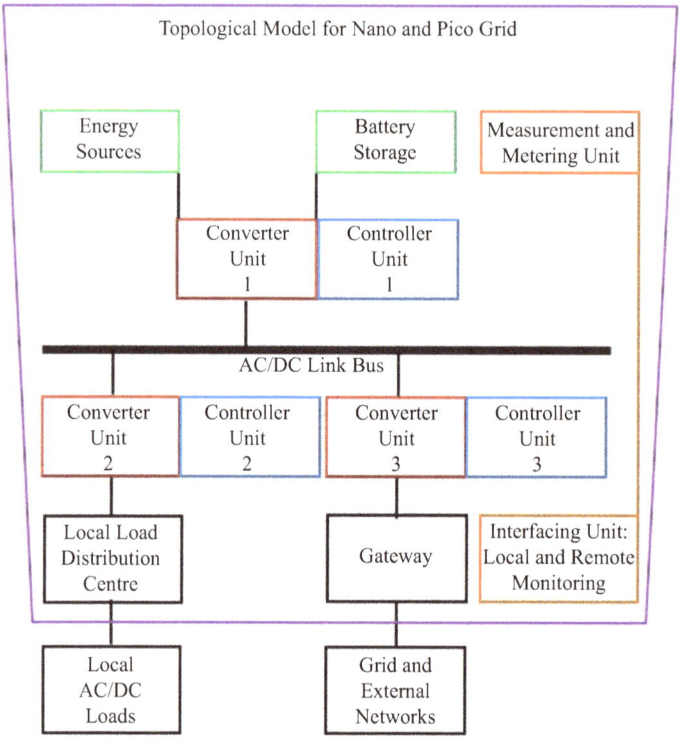

Figure 3.36 Generalized internal structure of NG or PG

The main features of these sub-members are described below.

(a) *Sources*: Sources provide input to grids. Three types of sources may exist renewable sources, non-renewable sources, and battery storage.

(b) *Converter unit*: It is used for voltage conversion. Based on utility converter, units may be divided into three categories:
- Converter unit 1 will be connected with power resources and will feed power to the link bus.
- Converter unit 2 will take power from the link bus and feed to local loads.
- Converter unit 3 will take power from the link bus and feed power to the grid through a gate. It may also take power from the grid and feed to a bus when local power resources are not available or insufficient to feed local loads. Thus, converter of unit 3 may be bidirectional in operation.

Based on AC and DC, converters may be of the following categories:

(a) AC–DC.
(b) DC–AC.
(c) AC–AC.
(d) DC–DC.

(c) *Controller unit*: Control unit controls the operation of each converter. Thus, each converter should individual control unit. However, all control units should work keeping coordination with each other.
(d) *Link bus*: Link bus crates the link between sources through converter with local loads and grids. It normally takes power from energy resources and acts as a source of power to local loads and the grid. If necessary, it can take power from the external grid. Link bus may be AC type or DC type.
(e) *Local load distribution center*: It consists of many switches or isolators and breakers. It connects or disconnects load with or from the system.
(f) *Gateway*: It creates the path for control and energy management between the system (NPG) and the external grid.
(g) *Measurement and metering unit*: It consists of different sensors, meters, etc. This unit helps to measure electrical parameters which are then used for monitoring and controlling purposes.
(h) *Interfacing unit for local and remote monitoring*: This unit provides the interfacing system for remote monitoring and remote control. It may consist of IoT devices for this purpose.

3.20.8.4 Hexagonal coordination for NPG

Excluding energy resources and link bus, six units are important for a typical NPG system. These are as follows:

(a) sources,
(b) converter unit,
(c) controller unit,
(d) link bus,
(e) local load distribution center,
(f) gateway,
(g) measurement and metering unit,
(h) interfacing unit for local and remote monitoring.

All these sub-members of NPG systems should have communication between them. Hexagonal coordination for NPG is shown in Figure 3.37 and corresponding nodal coordination is shown in Figure 3.38.

3.20.8.5 Battery storage connection

Different connections for the battery storage unit are shown in Figure 3.39. A battery storage unit may be connected in three major ways as follows:

(a) It may be connected directly with sources like PV strings, etc.
(b) It may be connected through a charge controller with sources.
(c) It may be connected through a converter and charge controller with the source.

3.21 Smart metering, measurement, and monitoring unit

Measurement and monitoring units incorporate two types of subunits as shown in Figure 3.40. One subunit involves measuring and metering different parameters.

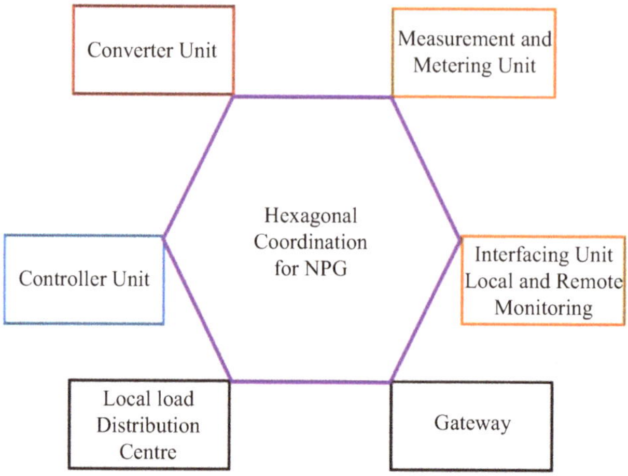

Figure 3.37 Hexagonal coordination network for NPG

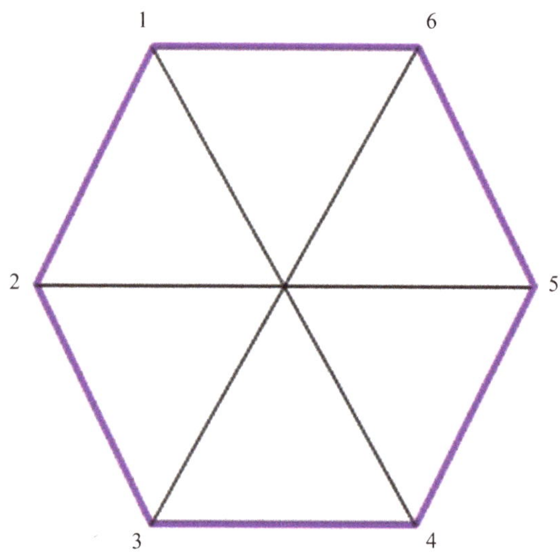

Figure 3.38 Nodal structure of hexagonal coordination for NPG

Sensors and conventional meters are involved. Along with them, smart meters are used in advanced type grid systems. Smart meters help in the best features of different components of energy management. Another subunit involves local and remote monitoring of different electrical and non-electrical parameters. Operation of this unit links with measuring and metering unit. It also involves IoT-based components for remote monitoring and control. The main purpose of this subunit includes the following aspects:

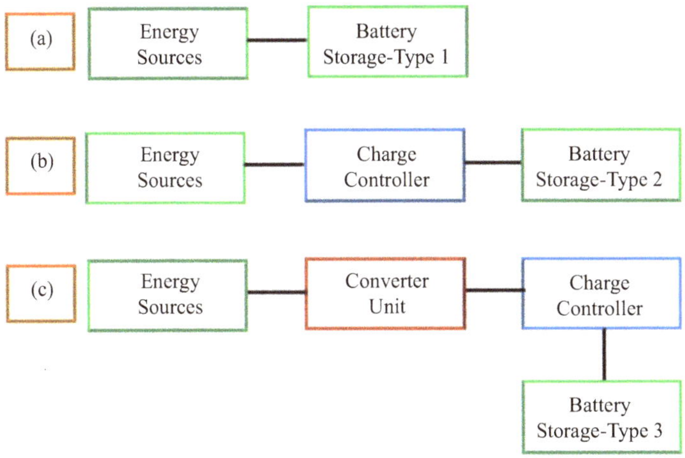

Figure 3.39 Different types of connection of battery storage units with energy sources: (a) type 1, (b) type 2, and (c) type 3

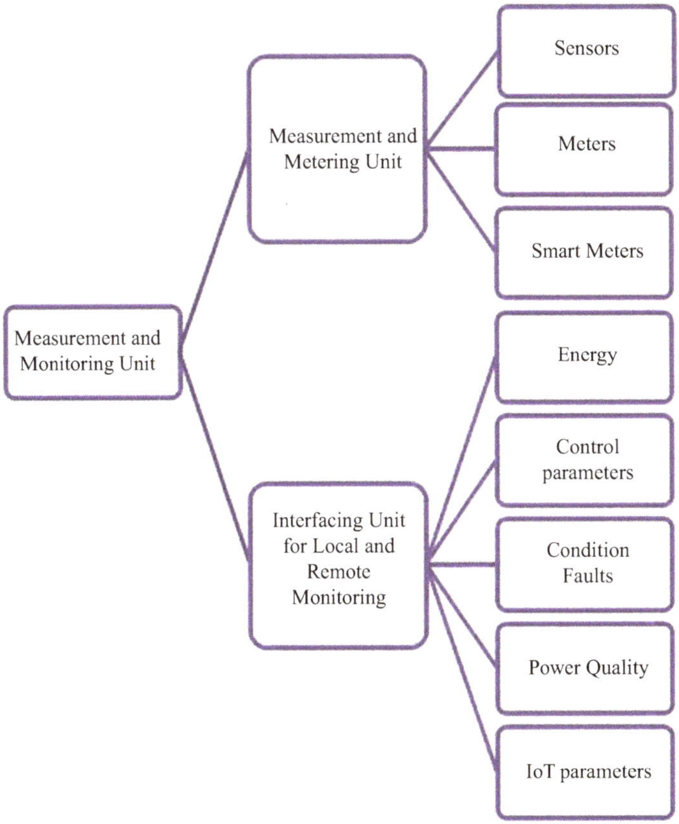

Figure 3.40 Different components of measuring and monitoring unit

- to monitor the performance of the grid;
- to monitor operations of different units of NPG;
- to monitor the health of NGP;
- to monitors conditions of different resources;
- to monitor faults, if any;
- to monitor control parameters;
- to monitor energy parameters and present capacity.

3.22 Power flow in AC multi-bus mesh and ring system

3.22.1 Load flow

With the increase in the number and diversity of loads, power system networks are becoming more and more complex. To mention the quality, reliability, and continuity of the power system, it has become essential to assess the flow of power in the different parts of the system. The study of the flow of active and reactive power along with voltages and their angles is commonly known as a load flow system. Load flow analysis is normally carried out with the help of the admittance matrix (made of system parameters) of the system. As the power system network is large and complicated, often, the number of unknown variables is higher than that of known variables which makes the computation difficult to get reach a solution. To cope with this, iterative methods are followed in load flow analysis. Different approaches useful for load flow analysis have been discussed in the following sections.

3.22.2 Power and admittance matrix

In load flow analysis, most of the parameters are considered complex. One important point that should be noted is that in normal circuit theory analysis, most of the sources are considered either voltage source or current source. But, in load flow analysis, all sources in the power system network are considered as a source of power. This power is complex and consists of real power or active power and imaginary power or reactive power. Mathematically, the power can be expressed as follows:

$$S = P + jQ \tag{3.20}$$

Voltage in a power system network can be expressed in terms of current as follows:

$$[V] = [Z][I]$$

Therefore, current can be expressed in terms of voltage as follows:

$$[I] = [Y][V]$$

$[Y]$ is known as the admittance matrix of a network. As in a power system network, the number of the known quantities is few and insufficient for a solution, to solve the load flow-related problems, iterative methods are used.

3.22.3 Classification of buses

Depending upon the use and connections, buses used in power system network can be classified as follows:

(a) *Generator bus or voltage-controlled buys*: bus which is connected with generators is known as generator bus. For such buses, active power and voltage are initially known in load flow analysis and for this reason, such bus is called P-V bus. As voltage is maintained at a constant value, sometimes they are called a regulated bus.
(b) *Load bus*: bus which is connected with load is known as load bus. For such buses, active power and reactive power (connected) are initially known in load flow analysis.
(c) *Slack bus or swing bus*: for such buses voltage and angle (δ) are initially known in load flow analysis. Swing bus or slack bus are normally connected to large sources of power (either large unit or grid which can maintain voltage and angle (δ) at a constant level).

3.22.4 Application of Kirchhoff's current law (KCL)

The bus may be connected to generators, loads, or transmission lines. Consider a simple case as shown in Figure 3.41, where a bus is connected with one generator and one load. Therefore, the generator current can be written as

$$I = I_1 \tag{3.21}$$

Now, I_1 can be expressed as

$$I_1 = Y_1 V_1 \tag{3.22}$$

where $Y_1$is the admittance of the connected load and $V_1$is the voltage of the bus.

Now, suppose there are n number of loads connected to the bus as shown in Figure 3.42 and those loads have admittance $Y_1, Y_2, Y_3, \ldots, Y_n$, respectively. Therefore, the current drawn from the generator can be written as

$$I = I_1 + I_2 + I_3 + \ldots + I_n \tag{3.23}$$

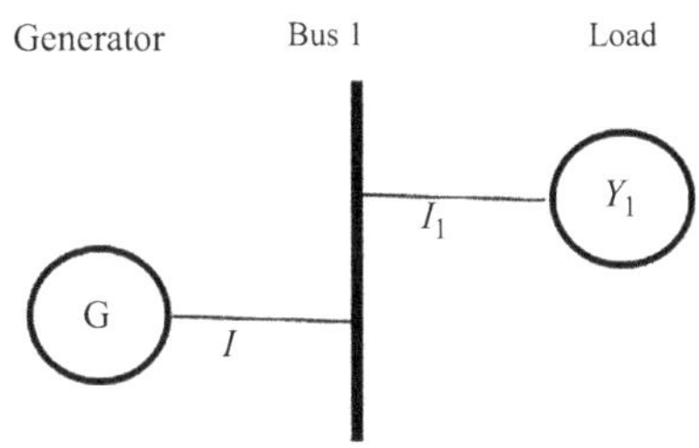

Figure 3.41 Bus connected with single generator and load

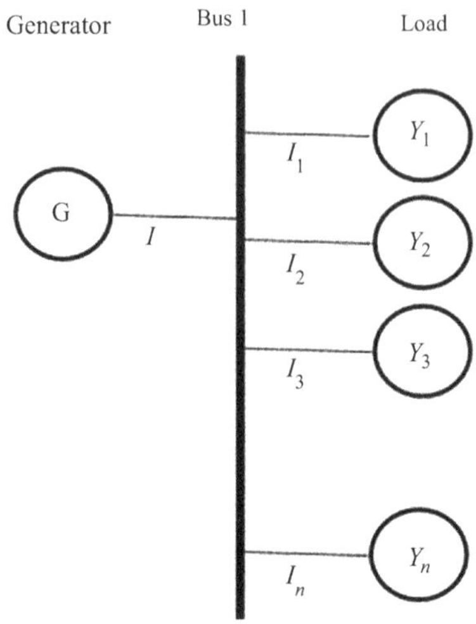

Figure 3.42 Bus connected with single generator and "n" is the number of loads

Now,

$$I_1 = Y_1 V_1 \tag{3.24}$$

$$I_2 = Y_2 V_1 \tag{3.25}$$

$$I_3 = Y_3 V_1 \tag{3.26}$$

$$I_n = Y_n V_1 \tag{3.27}$$

Therefore,

$$I = I_1 + I_2 + I_3 + \ldots + I_n \tag{3.28}$$

$$I = Y_1 V_1 + Y_2 V_1 + Y_3 V_1 + \ldots + Y_n V_1 \tag{3.29}$$

Let us consider a transmission line is connected with the bus. Admittance of the transmission line is Y_{1i} and the voltage level of receiving end is V_i Figure 3.43 Therefore, current drawn from the generator can be expressed as

$$I = Y_1 V_1 + Y_2 V_1 + Y_3 V_1 + \ldots + Y_n V_1 + Y_{1i} V_i \tag{3.30}$$

3.22.5 Methods of load flow solutions

Following iterative methods are used for load flow solutions:

(a) Gauss–Seidel method.
(b) Newton–Raphson method.
(c) Rangi–Kurta method.
(d) Fast decoupled method.

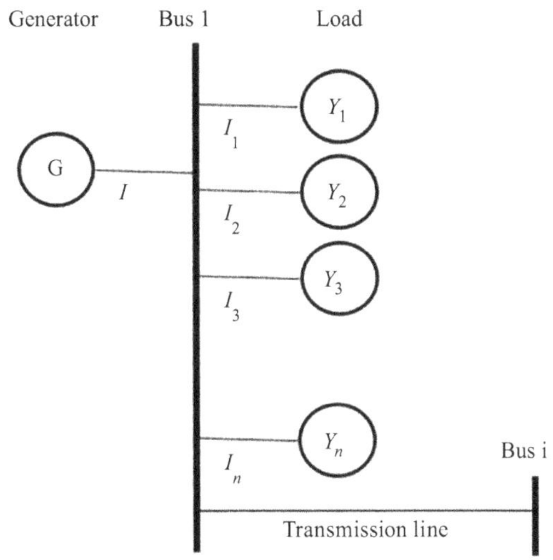

Figure 3.43 Bus connected with single generator, one transmission line, and "n" is the number of loads

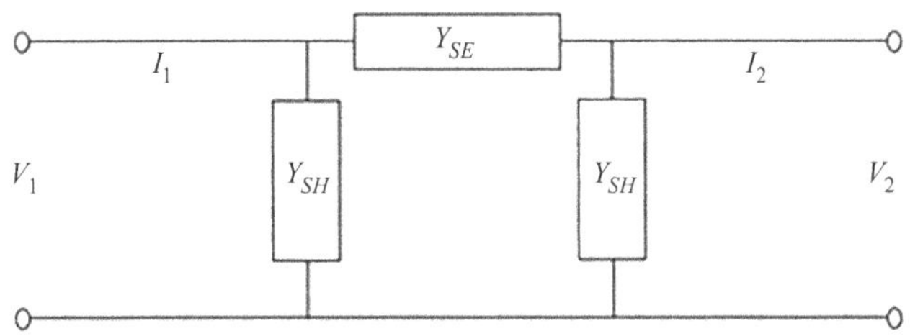

Figure 3.44 Short transmission line

3.22.6 Admittance matrix of a transmission line

Consider the Pi (π) model of a transmission line as shown in Figure 3.44 where Y_{SE} is the series admittance made of line resistance and inductance and Y_{SH} is the shunt admittance made of shunt line capacitance.

The admittance matrix of this network can be written as

$$[Y] = \begin{bmatrix} Y_{11} & Y_{12} \\ Y_{21} & Y_{22} \end{bmatrix} = \begin{bmatrix} Y_{SE} + Y_{SH} & -Y_{SH} \\ -Y_{SH} & Y_{SE} + Y_{SH} \end{bmatrix} \quad (3.31)$$

Therefore, currents can be expressed as follows:

$$\begin{bmatrix} I_1 \\ I_2 \end{bmatrix} = \begin{bmatrix} Y_{SE} + Y_{SH} & -Y_{SH} \\ -Y_{SH} & Y_{SE} + Y_{SH} \end{bmatrix} \begin{bmatrix} V_1 \\ V_2 \end{bmatrix} \quad (3.32)$$

3.22.7 Admittance matrix of three bus network (all buses are connected to load)

Consider a three bus system as shown in Figure 3.45.

From bus 1,

$$Y_{12}(V_1 - V_2) + Y_{31}(V_1 - V_3) = i_1 \tag{3.33}$$

From bus 2,

$$Y_{12}(V_1 - V_2) + Y_{23}(V_2 - V_3) = i_2 \tag{3.34}$$

From bus 3,

$$Y_{23}(V_3 - V_2) + Y_{31}(V_3 - V_1) = i_3 \tag{3.35}$$

From the above three equations,

$$i_1 = (Y_{12} + Y_{31})V_1 + (-Y_{12})V_2 + (-Y_{31})V_3 \tag{3.36}$$

$$i_2 = (-Y_{12})V_1 + (Y_{12} + Y_{23})V_2 + (-Y_{23})V_3 \tag{3.37}$$

$$i_3 = (-Y_{31})V_1 + (-Y_{23})V_2 + (Y_{23} + Y_{31})V_3 \tag{3.38}$$

Therefore, in the matrix form,

$$\begin{bmatrix} i_1 \\ i_2 \\ i_3 \end{bmatrix} = \begin{bmatrix} (Y_{12} + Y_{31}) & (-Y_{12}) & (-Y_{31}) \\ (-Y_{12}) & (Y_{12} + Y_{23}) & (-Y_{23}) \\ (-Y_{31}) & (-Y_{23}) & (Y_{23} + Y_{31}) \end{bmatrix} \begin{bmatrix} V_1 \\ V_2 \\ V_3 \end{bmatrix} \tag{3.39}$$

Let,

$$[Y] = \begin{bmatrix} Y_{11} & Y_{12} & Y_{13} \\ Y_{21} & Y_{22} & Y_{23} \\ Y_{31} & Y_{32} & Y_{33} \end{bmatrix} = \begin{bmatrix} (Y_{12} + Y_{31}) & (-Y_{12}) & (-Y_{31}) \\ (-Y_{12}) & (Y_{12} + Y_{23}) & (-Y_{23}) \\ (-Y_{31}) & (-Y_{23}) & (Y_{23} + Y_{31}) \end{bmatrix} \tag{3.40}$$

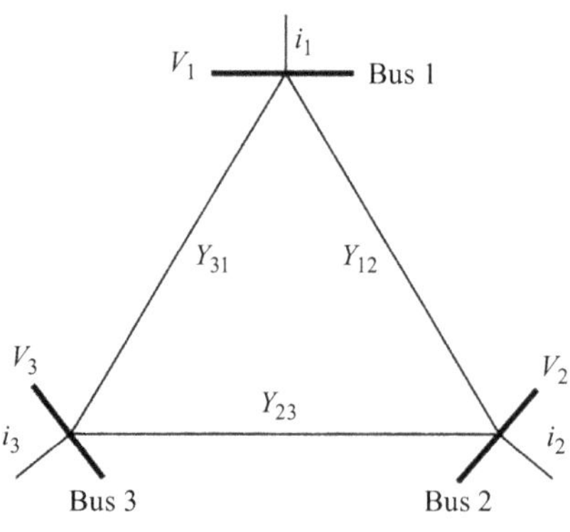

Figure 3.45 A three bus power system network

$[Y]$ is known as the admittance matrix of the three bus networks.

3.22.8 Admittance matrix of three bus network (all bus is connected to its load)

Consider a three bus system as shown in Figure 3.46.

From bus 1,

$$Y_{11}V_1 + Y_{12}(V_1 - V_2) + Y_{31}(V_1 - V_3) = i_1 \tag{3.41}$$

From bus 2,

$$Y_{22}V_2 + Y_{12}(V_1 - V_2) + Y_{23}(V_2 - V_3) = i_2 \tag{3.42}$$

From bus 3,

$$Y_{33}V_3 + Y_{23}(V_3 - V_2) + Y_{31}(V_3 - V_1) = i_3 \tag{3.43}$$

From the above three equations,

$$i_1 = (Y_{11} + Y_{12} + Y_{31})V_1 + (-Y_{12})V_2 + (-Y_{31})V_3 \tag{3.44}$$

$$i_2 = (-Y_{12})V_1 + (Y_{22} + Y_{12} + Y_{23})V_2 + (-Y_{23})V_3 \tag{3.45}$$

$$i_3 = (-Y_{31})V_1 + (-Y_{23})V_2 + (Y_{33} + Y_{23} + Y_{31})V_3 \tag{3.46}$$

Therefore, in the matrix form,

$$\begin{bmatrix} i_1 \\ i_2 \\ i_3 \end{bmatrix} = \begin{bmatrix} (Y_{11} + Y_{12} + Y_{31}) & (-Y_{12}) & (-Y_{31}) \\ (-Y_{12}) & (Y_{22} + Y_{12} + Y_{23}) & (-Y_{23}) \\ (-Y_{31}) & (-Y_{23}) & (Y_{33} + Y_{23} + Y_{31}) \end{bmatrix} \begin{bmatrix} V_1 \\ V_2 \\ V_3 \end{bmatrix} \tag{3.47}$$

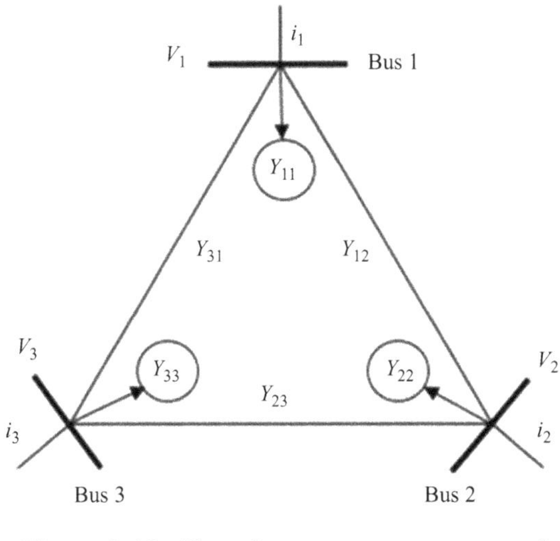

Figure 3.46 Three bus power system network

Let,

$$[Y]=\begin{bmatrix} Y_{11} & Y_{12} & Y_{13} \\ Y_{21} & Y_{22} & Y_{23} \\ Y_{31} & Y_{32} & Y_{33} \end{bmatrix}$$

$$=\begin{bmatrix} (Y_{11}+Y_{12}+Y_{31}) & (-Y_{12}) & (-Y_{31}) \\ (-Y_{12}) & (Y_{22}+Y_{12}+Y_{23}) & (-Y_{23}) \\ (-Y_{31}) & (-Y_{23}) & (Y_{33}+Y_{23}+Y_{31}) \end{bmatrix} \tag{3.48}$$

$[Y]$ is known as the admittance matrix of the three bus networks.

3.22.9 Admittance matrix of four bus network

Consider a four bus system as shown in Figure 3.47.

From bus 1,

$$Y_{12}(V_1 - V_2) + Y_{41}(V_1 - V_4) = i_1 \tag{3.49}$$

From bus 2,

$$Y_{12}(V_2 - V_1) + Y_{23}(V_2 - V_3) = i_2 \tag{3.50}$$

From bus 3,

$$Y_{23}(V_3 - V_2) + Y_{34}(V_3 - V_4) = i_3 \tag{3.51}$$

From bus 4,

$$Y_{34}(V_4 - V_3) + Y_{41}(V_4 - V_1) = i_4 \tag{3.52}$$

From the above four equations,

$$i_1 = (Y_{12} + Y_{41})V_1 + (-Y_{12})V_2 + (-Y_{41})V_4 \tag{3.53}$$

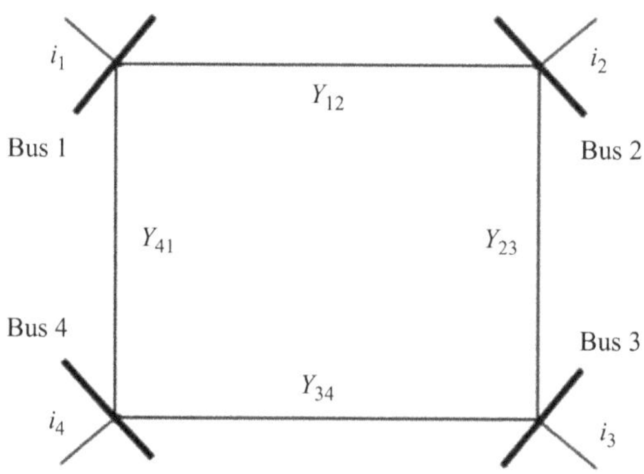

Figure 3.47 Four bus power system network

$$i_2 = (-Y_{12})V_1 + (Y_{12} + Y_{23})V_2 + (-Y_{23})V_3 \tag{3.54}$$

$$i_3 = (-Y_{23})V_2 + (Y_{23} + Y_{34})V_3 + (-Y_{34})V_4 \tag{3.55}$$

$$i_4 = (-Y_{41})V_1 + (-Y_{34})V_3 + (Y_{34} + Y_{41})V_4 \tag{3.56}$$

Therefore, in the matrix form,

$$\begin{bmatrix} i_1 \\ i_2 \\ i_3 \\ i_4 \end{bmatrix} = \begin{bmatrix} (Y_{12}+Y_{41}) & (-Y_{12}) & 0 & (-Y_{41}) \\ (-Y_{12}) & (Y_{12}+Y_{23}) & (-Y_{23}) & 0 \\ 0 & (-Y_{23}) & (Y_{23}+Y_{34}) & (-Y_{34}) \\ (-Y_{41}) & 0 & (-Y_{34}) & (Y_{34}+Y_{41}) \end{bmatrix} \begin{bmatrix} V_1 \\ V_2 \\ V_3 \\ V_4 \end{bmatrix} \tag{3.57}$$

Let,

$$[Y] = \begin{bmatrix} Y_{11} & Y_{12} & Y_{13} & Y_{14} \\ Y_{21} & Y_{22} & Y_{23} & Y_{24} \\ Y_{31} & Y_{32} & Y_{33} & Y_{34} \\ Y_{41} & Y_{42} & Y_{43} & Y_{44} \end{bmatrix}$$

$$= \begin{bmatrix} (Y_{12}+Y_{41}) & (-Y_{12}) & 0 & (-Y_{41}) \\ (-Y_{12}) & (Y_{12}+Y_{23}) & (-Y_{23}) & 0 \\ 0 & (-Y_{23}) & (Y_{23}+Y_{34}) & (-Y_{34}) \\ (-Y_{41}) & 0 & (-Y_{34}) & (Y_{34}+Y_{41}) \end{bmatrix} \tag{3.58}$$

$[Y]$ is known as the admittance matrix of the four bus networks.

3.22.10 Admittance matrix of four bus networks (all buses are connected to load)

Consider a four bus system as shown in Figure 3.48.

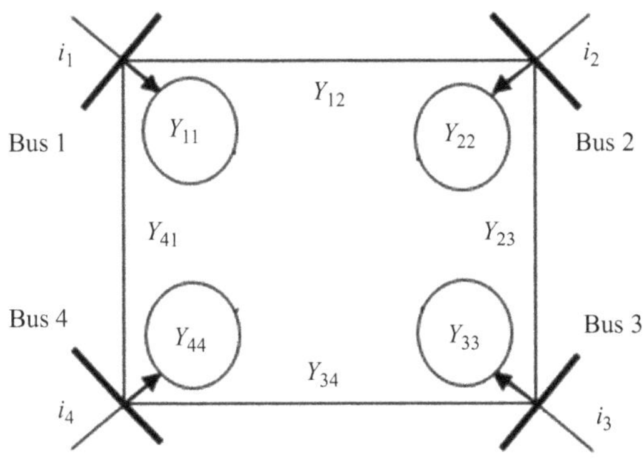

Figure 3.48 Four bus power system network

From bus 1,

$$Y_{11}V_1 + Y_{12}(V_1 - V_2) + Y_{41}(V_1 - V_4) = i_1 \tag{3.59}$$

From bus 2,

$$Y_{22}V_2 + Y_{12}(V_2 - V_1) + Y_{23}(V_2 - V_3) = i_2 \tag{3.60}$$

From bus 3,

$$Y_{33}V_3 + Y_{23}(V_3 - V_2) + Y_{34}(V_3 - V_4) = i_3 \tag{3.61}$$

From bus 4,

$$Y_{44}V_4 + Y_{34}(V_4 - V_3) + Y_{41}(V_4 - V_1) = i_4 \tag{3.62}$$

From the above four equations,

$$i_1 = (Y_{11} + Y_{12} + Y_{41})V_1 + (-Y_{12})V_2 + (-Y_{41})V_4 \tag{3.63}$$

$$i_2 = (-Y_{12})V_1 + (Y_{22} + Y_{12} + Y_{23})V_2 + (-Y_{23})V_3 \tag{3.64}$$

$$i_3 = (-Y_{23})V_2 + (Y_{33} + Y_{23} + Y_{34})V_3 + (-Y_{34})V_4 \tag{3.65}$$

$$i_4 = (-Y_{41})V_1 + (-Y_{34})V_3 + (Y_{44} + Y_{34} + Y_{41})V_4 \tag{3.66}$$

Therefore, in the matrix form,

$$\begin{bmatrix} i_1 \\ i_2 \\ i_3 \\ i_4 \end{bmatrix} = \begin{bmatrix} (Y_{11}+Y_{12}+Y_{41}) & (-Y_{12}) & 0 & (-Y_{41}) \\ (-Y_{12}) & (Y_{22}+Y_{12}+Y_{23}) & (-Y_{23}) & 0 \\ 0 & (-Y_{23}) & (Y_{33}+Y_{23}+Y_{34}) & (-Y_{34}) \\ (-Y_{41}) & 0 & (-Y_{34}) & (Y_{44}+Y_{34}+Y_{41}) \end{bmatrix} \begin{bmatrix} V_1 \\ V_2 \\ V_3 \\ V_4 \end{bmatrix} \tag{3.67}$$

Let,

$$[Y] = \begin{bmatrix} Y_{11} & Y_{12} & Y_{13} & Y_{14} \\ Y_{21} & Y_{22} & Y_{23} & Y_{24} \\ Y_{31} & Y_{32} & Y_{33} & Y_{34} \\ Y_{41} & Y_{42} & Y_{43} & Y_{44} \end{bmatrix}$$

$$= \begin{bmatrix} (Y_{11}+Y_{12}+Y_{41}) & (-Y_{12}) & 0 & (-Y_{41}) \\ (-Y_{12}) & (Y_{22}+Y_{12}+Y_{23}) & (-Y_{23}) & 0 \\ 0 & (-Y_{23}) & (Y_{33}+Y_{23}+Y_{34}) & (-Y_{34}) \\ (-Y_{41}) & 0 & (-Y_{34}) & (Y_{44}+Y_{34}+Y_{41}) \end{bmatrix} \tag{3.68}$$

$[Y]$is known as the admittance matrix of the four bus networks.

3.22.11 Generalized structure of admittance matrix

For a power system network having "n" number of buses, the admittance matrix can be written as follows:

$$[Y]_{n\times n} = \begin{bmatrix} Y_{11} & \dots & Y_{1n} \\ \vdots & \ddots & \vdots \\ Y_{n1} & \dots & Y_{nn} \end{bmatrix} \tag{3.69}$$

Current can be expressed in terms of voltage as follows:

$$\begin{bmatrix} I_1 \\ \vdots \\ I_2 \end{bmatrix} = \begin{bmatrix} Y_{11} & \dots & Y_{1n} \\ \vdots & \ddots & \vdots \\ Y_{n1} & \dots & Y_{nn} \end{bmatrix} \begin{bmatrix} V_1 \\ \vdots \\ V_2 \end{bmatrix} \tag{3.70}$$

3.22.12 Parameters and variables in load flow analysis

Elements of the admittance matrix are used as parameters in load flow analysis. Variables used in load flow analysis are voltage (V), active power (P), reactive power (Q), and angle (δ). Among four variables, some are known to us and some are unknown to us as presented in Table 3.5.

3.22.13 Two-port network analysis and load flow analysis

In two-port analysis, voltages are expressed in terms of currents by impedance matrix [Z], currents are expressed in terms of voltages by admittance matrix [Y], voltage and current of sending end are expressed in terms of voltage and current of receiving end by transmission parameters (A, B, C, and D) and voltage and current of opposite sides are expressed in terms of remaining voltage and current by hybrid parameters [h]. Thus, more stress is given to find out either voltage or current or both. No attempt is taken to determine active power (P), reactive power (Q), and angle (δ).

However, in load flow analysis, attention is focused mainly on determining voltage, active power (P), reactive power (Q), and angle (δ).

The voltage at a particular bus

Current in the ith bus can be written as follows:

$$I_i = \sum_{j=1}^{n} Y_{ij}V_i = Y_{ii}V_i + \sum_{\substack{j=1 \\ j\neq I}}^{n} Y_{ij}V_j \tag{3.71}$$

Table 3.5 Classification of buses

Status	Generator bus	Load bus	Swing or slack bus
Known	P and V	P and Q	V and δ
Unknown	Q and δ	V and δ	P and Q

$$Y_{ii}V_i = I_i - \sum_{\substack{j=1 \\ j \neq I}}^{n} Y_{ij}V_j \tag{3.72}$$

$$V_i = \frac{1}{Y_{ii}}\left[I_i - \sum_{\substack{j=1 \\ j \neq I}}^{n} Y_{ij}V_j \right] \tag{3.73}$$

Now power can be expressed as follows:

$$V_i^* I_i = P_i - jQ_i \tag{3.74}$$

$$I_i = \frac{P_i - jQ_i}{V_i^*} \tag{3.75}$$

Therefore,

$$V_i = \frac{1}{Y_{ii}}\left[I_i - \sum_{\substack{j=1 \\ j \neq I}}^{n} Y_{ij}V_j \right] = \frac{1}{Y_{ii}}\left[\frac{P_i - jQ_i}{V_i^*} - \sum_{\substack{j=1 \\ j \neq I}}^{n} Y_{ij}V_j \right] \tag{3.76}$$

Iterative approach

In the above equation, the voltage of the bus "i" is expressed in terms of its complex conjugate and voltages of other buses with the help of active power, reactive power, and admittance parameters. As in load flow analysis, most of the parameters and variables are unknown to us; an initial assumption is needed for some bus voltages. Then bus voltage of each bus is calculated by the above equation and the results are compared with the assumed values. Initially, difference (ΔV) will be there and the voltages of the right-hand side of the equation are replaced by the calculated voltage. Then again voltage of each bus is determined by the above equation. If the difference is there, the process is continued. Thus, this is nothing but a problem of iteration and after each level of iteration convergence is testified in terms of difference (ΔV). Ideally, the difference (ΔV) should become "zero"; however, in practice, the iterative method for finding out the voltage is continued until and unless the difference is less than a certain predefined value $(\in)$.

3.22.14 Gauss's iterative method without considering voltage-controlled bus

3.22.14.1 Known parameters

Consider a network consisting of one slack bus and another load (P, Q) bus. Admittance parameters are to be determined using the previous approach. Admittance parameters and available powers are to be noted. Also, note the number of buses (n). Let there be one slack bus and the remaining buses are load bus (P, Q bus) where active and reactive powers are known. There is no voltage-controlled bus.

3.22.14.2 Slack bus voltage and initial assumption of other bus voltages

Specify voltage corresponding to the slack bus, voltage is normally assumed as "$(a + j0)$"; "a" may be considered as "1" if it is connected as a base. Let "k" is the number of iterations and also let initially $k = 0$. Assume other voltages (V_i) for both $j = i$ and $j \neq i$ and denote them as $\left(V_i^k\right)$.

For $k = 0$,

$$V_i^k = V_i^0 \tag{3.77}$$

3.22.14.3 Convergence criterion

After each iteration error or difference (ΔV) between V_i^k and V_i^{k+1} is determined. Set a convergence criterion $(\in)$, i.e., $\Delta V_{\max} \leq \in$

Determination of bus voltages

Determine voltage of each bus (except slack bus voltage is known and constant) using the following equation by varying "j" and denoting them as (V_i^{k+1}):

$$V_i^{k+1} = \frac{1}{Y_{ii}}\left[\frac{P_i - Q_i}{V_i^k *} - \sum_{\substack{j=1 \\ j \neq I}}^{n} Y_{ij} V_j\right] \tag{3.78}$$

As $k = 0$,

$$V_i^{k+1} = V_i^{0+1} = V_i^1. \tag{3.79}$$

Determine V_i^1 for each bus by varying "I" from "1" to "n" except for the slack bus:

$$V_i^1 = \frac{1}{Y_{ii}}\left[\frac{P_i - Q_i}{V_i^0 *} - \sum_{\substack{j=1 \\ j \neq I}}^{n} Y_{ij} V_j\right] \tag{3.80}$$

Determine ΔV for each bus. Note maximum difference $\Delta V_{\max}$. Test convergence, i.e., see whether $\Delta V_{\max} \leq \in$? If no, increment by k by "1" and replace V_i^k by V_i^{k+1} for each bus and again determine bus voltages. If yes, the results indicate that the voltage has been obtained and therefore go for load flow determination.

Load flow assessment

After calculating bus voltages, other parameters like active power, reactive power, angle, etc. are determined.

3.22.14.4 Algorithm

1. Start
2. Take the admittance parameters and available powers
3. Assume voltage V_i^0 for both $i = j$ and $i \neq j$; voltage corresponding to the slack bus is normally assumed as $(a + j0) = (1 + j0)$
4. Set convergence criterion ($\in$) i.e. $\Delta V_{max} \leq \in$.
5. Then determine the voltage of each bus (except slack bus as the voltage of the slack bus is known and constant) using the following equation by varying "*j*" and denoting them as V_i^{k+1}

$$V_i^{k+1} = \frac{1}{Y_{ii}} \left[\frac{P_i - jQ_i}{V_i^k *} - \sum_{\substack{j=1 \\ j \neq I}}^{n} Y_{ij} V_j \right]$$

6. Determine V_i^1 for each bus by varying "*i*" from "1" to "*n*" except for the slack bus.
7. Determine difference $\Delta V = V_i^{k+1} - V_i^k$ for each value of "*i*".
8. Note the maximum difference ΔV_{max}.
9. Test convergence i.e. see whether $\Delta V_{max} \leq \in$?
10. If no, increment by k by "1" and replace V_i^k by V_i^{k+1} for each bus and again determine bus voltages.
11. If yes, the result indicates that voltage is obtained. Now go for load flow determination.
12. End.

3.22.15 *Explanation of Gauss–Seidel method in three bus system*

A three bus network can be represented by an admittance matrix of order $(n \times n = 3 \times 3)$. The admittance matrix can be written as follows:

$$[Y] = \begin{bmatrix} Y_{11} & Y_{12} & Y_{13} \\ Y_{21} & Y_{22} & Y_{23} \\ Y_{31} & Y_{32} & Y_{33} \end{bmatrix} \tag{3.81}$$

In this maximum number of the bus is n and let bus-1 be the slack bus or swing bus. The voltage of slack bus is known to us, say, it is V_1 (say). Therefore, in load flow analysis, "*i*" and "*j*" will vary from 1 to 3. Set the convergence criterion $\Delta V_{max} \leq \in$.

In the first iteration,

$$k = 0 \tag{3.82}$$

Voltages of the other two buses are assumed as follows:

$$V_2{}^k = V_2{}^0 \tag{3.83}$$

$$V_3{}^k = V_3{}^0 \tag{3.84}$$

Therefore,

$$V_i^{k+1} = \frac{1}{Y_{ii}}\left[\frac{P_i - jQ_i}{V_i^k *} - \sum_{\substack{j=1 \\ j \neq I}}^{n} Y_{ij}V_j\right] \tag{3.85}$$

As $k = 0, V_i^{k+1} = V_i^{0+1} = V_i{}^1$. Determine $V_i{}^1$ for each bus by varying "i" from "2" to "3" except for the slack bus:

$$V_2{}^1 = \frac{1}{Y_{ii}}\left[\frac{P_i - jQ_i}{V_2^0 *} - \left\{Y_{21}V_1 + Y_{23}V_3{}^0\right\}\right] \tag{3.86}$$

$$V_3{}^1 = \frac{1}{Y_{ii}}\left[\frac{P_i - jQ_i}{V_3^0 *} - \left\{Y_{31}V_1 + Y_{32}V_2{}^0\right\}\right] \tag{3.87}$$

Then errors are to be calculated as follows:

$$\Delta V_2 = V_2{}^1 - V_2{}^0 \tag{3.88}$$

$$\Delta V_3 = V_3{}^1 - V_3{}^0 \tag{3.89}$$

From ΔV_2 and ΔV_3, the maximum error $\Delta V_{\max}$ is to be determined and the convergence criterion is to be tested, i.e. $\Delta V_{\max} \leq \in$?

If yes, the solution of bus voltages is reached; otherwise, the iteration is to be replaced by incrementing k until and unless the convergence criterion is satisfied.

3.22.16 Explanation of Gauss–Seidel method in four bus system

A three bus network can be represented by an admittance matrix of order $(n \times n = 4 \times 4)$. The admittance matrix can be written as follows:

$$[Y] = \begin{bmatrix} Y_{11} & Y_{12} & Y_{13} & Y_{14} \\ Y_{21} & Y_{22} & Y_{23} & Y_{24} \\ Y_{31} & Y_{32} & Y_{33} & Y_{34} \\ Y_{41} & Y_{42} & Y_{43} & Y_{44} \end{bmatrix} \tag{3.90}$$

In this maximum, the number of the bus is n and let bus-1 be the slack bus or swing bus. The voltage of slack bus is known to us, say, it is V_1 (say). Therefore, in load flow analysis, "i" and "j" will vary from 1 to 4. Set the convergence criterion $\Delta V_{\max} \leq \in$.

In the first iteration,

$$k = 0 \tag{3.91}$$

Voltages of the other two buses are assumed as follows:

$$V_2{}^k = V_2{}^0 \tag{3.92}$$

$$V_3{}^k = V_3{}^0 \tag{3.93}$$

$$V_4{}^k = V_4{}^0 \tag{3.94}$$

Therefore,

$$V_i^{k+1} = \frac{1}{Y_{ii}}\left[\frac{P_i - jQ_i}{V_i^k *} - \sum_{\substack{j=1 \\ j \neq I}}^{n} Y_{ij}V_j\right] \tag{3.95}$$

As $k = 0, V_i^{k+1} = V_i^{0+1} = V_i{}^1$. Determine $V_i{}^1$ for each bus by varying "i" from "2" to "3" except for the slack bus:

$$V_2{}^1 = \frac{1}{Y_{ii}}\left[\frac{P_i - jQ_i}{V_2^0 *} - \left\{Y_{21}V_1 + Y_{23}V_3{}^2 + Y_{24}V_4{}^2\right\}\right] \tag{3.96}$$

$$V_3{}^1 = \frac{1}{Y_{ii}}\left[\frac{P_i - jQ_i}{V_3^0 *} - \left\{Y_{31}V_1 + Y_{32}V_2{}^0\right\}\right] \tag{3.97}$$

$$V_4{}^1 = \frac{1}{Y_{ii}}\left[\frac{P_i - jQ_i}{V_4^0 *} - \left\{Y_{41}V_1 + Y_{42}V_2{}^0 + Y_{43}V_3{}^0\right\}\right] \tag{3.98}$$

Then errors are to be calculated as follows:

$$\Delta V_2 = V_2{}^1 - V_2{}^0 \tag{3.99}$$

$$\Delta V_3 = V_3{}^1 - V_3{}^0 \tag{3.100}$$

$$\Delta V_4 = V_4{}^1 - V_4{}^0 \tag{3.101}$$

From $\Delta V_2$and ΔV_3, ΔV_4 maximum error $\Delta V_{\max}$ is to be determined and the convergence criterion is to be tested, i.e. $\Delta V_{\max} \leq \in$?

If yes, the solution of bus voltages is reached; otherwise, the iteration is to be replaced by incrementing k until and unless the convergence criterion is satisfied.

3.22.17 Gauss's iterative method considering generator bus (P, V bus)

3.22.17.1 Known parameters

Consider a network consisting of one slack bus, load (P, Q) bus, and some generator bus (P, V bus). Admittance parameters are to be determined using the previous approach. Admittance parameters and available powers are to be noted. Note the number of buses (n). Let there be one slack bus, some load buses where active power and reactive power are known and some generator buses where active power is known but reactive power is unknown. In such cases, reactive power is to be calculated for generator buses and considered in the load flow solution.

Power in a bus can be written as

$$P_i - Q_i = V_i^* I_i \tag{3.102}$$

$$\text{As } I_i = \sum_{j=1}^{n} Y_{ij} V_i \tag{3.103}$$

$$P_i - Q_i = V_i^* \sum_{j=1}^{n} Y_{ij} V_i \tag{3.104}$$

Let,

$$V_i = e_i + jf_i \tag{3.105}$$

$$Y_{ij} = G_{ij} - jB_{ij} \tag{3.106}$$

Therefore,
Power can be expressed as follows:

$$P_i - Q_i = V_i^* \sum_{j=1}^{n} Y_{ij} V_i \tag{3.107}$$

$$P_i - Q_i = (e_i + jf_i) \sum_{j=1}^{n} \left(G_{ij} - jB_{ij}\right)(e_i + jf_i) \tag{3.108}$$

$$P_i = \sum_{j=1}^{n} \left\{ e_i\left(e_i G_{ij} + f_i B_{ij}\right) + f_i\left(f_i G_{ij} - e_i B_{ij}\right) \right\} \tag{3.109}$$

$$Q_i = \sum_{j=1}^{n} \left\{ f_i\left(e_i G_{ij} + f_i B_{ij}\right) + e_i\left(f_i G_{ij} - e_i B_{ij}\right) \right\} \tag{3.110}$$

$$|V_i|^2 = e_i^2 + f_i^2 \tag{3.111}$$

Therefore, in load flow analysis by the Gauss–Seidel method, the reactive power of the generator bus is first to be calculated by the above equations. Reactive power is to be checked whether it is in between certain limits. The remaining part is the same as that of the Gauss method which is described previously.

3.22.17.2 Slack bus voltage and initial assumption of other bus voltages

Specify, the voltage corresponding to the slack bus, voltage is normally assumed as "$(a + j0)$"; "a" may be considered as "1" if it is connected as a base. Let "k" be the number of iterations and also let initially $k = 0$. Assume other voltages (V_i) for both $j = i$ and $j \neq i$ and denote them as $\left(V_i^k\right)$.

For $k = 0$,

$$V_i^k = V_i^0 \tag{3.112}$$

3.22.17.3 Convergence criterion

After each iteration error or difference (ΔV) between V_i^k and V_i^{k+1} is determined. Set a convergence criterion $(\in)$, i.e., $\Delta V_{\max} \leq \in$. Another convergence criterion is based on the reactive power as follows:

$$\Delta Q = Q_i^k - Q_i^{k+1} \tag{3.113}$$

$$\Delta Q_{\max} \leq \in \tag{3.114}$$

Additional convergence criteria can be set based on active power as follows:

$$\Delta P = P_i^k - P_i^{k+1} \tag{3.115}$$

$$\Delta P_{\max} \leq \in \tag{3.116}$$

Determination of reactive power for generator bus (P, V bus)

As the reactive power of a generator bus is unknown, it is first to be determined and checked whether it is within the limit by the following equation:

$$Q_i = \sum_{j=1}^{n} \left\{ f_i\left(e_i G_{ij} + f_i B_{ij}\right) + e_i\left(f_i G_{ij} - e_i B_{ij}\right) \right\} \tag{3.117}$$

3.22.17.4 Determination of bus voltages

Determine voltage of each bus (except slack bus voltage is known and constant) using the following equation by varying "j" and denote them as (V_i^{k+1}):

$$V_i^{k+1} = \frac{1}{Y_{ii}} \left[\frac{P_i - Q_i}{V_i^{k}*} - \sum_{\substack{j=1 \\ j \neq I}}^{n} Y_{ij} V_j \right] \tag{3.118}$$

As $k = 0$,

$$V_i^{k+1} = V_i^{0+1} = V_i^1 \tag{3.119}$$

Determine V_i^1 for each bus by varying "*i*" from "1" to "*n*" except for the slack bus:

$$V_i^1 = \frac{1}{Y_{ii}}\left[\frac{P_i - Q_i}{V_i^0 *} - \sum_{\substack{j=1\\ j\neq I}}^{n} Y_{ij}V_j\right] \quad (3.120)$$

Determine ΔV for each bus. Note the maximum difference $\Delta V_{\max}$. Test convergence i.e. see whether $\Delta V_{\max} \leq \in$? If no, increment by *k* by "1" and replace V_i^k by V_i^{k+1} for each bus and again determine bus voltages. If yes, the results indicate that voltage is obtained and then go for load flow determination.

Load flow assessment

After calculating bus voltages, other parameters like active power, reactive power, angle, etc. are determined.

3.22.17.5 Algorithm

1. Start.
2. Take the admittance parameters and available powers.
3. As the reactive power of a generator bus is unknown, it is first to be determined and checked whether it is within the limit by the following equation:

$$Q_i = \sum_{j=1}^{n} \left\{f_i\left(e_i G_{ij} + f_i B_{ij}\right) + e_i\left(f_i G_{ij} - e_i B_{ij}\right)\right\}$$

4. Assume voltage V_i^0 for both $i = j$ and $i \neq j$; voltage corresponding to the slack bus is normally assumed as $(a + j0) = (1 + j0)$.
5. Set convergence criterion ($\in$) i.e. $\Delta V_{\max} \leq \in$.
6. Then determine the voltage of each bus (except slack bus as the voltage of slack bus is known and constant) using the following equation by varying "*j*" and denote them as V_i^{k+1}

$$V_i^{k+1} = \frac{1}{Y_{ii}}\left[\frac{P_i - jQ_i}{V_i^k *} - \sum_{\substack{j=1\\ j\neq I}}^{n} Y_{ij}V_j\right]$$

7. Determine V_i^1 for each bus by varying "*i*" from "1" to "*n*" except for the slack bus.
8. Determine difference $\Delta V = V_i^{k+1} - V_i^k$ for each value of """.
9. Note the maximum difference $\Delta V_{\max}$.
10. Test convergence i.e. see whether $\Delta V_{\max} \leq \in$?
11. If no, increment by *k* by "1" and replace V_i^k by V_i^{k+1} for each bus and again determine bus voltages and follow steps 6–10.

12. If yes, the results indicate that voltage is obtained and then go for load flow determination.
13. End.

3.22.18 Advantage of Gauss–Seidel method over Gauss method

The main advantage of the Gauss–Seidel method over the Gauss method is that the Gauss–Seidel method is capable of giving load flow solution considering both loads (P, Q are known) bus and generator bus (P, V bus) or voltage controlled bus where reactive power Q is unknown to us. Other features of the Gauss method are applicable in the Gauss–Seidel method.

3.22.19 Limitation of Gauss–Seidel method

- The method needs a large number of iteration.
- Sometimes it becomes difficult to reach a convergence criterion.

3.22.20 Acceleration factor

Because of a large number of iteration, the time of execution becomes high. To speed up the execution process, sometimes acceleration factor is introduced. It is defined as follows:

$$V_{i(\text{accelerated})}^{k+1} = V_i^k + \sigma \Delta V \tag{3.121}$$

σ is called the acceleration factor; normally $1.4 \leq \sigma \leq 1.6$.

3.22.21 Advantage of Gauss–Seidel method

The advantages of the Gauss–Seidel method are as follows:

(a) The method is very simple.
(b) The number of steps for the iterative loop is small.
(c) It needs less computing time.

3.22.22 Newton–Raphson method

Newton–Raphson method is based on Taylor series expansion. In this method, a set of non-linear equations is expressed in terms of a set of linear equations, and approximation is made up to first order. Let x and y be two sets. Each depends on all elements of y.

Mathematically,

$$y = f(x) \tag{3.122}$$

i.e.,

$$y_i = f_i(x_1, x_2, \ldots\ldots\ldots\ldots x_n) \tag{3.123}$$

Therefore,

$$y_1 = f_1(x_1, x_2, \ldots\ldots\ldots\ldots x_n) \tag{3.124}$$

$$y_2 = f_2(x_1, x_2, \ldots\ldots\ldots\ldots x_n) \tag{3.125}$$

$$y_3 = f_3(x_1, x_2, \ldots\ldots\ldots\ldots x_n) \tag{3.126}$$

—

$$y_n = f_n(x_1, x_2, \ldots\ldots\ldots\ldots x_n) \tag{3.127}$$

Let initially,

$$x_i = x_i^0$$

Then,

$$y_i = f_i\left(x_1{}^0, x_2{}^0, \ldots\ldots\ldots\ldots x_n{}^0\right) \tag{3.128}$$

Therefore,

$$y_1 = f_1\left(x_1{}^0, x_2{}^0, \ldots\ldots\ldots\ldots x_n{}^0\right) \tag{3.129}$$

$$y_2 = f_2\left(x_1{}^0, x_2{}^0, \ldots\ldots\ldots\ldots x_n{}^0\right) \tag{3.130}$$

$$y_3 = f_3\left(x_1{}^0, x_2{}^0, \ldots\ldots\ldots\ldots x_n{}^0\right) \tag{3.131}$$

...

$$y_n = f_n\left(x_1{}^0, x_2{}^0, \ldots\ldots\ldots\ldots x_n{}^0\right) \tag{3.132}$$

Now, consider x is changed slightly and the new value of x is as follows:

$$x_i = x_i^0 + \Delta x_i \tag{3.133}$$

Hence,

$$y_i = f_i\left(x_1{}^0 + \Delta x_1, x_2{}^0 + \Delta x_2, \ldots\ldots\ldots\ldots x_n{}^0 + \Delta x_n\right) \tag{3.134}$$

Therefore,

$$y_1 = f_1\left(x_1{}^0 + \Delta x_1, x_2{}^0 + \Delta x_2, \ldots\ldots\ldots\ldots x_n{}^0 + \Delta x_n\right) \tag{3.135}$$

$$y_2 = f_2\left(x_1{}^0 + \Delta x_1, x_2{}^0 + \Delta x_2, \ldots\ldots\ldots\ldots x_n{}^0 + \Delta x_n\right) \tag{3.136}$$

$$y_3 = f_3\left(x_1{}^0 + \Delta x_1, x_2{}^0 + \Delta x_2, \ldots\ldots\ldots\ldots x_n{}^0 + \Delta x_n\right) \tag{3.137}$$

...

$$y_n = f_n\left(x_1{}^0 + \Delta x_1, x_2{}^0 + \Delta x_2, \ldots\ldots\ldots\ldots x_n{}^0 + \Delta x_n\right) \tag{3.138}$$

The above equation can be expanded in Taylor's series with the first-order approximation as follows:

$$y_i = f_i\left(x_1{}^0, x_2{}^0, \ldots\ldots\ldots\ldots x_n{}^0\right) + \left[\Delta x_1 \frac{\partial f_i}{\partial x_1}\right]_0 + \left[\Delta x_2 \frac{\partial f_i}{\partial x_2}\right]_0 + \ldots + \left[\Delta x_n \frac{\partial f_i}{\partial x_n}\right]_0 \tag{3.139}$$

Therefore,

$$y_1 = f_1\left(x_1{}^0, x_2{}^0, \ldots\ldots\ldots\ldots x_n{}^0\right) + \left[\Delta x_1 \frac{\partial f_1}{\partial x_1}\right]_0 + \left[\Delta x_2 \frac{\partial f_1}{\partial x_2}\right]_0 + \ldots + \left[\Delta x_n \frac{\partial f_1}{\partial x_n}\right]_0 \tag{3.140}$$

$$y_2 = f_2\left(x_1{}^0, x_2{}^0, \ldots\ldots\ldots\ldots x_n{}^0\right) + \left[\Delta x_1 \frac{\partial f_2}{\partial x_1}\right]_0 + \left[\Delta x_2 \frac{\partial f_2}{\partial x_2}\right]_0 + \ldots + \left[\Delta x_n \frac{\partial f_2}{\partial x_n}\right]_0 \tag{3.141}$$

$$y_i = f_i\left(x_1{}^0, x_2{}^0, \ldots\ldots\ldots\ldots x_n{}^0\right) + \left[\Delta x_1 \frac{\partial f_3}{\partial x_1}\right]_0 + \left[\Delta x_2 \frac{\partial f_3}{\partial x_2}\right]_0 + \ldots + \left[\Delta x_n \frac{\partial f_3}{\partial x_n}\right]_0 \tag{3.142}$$

$$y_n = f_n\left(x_1{}^0, x_2{}^0, \ldots\ldots\ldots\ldots x_n{}^0\right) + \left[\Delta x_1 \frac{\partial f_n}{\partial x_1}\right]_0 + \left[\Delta x_2 \frac{\partial f_n}{\partial x_2}\right]_0 + \ldots + \left[\Delta x_n \frac{\partial f_n}{\partial x_n}\right]_0 \tag{3.143}$$

Therefore, the difference of y due to a small change of x can be written as follows:

$$y_i - f_i\left(x_1{}^0, x_2{}^0, \ldots\ldots\ldots\ldots x_n{}^0\right) = \left[\Delta x_1 \frac{\partial f_i}{\partial x_1}\right]_0 + \left[\Delta x_2 \frac{\partial f_i}{\partial x_2}\right]_0 + \ldots + \left[\Delta x_n \frac{\partial f_i}{\partial x_n}\right]_0 \tag{3.144}$$

Therefore,

$$y_1 - f_1\left(x_1{}^0, x_2{}^0, \ldots\ldots\ldots\ldots x_n{}^0\right) = \left[\Delta x_1 \frac{\partial f_1}{\partial x_1}\right]_0 + \left[\Delta x_2 \frac{\partial f_1}{\partial x_2}\right]_0 + \ldots + \left[\Delta x_n \frac{\partial f_1}{\partial x_n}\right]_0 \tag{3.145}$$

$$y_2 - f_2\left(x_1{}^0, x_2{}^0, \ldots\ldots\ldots\ldots x_n{}^0\right) = \left[\Delta x_1 \frac{\partial f_2}{\partial x_1}\right]_0 + \left[\Delta x_2 \frac{\partial f_2}{\partial x_2}\right]_0 + \ldots + \left[\Delta x_n \frac{\partial f_2}{\partial x_n}\right]_0 \tag{3.146}$$

$$y_i - f_i\left(x_1{}^0, x_2{}^0, \ldots\ldots\ldots\ldots x_n{}^0\right) = \left[\Delta x_1 \frac{\partial f_3}{\partial x_1}\right]_0 + \left[\Delta x_2 \frac{\partial f_3}{\partial x_2}\right]_0 + \ldots + \left[\Delta x_n \frac{\partial f_3}{\partial x_n}\right]_0 \tag{3.147}$$

$$y_n - f_n\left(x_1{}^0, x_2{}^0, \ldots\ldots\ldots\ldots x_n{}^0\right) = \left[\Delta x_1 \frac{\partial f_n}{\partial x_1}\right]_0 + \left[\Delta x_2 \frac{\partial f_n}{\partial x_2}\right]_0 + \ldots + \left[\Delta x_n \frac{\partial f_n}{\partial x_n}\right]_0 \tag{3.148}$$

In the matrix form, the above equation can be written as follows:

$$\begin{bmatrix} y_1 - f_1(x_1{}^0, x_2{}^0, \ldots\ldots\ldots\ldots x_n{}^0) \\ \vdots \\ y_n - f_n(x_1{}^0, x_2{}^0, \ldots\ldots\ldots\ldots x_n{}^0) \end{bmatrix} = \begin{bmatrix} \left[\frac{\partial f_1}{\partial x_1}\right]_0 & \cdots & \left[\frac{\partial f_1}{\partial x_n}\right]_0 \\ \vdots & \ddots & \vdots \\ \left[\frac{\partial f_n}{\partial x_1}\right]_0 & \cdots & \left[\frac{\partial f_n}{\partial x_n}\right]_0 \end{bmatrix} \begin{bmatrix} \Delta x_1 \\ \vdots \\ \Delta x_n \end{bmatrix} \tag{3.149}$$

$$\text{or, } A = JB \tag{3.150}$$

where,

$$A = \begin{bmatrix} y_1 - f_1(x_1{}^0, x_2{}^0, \ldots\ldots\ldots\ldots x_n{}^0) \\ \vdots \\ y_n - f_n(x_1{}^0, x_2{}^0, \ldots\ldots\ldots\ldots x_n{}^0) \end{bmatrix} \tag{3.151}$$

$$B = \begin{bmatrix} \Delta x_1 \\ \vdots \\ \Delta x_n \end{bmatrix} \tag{3.152}$$

$$J = \begin{bmatrix} \left[\frac{\partial f_1}{\partial x_1}\right]_0 & \cdots & \left[\frac{\partial f_1}{\partial x_n}\right]_0 \\ \vdots & \ddots & \vdots \\ \left[\frac{\partial f_n}{\partial x_1}\right]_0 & \cdots & \left[\frac{\partial f_n}{\partial x_n}\right]_0 \end{bmatrix} \tag{3.153}$$

J is called the Jacobian matrix

3.22.23 Newton–Raphson method in the load flow solution

In load flow problems, four variables are most important. They are P, Q, V, and δ. In most cases, two are known and the remaining two are to be determined. Thus, P and Q can be expressed in terms of V and δ.

With the help of the above Newton–Raphson approach, P can be written in terms of V as follows:

$$\begin{bmatrix} \Delta P_1 \\ \vdots \\ \Delta P_n \end{bmatrix} = \begin{bmatrix} \left[\frac{\partial P_1}{\partial V_1}\right]_0 & \cdots & \left[\frac{\partial P_1}{\partial V_n}\right]_0 \\ \vdots & \ddots & \vdots \\ \left[\frac{\partial P_n}{\partial V_1}\right]_0 & \cdots & \left[\frac{\partial P_n}{\partial V_n}\right]_0 \end{bmatrix} \begin{bmatrix} \Delta V_1 \\ \vdots \\ \Delta V_n \end{bmatrix} \tag{3.154}$$

With the help of the above Newton–Raphson approach, P can be written in terms of δ as follows:

$$\begin{bmatrix} \Delta P_1 \\ \vdots \\ \Delta P_n \end{bmatrix} = \begin{bmatrix} \left[\frac{\partial P_1}{\partial \delta_1}\right]_0 & \cdots & \left[\frac{\partial P_1}{\partial \delta_n}\right]_0 \\ \vdots & \ddots & \vdots \\ \left[\frac{\partial P_n}{\partial \delta_1}\right]_0 & \cdots & \left[\frac{\partial P_n}{\partial \delta_n}\right]_0 \end{bmatrix} \begin{bmatrix} \Delta \delta_1 \\ \vdots \\ \Delta \delta_n \end{bmatrix} \tag{3.155}$$

Combining these above two equations,

$$\begin{bmatrix} \Delta P_1 \\ \vdots \\ \Delta P_n \end{bmatrix} = \begin{bmatrix} \left[\frac{\partial P_1}{\partial \delta_1}\right]_0 & \cdots & \left[\frac{\partial P_1}{\partial \delta_n}\right]_0 & \left[\frac{\partial P_1}{\partial V_1}\right]_0 & \cdots & \left[\frac{\partial P_1}{\partial V_n}\right]_0 \\ \vdots & \ddots & \vdots & \vdots & \ddots & \vdots \\ \left[\frac{\partial P_n}{\partial \delta_1}\right]_0 & \cdots & \left[\frac{\partial P_n}{\partial \delta_n}\right]_0 & \left[\frac{\partial P_n}{\partial V_1}\right]_0 & \cdots & \left[\frac{\partial P_n}{\partial V_n}\right]_0 \end{bmatrix} \begin{bmatrix} \Delta \delta_1 \\ \vdots \\ \Delta \delta_n \\ \Delta V_1 \\ \vdots \\ \Delta V_n \end{bmatrix} \tag{3.156}$$

Similarly, with the help of the above Newton–Raphson approach, Q can be written in terms of V as follows:

$$\begin{bmatrix} \Delta Q_1 \\ \vdots \\ \Delta Q_n \end{bmatrix} = \begin{bmatrix} \left[\frac{\partial Q_1}{\partial V_1}\right]_0 & \cdots & \left[\frac{\partial Q_1}{\partial V_n}\right]_0 \\ \vdots & \ddots & \vdots \\ \left[\frac{\partial Q_n}{\partial V_1}\right]_0 & \cdots & \left[\frac{\partial Q_n}{\partial V_n}\right]_0 \end{bmatrix} \begin{bmatrix} \Delta V_1 \\ \vdots \\ \Delta V_n \end{bmatrix} \tag{3.157}$$

With the help of the above Newton–Raphson approach, Q can be written in terms of δ as follows:

$$\begin{bmatrix} \Delta Q_1 \\ \vdots \\ \Delta Q_n \end{bmatrix} = \begin{bmatrix} \left[\frac{\partial Q_1}{\partial \delta_1}\right]_0 & \cdots & \left[\frac{\partial Q}{\partial \delta_n}\right]_0 \\ \vdots & \ddots & \vdots \\ \left[\frac{\partial Q_n}{\partial \delta_1}\right]_0 & \cdots & \left[\frac{\partial Q_n}{\partial \delta_n}\right]_0 \end{bmatrix} \begin{bmatrix} \Delta \delta_1 \\ \vdots \\ \Delta \delta_n \end{bmatrix} \tag{3.158}$$

Combining these above two equations,

$$\begin{bmatrix} \Delta Q_1 \\ \vdots \\ \Delta Q_n \end{bmatrix} = \begin{bmatrix} \begin{bmatrix} \left[\frac{\partial Q_1}{\partial \delta_1}\right]_0 & \cdots & \left[\frac{\partial Q}{\partial \delta_n}\right]_0 \\ \vdots & \ddots & \vdots \\ \left[\frac{\partial Q_n}{\partial \delta_1}\right]_0 & \cdots & \left[\frac{\partial Q_n}{\partial \delta_n}\right]_0 \end{bmatrix} & \begin{bmatrix} \left[\frac{\partial Q_1}{\partial V_1}\right]_0 & \cdots & \left[\frac{\partial Q_1}{\partial V_n}\right]_0 \\ \vdots & \ddots & \vdots \\ \left[\frac{\partial Q_n}{\partial V_1}\right]_0 & \cdots & \left[\frac{\partial Q_n}{\partial V_n}\right]_0 \end{bmatrix} \end{bmatrix} \begin{bmatrix} \Delta \delta_1 \\ \vdots \\ \Delta \delta_n \\ \Delta V_1 \\ \vdots \\ \Delta V_n \end{bmatrix} \tag{3.159}$$

Therefore, both active and reactive power can be expressed as follows:

$$\begin{bmatrix} \Delta P_1 \\ \vdots \\ \Delta P_n \\ \Delta Q_1 \\ \vdots \\ \Delta Q_n \end{bmatrix} \begin{bmatrix} \left[\frac{\partial P_1}{\partial \delta_1}\right]_0 & \cdots & \left[\frac{\partial P_1}{\partial \delta_n}\right]_0 & \left[\frac{\partial P_1}{\partial V_1}\right]_0 & \cdots & \left[\frac{\partial P_1}{\partial V_n}\right]_0 \\ \vdots & \ddots & \vdots & \vdots & \ddots & \vdots \\ \left[\frac{\partial P_n}{\partial \delta_1}\right]_0 & \cdots & \left[\frac{\partial P_n}{\partial \delta_n}\right]_0 & \left[\frac{\partial P_n}{\partial V_1}\right]_0 & \cdots & \left[\frac{\partial P_n}{\partial V_n}\right]_0 \\ \left[\frac{\partial Q_1}{\partial \delta_1}\right]_0 & \cdots & \left[\frac{\partial Q}{\partial \delta_n}\right]_0 & \left[\frac{\partial Q_1}{\partial V_1}\right]_0 & \cdots & \left[\frac{\partial Q_1}{\partial V_n}\right]_0 \\ \vdots & \ddots & \vdots & \vdots & \ddots & \vdots \\ \left[\frac{\partial Q_n}{\partial \delta_1}\right]_0 & \cdots & \left[\frac{\partial Q_n}{\partial \delta_n}\right]_0 & \left[\frac{\partial Q_n}{\partial V_1}\right]_0 & \cdots & \left[\frac{\partial Q_n}{\partial V_n}\right]_0 \end{bmatrix} \begin{bmatrix} \Delta \delta_1 \\ \vdots \\ \Delta \delta_n \\ \Delta V_1 \\ \vdots \\ \Delta V_n \end{bmatrix} \tag{3.160}$$

$$\begin{bmatrix} \Delta P \\ \Delta Q \end{bmatrix} \begin{bmatrix} J_1 & J_2 \\ J_3 & J_4 \end{bmatrix} \begin{bmatrix} \Delta\delta \\ \Delta V \end{bmatrix} \tag{3.161}$$

Therefore, in the Newton–Raphson method, convergence criteria are made on both active and reactive power as follows:

Reactive power:

$$\Delta Q = Q_i^k - Q_i^{k+1} \tag{3.162}$$

$$\Delta Q_{\max} \leq \in \tag{3.163}$$

Active power is as follows:

$$\Delta P = P_i^k - P_i^{k+1} \tag{3.164}$$

$$\Delta P_{\max} \leq \in \tag{3.165}$$

3.22.24 Fast decoupled method

In most real cases, it is observed that active power control is involved with only $\Delta\delta$ and reactive power control is involved with ΔV. In the fast decoupled method, the load flow solution can be achieved by considering the effects of ΔV in the determination of ΔQ and the effects of $\Delta\delta$ in the determination of ΔP. Thus, fast decoupled method reduces the time of execution for load flow solution.

Thus, the matrix equation obtained in the Newton–Raphson method can be modified as follows:

$$\begin{bmatrix} \Delta P \\ \Delta Q \end{bmatrix} \begin{bmatrix} J_1 & o \\ o & J_4 \end{bmatrix} \begin{bmatrix} \Delta\delta \\ \Delta V \end{bmatrix} \tag{3.166}$$

i.e.

$$\Delta P = J_1 \Delta\delta \tag{3.167}$$

$$\Delta Q = J_4 \Delta V \tag{3.168}$$

3.22.25 Sparse matrix

In a large power system network, most of the elements of the Jacobian matrix are zero. Such a matrix having most of the elements zero is known as the sparse matrix.

3.23 Load flow in DC network

DC load flow analysis is much simpler as the network is resistive. A simple load distribution network is shown in Figure 3.49. It has four buses connected with four resistive loads as shown in Figure 3.49(a). The distribution line is resistive and can be represented by equivalent resistance. Thus, in this distribution network, power is distributed among four loads. The current distribution is shown in Figure 3.49(b).

An equivalent circuit diagram is shown in Figure 3.49(c). The total load may be written as follows:

$$I = I_{L1} + I_1 \tag{3.169}$$

Now,

$$I_1 = I_{L2} + I_2$$

$$\therefore I = I_{L1} + I_1 = I_{L1} + I_{L2} + I_2 \tag{3.170}$$

Now,

$$I_2 = I_{L3} + I_3$$

$$\therefore I = I_{L1} + I_{L2} + I_2 = I_{L1} + I_{L2} + I_{L3} + I_3 \tag{3.171}$$

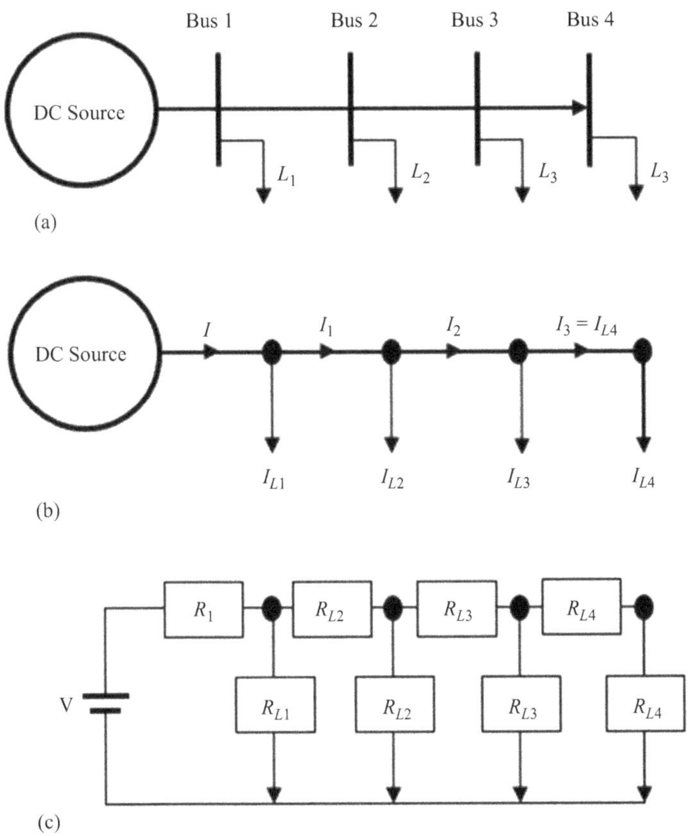

Figure 3.49 DC load distribution

Now,

$$I_3 = I_{L4}$$

$$\therefore I = I_{L1} + I_{L2} + I_{L3} + I_3 = I_{L1} + I_{L2} + I_{L3} + I_{L4} \quad (3.172)$$

Therefore, the power is distributed among 'n' number of loads, the total load may be written as follows:

$$\therefore I = I_{L1} + I_{L2} + I_{L3} + I_{L4} + \ldots + I_{Ln} \quad (3.173)$$

3.24 Summary

After discussing gradual changes of the grid structure, NG and PG were focused. Grid definitions have been provided for NG and PG have been provided. Then, components of NPG systems have been discussed. The discussion has revealed that converters are the most important unit of the internal section of NPG. Based on the type of NPG, different types of converters like DC–DC, DC–AC, and AC–DC converters are used. PWM-based choppers and inverters are widely used in NPG systems. Features of different choppers and inverters have been discussed. The transformer is used for connection with the external grid. Measurement units are important for monitoring and control operations. Considering all aspects, hexagonal topology is found very effective for NPG.

References

[1] B. Nordman and K. Christensen: 'Local power distribution with nanogrids', 2013 International Green Computing Conference Proceedings, Arlington, VA, 2013, pp. 1–8. DOI: 10.1109/IGCC.2013.6604464

[2] H. Samet, T. Ghanbari, E. Kazemzadeh, and E. Moshksar: 'Full-selective contribution of sources in DC nanogrids using a technique based on buck/boost converter', *IET Energy Systems Integration*, 1, no. 1, 2019, p. 14–22

[3] H. Samet, and E. Kazemzadeh, and T. Ghanbari: 'Realising full-selective sources contribution in AC nanogrids', *IET Energy Systems Integration*, 2, no. 1, 2020, p. 40–49.

[4] M. Heidari, T. Niknam, M. Zare, and S. Niknam: 'Integrated battery model in cost-effective operation and load management of grid-connected smart nanogrid', *IET Renewable Power Generation*, 13, no. 7, 2019, p. 1123–1131.

Further reading

[1] F. Asadi, K. Eguchi, and J. Hudgins, *Dynamics and Control of DC–DC Converters*, Morgan & Claypool, 2018.

[2] F. Asadi and J. Hudgins, *Computer Techniques for Dynamic Modeling of DC–DC Power Converters*, Morgan & Claypool, 2018.

[3] R. A. Dunlap, *Renewable Energy: Volume 1: Requirements and Sources*, Morgan & Claypool, 2020.

[4] M. Vilathgamuwa, D. Nayanasiri, and S. Gamini, *Power Electronics for Photovoltaic Power Systems*, Morgan & Claypool, 2015.

[5] S. Beeby and N. White, *Energy Harvesting for Autonomous Systems*, Artech, 2010.

[6] A. Phil Weicker, *Systems Approach to Lithium-Ion Battery Management*, Artech, 2013.

[7] Y. Chen, "Energy Sources," in *Energy Harvesting Communications: Principles and Theories*, Wiley, 2019, pp. 5–17, DOI: 10.1002/9781119383062.ch2.

[8] M. G. Simes and F. A. Farret, "Modeling Alternative Sources of Energy," in *Modeling Power Electronics and Interfacing Energy Conversion Systems*, IEEE, Wiley-IEEE Press, 2017, pp. 203–226, DOI: 10.1002/9781119058458.ch9.

[9] B. K. Bose, "Power Semiconductor Devices for Smart Grid and Renewable Energy Systems," in *Power Electronics in Renewable Energy Systems and Smart Grid: Technology and Applications*, IEEE, Wiley-IEEE Press, 2019, pp. 85–152, DOI: 10.1002/9781119515661.ch2.

[10] B. K. Bose, "Energy, Environment, Power Electronics, Renewable Energy Systems, and Smart Grid," in *Power Electronics in Renewable Energy Systems and Smart Grid: Technology and Applications*, IEEE, 2019, pp. 1–83, DOI: 10.1002/9781119515661.ch1.

[11] Q.-C. Zhong and T. Hornik, "Current Proportional – Integral Control," in *Control of Power Inverters in Renewable Energy and Smart Grid Integration*, IEEE, 2012, pp. 250–258, DOI: 10.1002/9781118481806.ch15.

[12] S. Rao, S. Katoch, V. Narayanaswamy, G. Muniraju, C. Tepedelenlioglu, A. Spanias, and P. Turaga, *Machine Learning for Solar Array Monitoring, Optimization, and Control*, Morgan & Claypool, 2020.

[13] H. Braun, M. Banavar, and A. Spanias, *Signal Processing for Solar Array Monitoring, Fault Detection, and Optimization*, Morgan & Claypool, 2012.

[14] P. R. Wolfe, "Economics of Solar Generation," in *The Solar Generation: Childhood and Adolescence of Terrestrial Photovoltaics*, IEEE, Wiley-IEEE Press, 2018, pp. 95–104, DOI: 10.1002/9781119425618.ch6.

[15] D. Grahame Holmes and T. A. Lipo, *Pulse Width Modulation for Power Converters: Principles and Practice*, Willey-IEEE Press, 2003.

[16] F. Asadi and J. Hudgins, *Robust Control of DC–DC Converters: The Kharitonov's Theorem Approach with MATLAB® Codes*, Morgan & Claypool, 2018.

Chapter 4
Operation and performance analysis

This chapter presents the operation and performance analysis of nanogrid and picogrid systems. At first, different control methods for nanogrids and picogrids have been described. Then some case studies on their operations have been presented. Then performance analysis has been carried out followed by a summary.

4.1 Introduction

The operation of nano and picogrids (NPGs) depends on grid configurations. Types of resources have a great influence on their operation. Availability and resource capacity become deciding factors for selecting modes of operation. The selection and operation of converter units are governed by the nature of resources, grid connectivity, and loads. Sensory units and measurement units keep in coordination with converters and control units. In most cases, discrete and numerical control logics are applied to operate NPG.

Performance analysis is normally done based on the outcome of the study on load management, implementation of power trading opportunity, reliability, and sustainability of the NPG system. Sensor-based supervisory monitoring is done before performance analysis. Bus voltages and waveforms at each intermediate stage of operation under different circumstances reveal the performance reports of each component of NPG.

4.2 Operation

The operation of NPG is interrelated with the operation of all sub-members of the NPG system. The operation of NPG varies with the type of NPG and depends largely on the operations of converters.

4.2.1 Open-loop operation

The open-loop operation is found very rare. In open-loop mode, rectifiers and inverters feed power to loads in an uncontrolled manner. The output voltage does not remain constant with the change of load. The output also varies with the change of different input variables. Thus, the use of such open-loop uncontrolled converters lacks in achieving consumers' satisfaction.

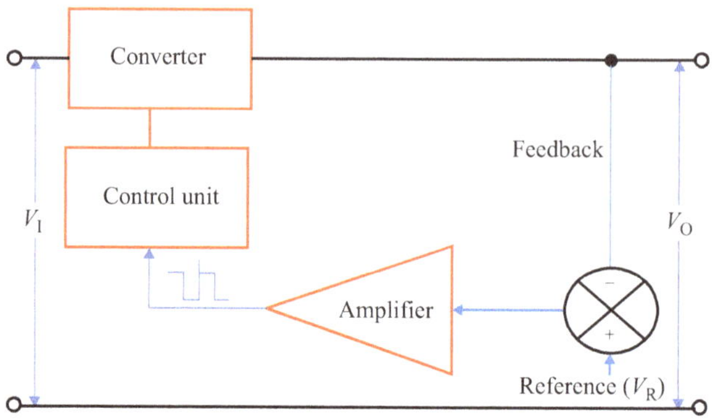

Figure 4.1 Closed-loop operations of converters

4.2.2 Closed-loop operation

Closed-loop operations are found in most cases of NPGs. The performance of closed-loop operation largely depends on the performance of converters. As converter output should synchronize with external grid (EG) connection and other loads, the operation is controlled based on feedback signals collected from the output. The closed-loop control system of the converter is shown in Figure 4.1. It collects output signals and processes them for feedback. With the help of a reference "set" value, an error signal is generated. Based on error signals, it operates. The operation involves with triggering and commutations of different semiconductor switches. "ON" and "OFF" sequence for different switching components in time scale decide the output of the converters. Switching operations are to be set by the type of switching devices and circuit configurations.

Output is judged based on the following main parameters of electrical signals:

- amplitude,
- frequency, and
- harmonics content.

The frequency and the amplitude of output signals decide the performance of the converter, while harmonics content decides the quality parameters of the output.

4.3 Classification of NPG control

Different control schemes are needed for NPG applications that are controlled and operated based on the availability of resources and load demands.

Broadly, control and operation schemes for NPG systems may be divided into two categories as follows:

(a) Supply-side management (SSM).
(b) Demand-side management.

4.3.1 Supply-side management (SSM)

In SSM, control operation considers the following things:

- type of resources e.g., solar PV, wind turbine, etc.;
- battery bank capacity;
- state of charge (SoC) of batteries;
- optimized source integration and storage control;
- manipulation or adjustment of loads.

Maximum power point tracking (MPPT) is a widely used control scheme that belongs to SSM.

4.3.2 Demand-side management (DSM)

In DSM, the control scheme considers the following issues:

- state of load demand;
- capacity of grids;
- integration of different NPGs;
- integration with the utility grid;
- control of load shading.

Demand feedback and real-time tariff-based different control schemes are common examples of demand-side load management. The main parameters of the performance index of control schemes belonging to DSM are

- demand response,
- dynamic demand, and
- energy efficiency.

4.4 MPPT in PV control system for NPG

Most of the PV-based NPG systems are incorporated with MPPT with the controller unit. This improves efficiency and increases the utilization of solar power. The operation of the NPG system greatly depends on the efficient performance of this tracking system along with the control logic. Different aspects of MPPT will be discussed in the following sub-sections.

4.4.1 Difficulties of using solar PV generation efficiently

There are few difficulties in solar PV-based electric power generation. These are as follows:

- Characteristics of PV cells depend on sunshine. Temperature and irradiance vary with sunshine and weather conditions. Thus, PV features change with irradiance, ambient temperature, and cell temperature.
- The generated voltage varies with temperature and irradiance. For example, if a typical PV module voltage at maximum power generation is around 17 V for

a measured cell temperature of 25 °C, it may drop to around 15 V on a very hot day and it may increase to 18 V on a very cold day.

- The solar cell can deliver maximum power at particular operating conditions near the knee point of *I–V* characteristics.
- Maximum power changes with irradiance and temperature.
- An open-loop PV-based supply system is not capable of extracting maximum power at variable temperature and irradiance.
- PV output voltage is not often directly compatible with battery/load voltage that needs for voltage conversion.

4.4.2 Importance of MPPT

Solar PV can deliver maximum power near the knee point of *I–V* characteristics. But, the operating point shifts with the change of temperature, irradiance, etc. Those are weather-dependent parameters. Therefore, if the voltage and current are not adjusted, maximum power cannot be extracted. MPPT system adjusts current concerning voltage in such a way that the operating point exists exactly at or near maximum power point in the characteristics resulting in maximum power generation. Therefore, MPPT is important for generating maximum power at different weather conditions influencing parameters like temperature, irradiance, etc.

4.4.3 Components of MPPT-based control system

The MPPT-based control system is shown in the block diagram in Figures 4.2 and 4.3. PV panel output is connected with the converter. The converter is of DC–DC

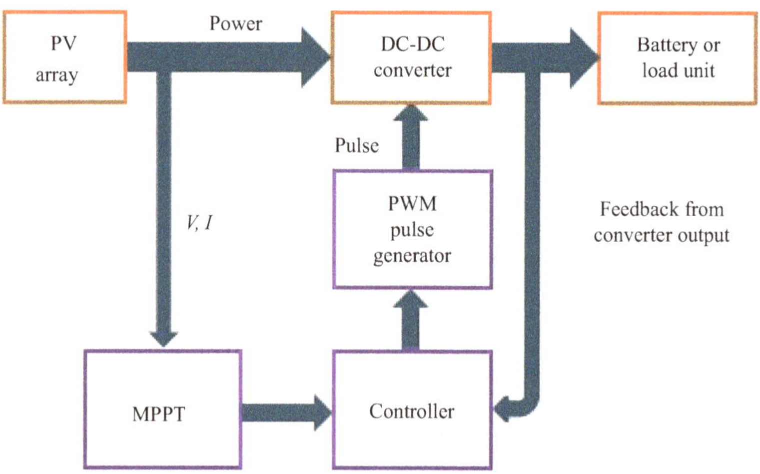

Figure 4.2 MPPT-based control system for NPG applications

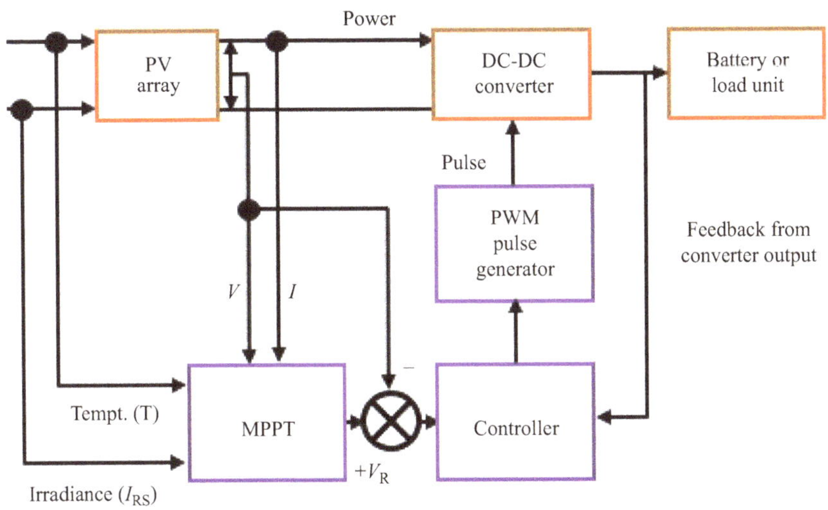

Figure 4.3 MPPT-based control system for NPG applications showing temperature and irradiance input

type. Based on battery or load voltage requirements, a boost or buck converter is chosen. The control system has three main parts as follows:

- MPPT.
- Controller for PWM-based pulse generation.
- PWM pulse generator or power driver.

4.4.4 Maximum power point tracker

The tracker aims for an operating point at or near the point of maximum power. It receives voltage and current signals from the PV panel for measuring output power. It analyses the power and current for the voltage variation and checks whether the operating point is located at or near the point of maximum power. If there is a deviation, it tries to understand how the nature of deviation is. Accordingly, it sends a signal to the controller to adjust the operating point.

4.4.5 Controller for PWM-based pulse generation

This control unit takes the output from MPPT and feedback signal from converter output. Based on the logic algorithm, it determines control signals for PWM. Different control logics have been introduced and research is going on to develop better controllers.

4.4.6 PWM pulse generator or power driver

The pulse generator generates pulse signals as per the control signal received from the control unit and sends them to the switching devices of the converter. Converter sends power generated by PV panel at a desired voltage level. It is also known as a PWM power driver.

4.5 MPPT solar charge controller

4.5.1 MPPT controller

An MPPT-based solar charge controller (SCC) is a solar PV-fed DC–DC converter with a maximum power tracking algorithm-based controller to maximize the charging current going into the battery. MPPT SCC is shown by a blue dotted box in a solar PV system in Figure 4.4. MPPT is DC to DC converter which operates by taking DC input from the PV module, changing it to AC, and converting it back to a different DC voltage and current to exactly match the PV module.

4.5.2 Main features of MPPT controller

The main features of the MPPT solar charge controller are as follows:

- MPPT-based charge controller well understands the characteristics of solar-PV resource i.e., *I–V* characteristics of PV module.
- MPPT-based charge controller extracts the maximum power from the solar module at variable weather conditions.
- MPPT-based charge controller forces the PV module to generate power at a voltage that is closer to a maximum power-point.
- MPPT-based charge controller can provide different voltage levels depending on the requirement of the battery.

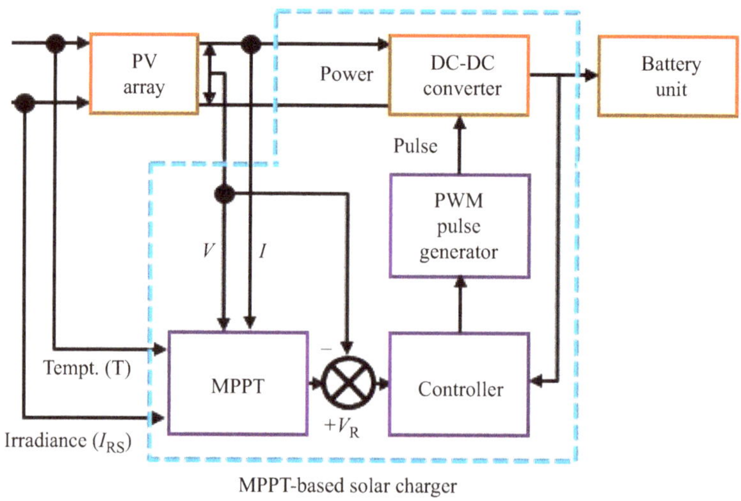

Figure 4.4 MPPT solar charge controller has been shown by blue dotted box in solar PV system

- MPPT-based charge controller is less complex and improves efficiency.
- MPPT-based charge controller can also be used to feed other loads applied to other loads.

4.5.3 MPPT with other controllers

Different types of the algorithm have been introduced and are in use for NPG application. Some control techniques are mentioned below:

- Conductance-based MPPT controller.
- PI controller.
- PID controller.
- Fuzzy-based controller.
- Genetic algorithm-based controller.
- Droop controller.

4.5.4 Incremental conductance-based MPPT controller

The incremental conductance-based MPPT controller is a modified version of Perturb and observes method. It works on the relative change of current concerning the relative change of voltage.

Let V is the output voltage of PV panel; I is the output current of PV panel; P is the output power of PV panel.

Therefore,

$$P = VI$$

The objective is to maximize P, i.e.,

$$\begin{aligned} \text{Max } P &= VI \\ \therefore \frac{dP}{dV} &= \frac{d(VI)}{dV} \\ &= V\frac{dI}{dV} + I\frac{dV}{dV} \\ &= V\frac{dI}{dV} + I \end{aligned} \tag{4.1}$$

When the power is maximum,

$$\begin{aligned} &\frac{dP}{dV} = 0 \\ &V\frac{dI}{dV} + I = 0 \\ &V\frac{dI}{dV} = -I \\ &\frac{dI}{dV} = -\frac{I}{V} \end{aligned} \tag{4.2}$$

Thus, the final inference is presented in Table 4.1.

Table 4.1 Incremental current and operating point

Observation	Inference
$\frac{\Delta I}{\Delta V} = -\frac{I}{V}$	The operating point is at the maximum power-point
$\frac{\Delta I}{\Delta V} > -\frac{I}{V}$	The operating point is at the left of the maximum power-point
$\frac{\Delta I}{\Delta V} < -\frac{I}{V}$	The operating point is at the right of maximum power-point

The reference signal of the control unit can be written by the following algorithm:

(i) Let i be variable and K be small constant for adjusting reference input V_r
(ii) Measure $V(i)$ and I (i)
(iii) Determine $\Delta V = V(i) - V(i-1)$ and $\Delta I = I(i) - I(i-1)$
(iv) If $\Delta V = 0$, then

if $\Delta I = 0$ then

$V(i) = V(i-1)$

$I(i) = I(i-1)$

else

if $\Delta I > 0$ then

$V_r = V_r + K$

$V(i) = V(i-1)$

$I(i) = I(i-1)$

else

$V_r = V_r - K$

$V(i) = V(i-1)$

$I(i) = I(i-1)$

if $\Delta I \neq 0$ then

if $\dfrac{\Delta I}{\Delta V} = -\dfrac{I}{V}$ then

$V(i) = V(i-1)$

$I(i) = I(i-1)$

else

if $\dfrac{\Delta I}{\Delta V} > -\dfrac{I}{V}$ then

$V_r = V_r + K$

$V(i) = V(i-1)$

$I(i) = I(i-1)$

else

$$V_r = V_r - K$$
$$V(i) = V(i-1)$$
$$I(i) = I(i-1)$$

4.5.5 Proportional-integration (PI) controller and proportional-integration-differentiation (PID) controller

The controller receives an output of MPPT and feedback signal from DC–DC converter output. The controller generates signals for the pulse-generator following PI logic or PID logic as shown in Figures 4.5 and 4.6, respectively. K_P, K_I, and K_D are constants for the control path for proportional (P) control, integration (I) control, and differentiation (D) control, respectively.

4.5.6 Fuzzy-based controller (FC) and genetic algorithm-based controller (GAC)

The controller receives an output of MPPT and feedback signal from DC–DC converter output. The operating point for maximum power is tracked and the

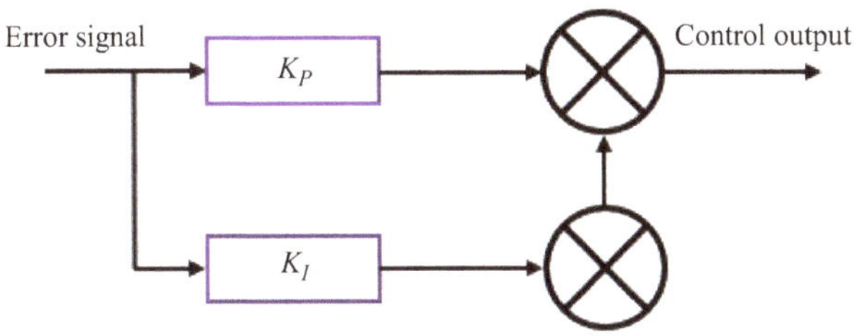

Figure 4.5 PI controller for NPG system

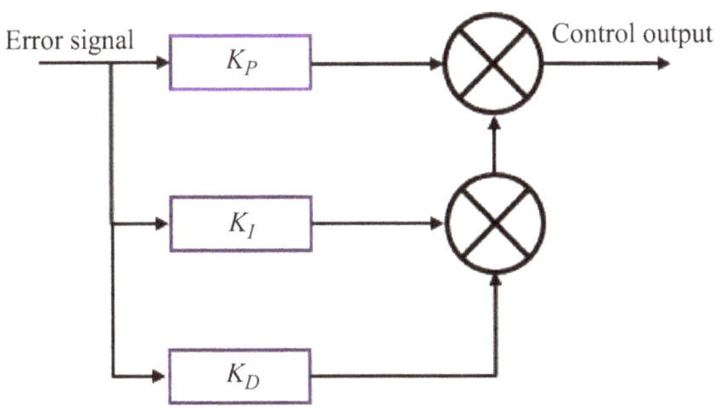

Figure 4.6 PID controller for NPG system

controller generates signals for pulse-generator based on Fuzzy logic and/or genetic algorithm (GA) as shown in Figure 4.7.

4.5.7 Voltage droop controller (VDC)

In VDC, droop in voltage profile is monitored. The controller receives the output of MPPT and feedback signal from the DC–DC converter output. Voltage drooping is monitored and the controller generates signals for the pulse-generator as shown in Figure 4.8.

4.6 Droop control

4.6.1 Voltage droop

Voltage droop refers to the loss that occurs between the supplier and the load at the time of power delivery. Let us understand voltage droop by simple Ohm's law.

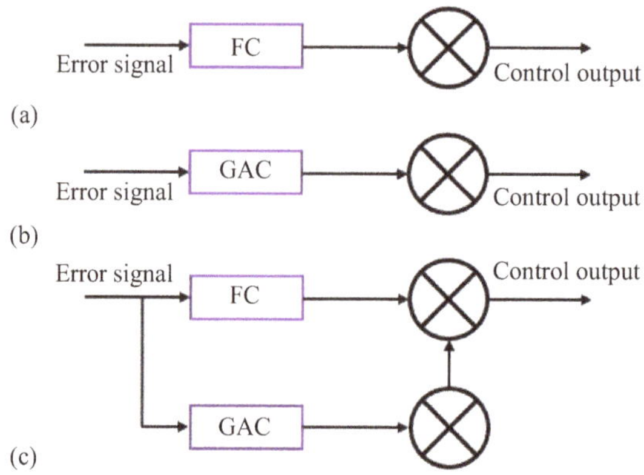

Figure 4.7 Fuzzy or GA-based controller for NPG system: (a) fuzzy logic-based controller, (b) GA-based controller, and (c) GA and fuzzy logic-based controller

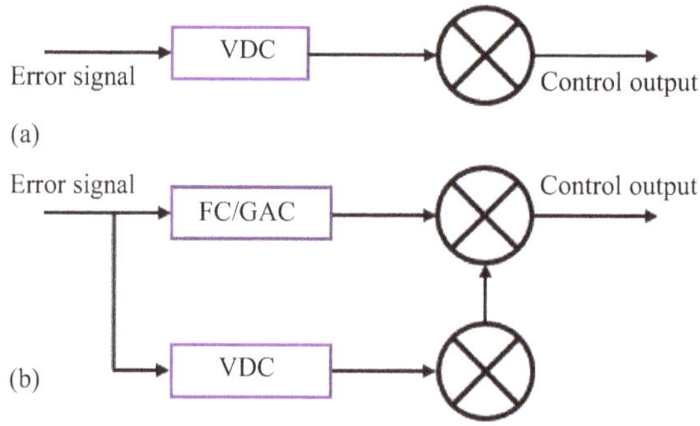

Figure 4.8 Droop controller for NPG system: (a) voltage droop controller and (b) voltage droop controller with fuzzy or GA-based controller

Consider a resistive circuit where a supply is connected with a load as shown in Figure 4.9.

Let

E is the supply voltage at any load; I is the current when the supply is delivering power to load unit; V is the voltage at load terminal; R is the resistance between regulated supply and load.

Using Kirchhoff's voltage law,

$$E = V + IR \tag{4.3}$$

or,

$$V = E - IR \tag{4.4}$$

In this circuit, following conditions may occur:

- no load,
- rated load,
- under load, and
- overload.

At no load,

$$I = 0,\ V = V_{NL}\ (say)$$
$$\therefore V = V_{NL} = E - IR = E - 0 = E$$

At rated load (I_R),

$$I = I_R,\ V = V_R\ (say)$$
$$\therefore V = V_R = E - I_R R$$
$$\therefore V < V_{NL}$$

At under load (I_{UL}),

$$I = I_{UL},\ \text{where } I_{UL} < I_R,$$
$$\therefore V = E - I_{UL} R$$
$$\therefore V < V_{NL} \text{ and } V > V_R$$

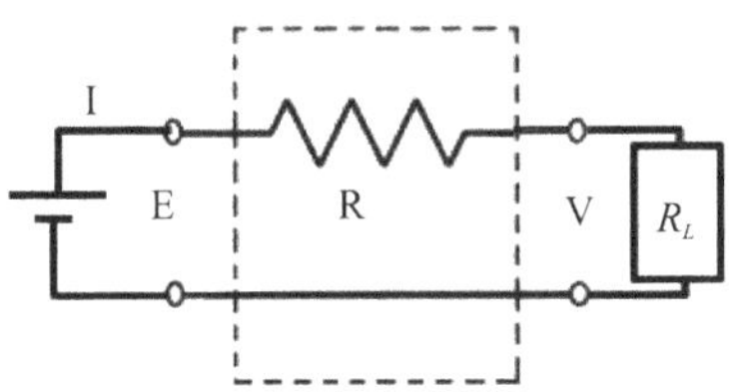

Figure 4.9 A resistive circuit with a supply connected with a load

At overload (I_{OL}),

$$I = I_{OL}, \text{ where } I_{OL} > I_R,$$
$$\therefore V = E - I_{OR}R$$
$$\therefore V < V_R < V_{NL}$$

All these cases are shown in Figure 4.10.

It is now clear that voltage droop occurs during load change that is decided by *IR* drop in a resistive circuit. Let us now see voltage droop concerning voltage (V_R) at rated load as shown in Figure 4.11. Let V_{min} and V_{max} are minimum voltage and maximum voltage, respectively. If load increases beyond rated value, droop increases, and voltage decreases. On the other hand, if load decreases below the rated value, droop decreases, and voltage increases up to no-load voltage V_{NL}.

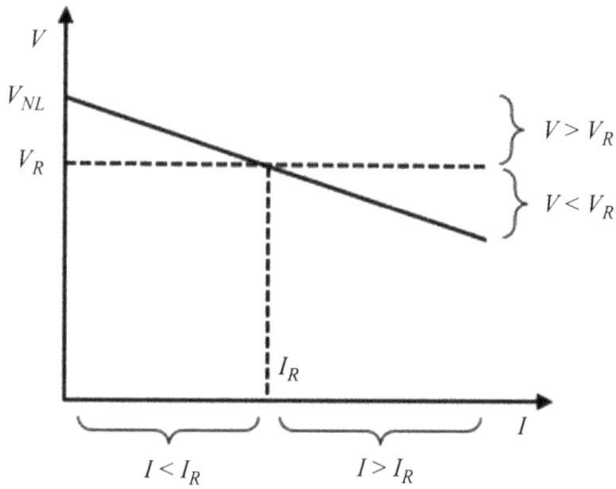

Figure 4.10 Load characteristics of the resistive circuit

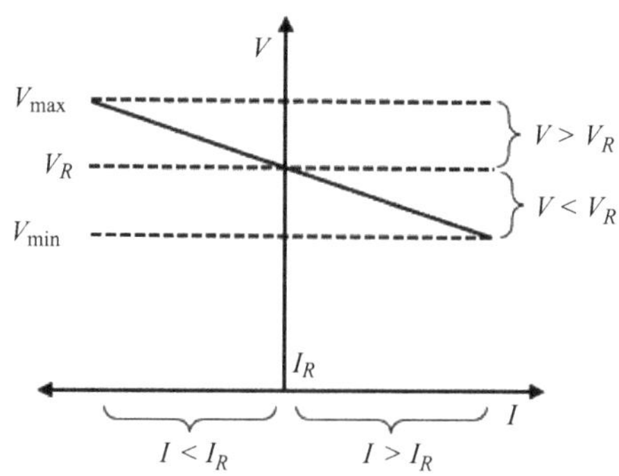

Figure 4.11 Droop characteristics of the resistive circuit

Therefore,

$$V_{\max} = V_{NL}$$

Concerning rated voltage, two extreme droops will be

$$\text{Positive droop} = V_{RL} - V_{\min}$$

$$\text{Negative droop} = V_{RL} - V_{\max} = -(V_{\max} - V_{RL})$$

Concerning maximum voltage, maximum voltage droop $(D_{\max})$ may be written as follows:

$$D_{\max} \quad = V_{\max} - V_{\min} \tag{4.5}$$

Therefore, half of the voltage droop (D_{half}) may be written as

$$D_{\text{half}} = \frac{V_{\max} - V_{\min}}{2} \tag{4.6}$$

In practice, droop may vary in time scale. An example has been shown for a system that goes through maximum droop and the then half droop as shown in Figure 4.12.

4.6.2 Droop control in AC network

Consider an AC network where a source is feeding power to a bus as shown in Figure 4.13.

Let

V_1 is the supply voltage at bus 1; V_2 is the voltage at bus2; I is the current.

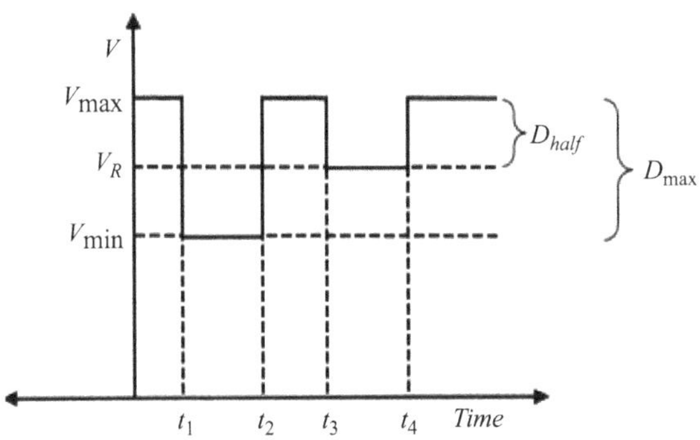

Figure 4.12 Example of maximum and half droops in a time scale

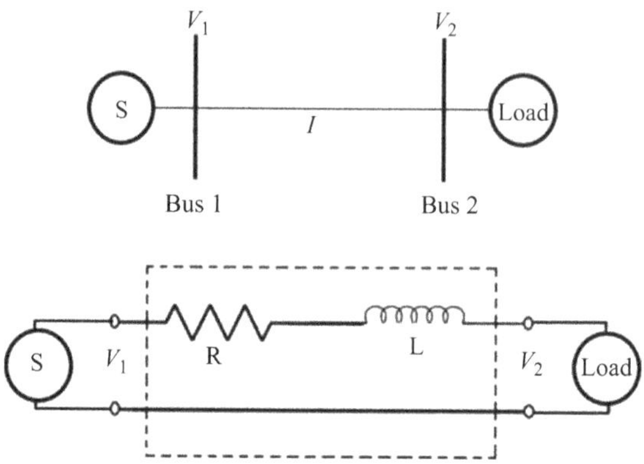

Figure 4.13 A network with an AC source connected with load and its equivalent circuit

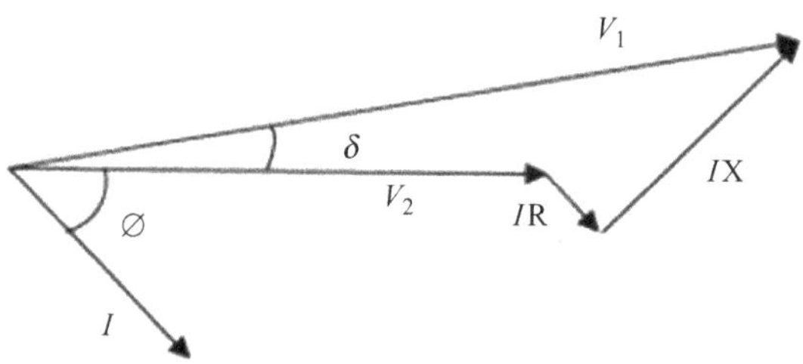

Figure 4.14 Phasor of the above AC network

The network is made of resistance and inductive reactance and is connected with inductive load. Therefore, the model is an inductive network.

Let

R is the line resistance; L is the line inductance; X is the line reactance.

The current flowing through the line is inductive, and lagging behind the voltage. Voltage drops are shown in Figure 4.14.

From the phasor, IR is the drop across line resistance and it is parallel to I, IX is the drop across inductance and it is perpendicular to I or IR. Sending end voltage V_1 is the vector sum of receiving end voltage V_2 , drop across the resistance (IR) and drop across the reactance (IX).

The phasor indicates that voltage at bus 1 is greater than the voltage at bus 2. Let

$$\varnothing = \angle V_1,\ I$$
$$\delta = \angle V_1,\ V_2$$

From the geometry of the phasor, sending end voltage can be written as follows:

$$V_1^2 = (V_2 \cos \varnothing + IR)^2 + (V_2 \sin \varnothing + IX)^2$$

$$V_1 = \sqrt{(V_2 \cos \varnothing + IR)^2 + (V_2 \sin \varnothing + IX)^2}$$

$$V_1 = \sqrt{V_2^2 + 2V_R IR \cos \varnothing + 2V_2 IX \sin \varnothing + (IR)^2 + (IX)^2} \tag{4.7}$$

$$V_1 \cong \sqrt{V_2^2 + 2V_R IR \cos \varnothing + 2V_2 IX \sin \varnothing}$$

$$V_1 \cong V_2 \left(1 + \frac{2IR \cos \varnothing + 2IX \sin \varnothing}{V_2}\right)^{\frac{1}{2}}$$

$$V_1 \cong V_2 \left(1 + \frac{IR\cos \varnothing + IX\sin \varnothing}{V_2}\right) \tag{4.8}$$

$$V_1 \cong V_2 + (IR\cos \varnothing + IX\sin \varnothing) \tag{4.9}$$

$$\therefore Reg\% = \frac{V_1 - V_2}{V_1} \times 100\%$$

$$\cong \frac{IR\cos \varnothing + IX\sin \varnothing}{V_1} \times 100\% \tag{4.10}$$

Active power flow through the circuit can be written as follows:

$$P = \frac{V_1}{R^2 + X^2}\{R(V_1 - V_2 \cos \delta) + XV_2 \sin \delta\} \tag{4.11}$$

Active power flow through the circuit can be written as follows:

$$Q = \frac{V_1}{R^2 + X^2}\{-RV_2 \cos \delta + X(V_1 - V_2 \cos \delta) + XV_2 \sin \delta\} \tag{4.12}$$

When reactance is very much greater than resistance, i.e., $X \gg R$, δ will be very small. Therefore, the following approximation can be made:

$$\sin \delta \approx 1$$
$$\sin \delta \approx 1$$

Then, active and reactive powers can be written as follows:

$$P = \frac{V_1 V_2}{X} \delta \tag{4.13}$$

$$Q = \frac{V_1^2 - V_1 V_2}{X} \tag{4.14}$$

Voltages and frequency or angular frequency are controlled to control the flow of active power (P) and reactive power (Q). Error signal generation in voltage droop controller is shown by a block diagram in Figure 4.15.

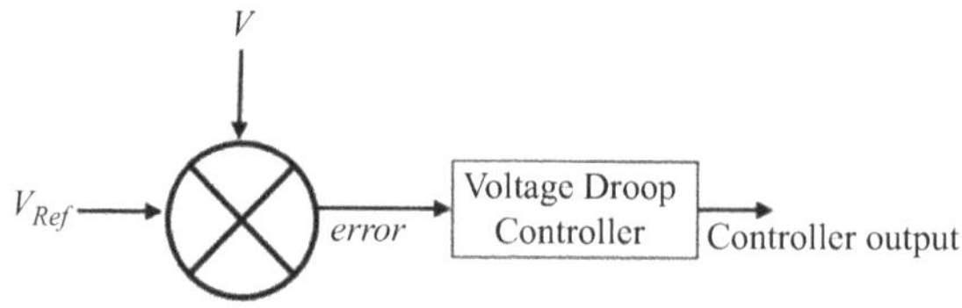

Figure 4.15 Error signal generation in voltage droop controller

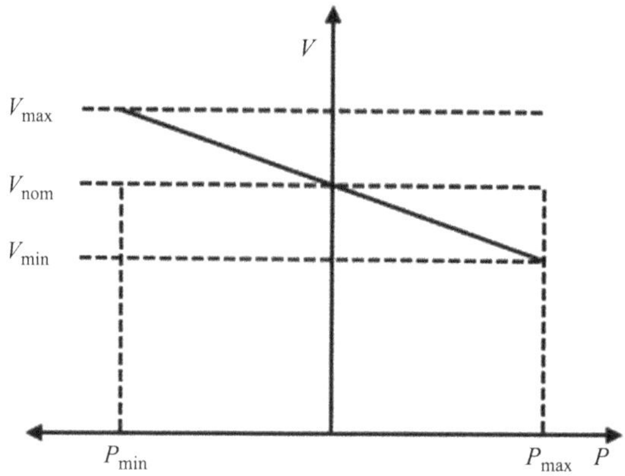

Figure 4.16 Active power versus voltage used in droop control

Active power voltage used in voltage droop control is shown in Figure 4.16. In droop control, the voltage can be written as follows:

$$V = V_0 + k_P(P_0 - P) \tag{4.15}$$

where V_0 is the set or rated voltage; P_0 is the set or rated active power;

$$k_P = \Delta V/Q_{\max}.$$

where $P_{\max}$is the maximum limit for flow of active power.

k_P is called the voltage droop coefficient for active power.

Reactive power–voltage used in voltage droop control is shown in Figure 4.17. In droop control, the voltage can be written as follows:

$$V = V_0 + k_Q(Q_0 - Q) \tag{4.16}$$

where

V_0 is the set or rated voltage; Q_0 is the set or rated reactive power; k_Q

$= \Delta V/Q_{\max}.$

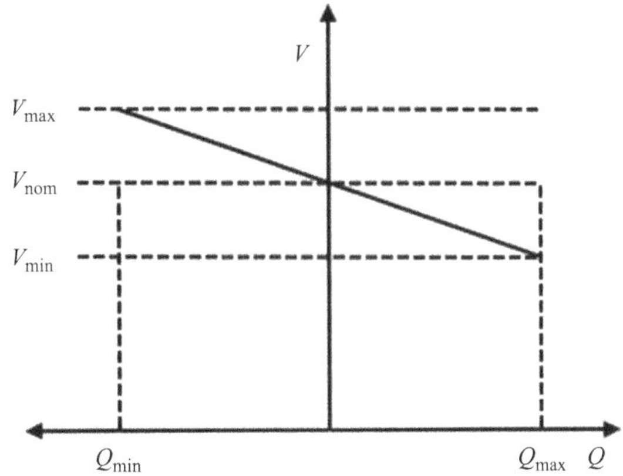

Figure 4.17 Reactive power versus voltage used in droop control

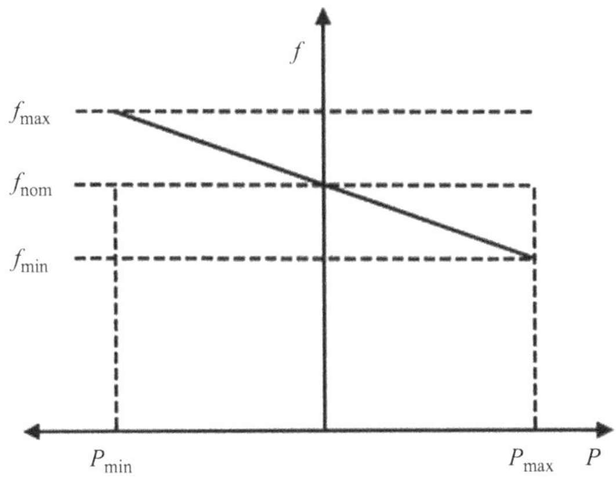

Figure 4.18 Active power versus frequency used in droop control

where Q_{max} is the maximum limit for flow of reactive power. k_Q is called the voltage droop coefficient for reactive power.

During control, frequency, angular frequency, and load angle are related by

$$f_n = \frac{\omega_n}{2\pi} = \frac{d\delta_n}{dt} \tag{4.17}$$

Active power–frequency used in droop control is shown in Figure 4.18. In droop control, the frequency can be written as follows:

$$f = f_0 + k_{fP}(P_0 - P) \tag{4.18}$$

where f_0 is the set or rated frequency; P_0 is the set or rated active power; $k_{fP} = \frac{\Delta f}{P_{max}}$. where P_{max} is the maximum limit for flow of active power. k_{fP} is called the frequency droop coefficient for active power.

The active power–angular frequency used in droop control is shown in Figure 4.19. In droop control, the angular frequency can be written as follows:

$$\omega = \omega_0 + k_{\omega P}(P_0 - P) \tag{4.19}$$

where

ω_0 is the set or rated angular frequency; P_0 is the set or rated active power; $k_{\omega P} = \frac{\Delta \omega}{P_{max}}$.

where P_{max} is the maximum limit for flow of active power. $k_{\omega P}$ is called the frequency droop coefficient for active power.

4.6.3 Traditional droop control in AC network

Traditional droop control involves frequency and voltage control for control of active and reactive powers, respectively.

$$\omega = \omega_0 - f_{dP}P \tag{4.20}$$

where f_{dP} is the droop factor for active power, defined as

$$f_{dP} = \frac{\omega_0 - \omega}{P} \tag{4.21}$$

$$V = V_0 - f_{dQ}Q \tag{4.22}$$

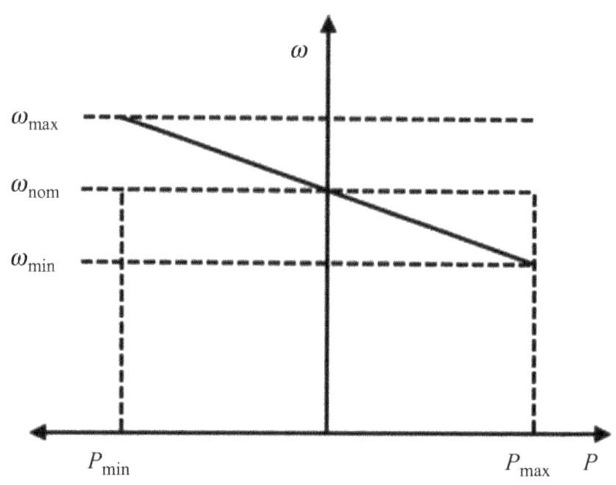

Figure 4.19 Active power versus angular frequency used in droop control

where f_{dQ} is the droop factor for reactive power, defined as

$$f_{dQ} = \frac{V_0 - V}{Q} \tag{4.23}$$

4.6.4 Droop control in DC network

Droop control in the DC network can be done by controlling voltage with the change of either power or current.

Power–voltage used in droop control is shown in Figure 4.20. In droop control, power can be expressed as follows:

$$P_{DCi} = P_{DC0} - \frac{V_{DCi} - V_{DC0}}{k_{DCP}} \tag{4.24}$$

where

P_{DCi} is the power at ith step; P_{DC0} is the power previous step;
V_{DCi} is the voltage at i th step; V_{DC0} is the voltage previous step;

$$k_D = \frac{V_{DCi} - V_{DC0}}{P_{DCi} - P_{DC0}}$$

k_{DCP} is known as the DC voltage droop constant in terms of power (it is less than 1).

Direct current–voltage used in droop control is shown in Figure 4.21. In droop control, current can be expressed as follows:

$$I_{DCi} = I_{DC0} - \frac{V_{DCi} - V_{DC0}}{k_{DCP}} \tag{4.25}$$

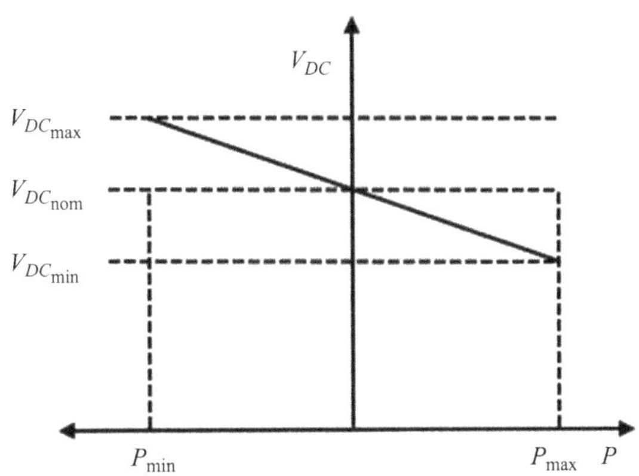

Figure 4.20 Power versus DC voltage in DC network used in droop control

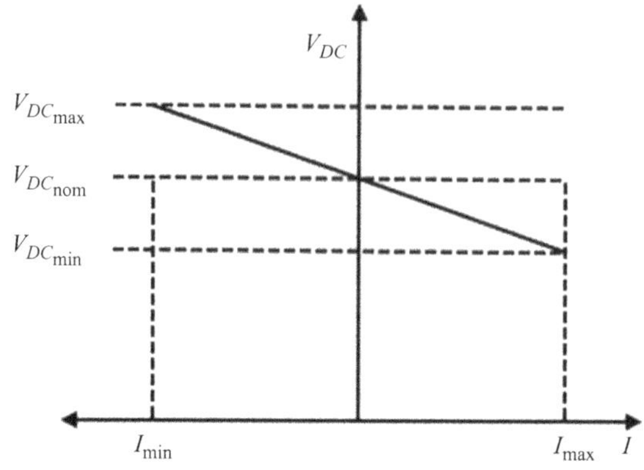

Figure 4.21 Current versus voltage in DC network used in droop control

where I_{DCi} is the power at *i*th step; I_{DC0} is the power previous step; I_{DCi} is the voltage at *i*th step; I_{DC0} is the voltage previous step;

$$k_{DCI} = \frac{V_{DCi} - V_{DC0}}{P_{DCi} - P_{DC0}}$$

k_{DCI} is known as the DC voltage droop constant in terms of current (it is less than 1).

4.6.5 Main features of droop control

The main features of the droop control technique are as follows:

- It is easy to implement.
- It depends much on output feedback and does not rely on communication.
- It can be used between inverter and grid or load and between chopper and load.
- Filter effects can be included.
- Droop gain may be controlled.
- The filter time constant may be controlled.
- Distributed control is possible in the grid network.

4.7 Virtual synchronous machine (VSM)-based control

An alternate approach to droop control is a VSM. It is an emerging control technique used for microgrid and nanogrid controls. VSM refers to static synchronous machines that consider the effect of inertial of the conventional synchronous machine along with active and reactive power. Both current source and voltage source-based VSM have been introduced. The damping effect of VSM based control system can be considered equivalent to droop gain and filter time constant of the droop control system. There are various types of models for VSM.

Models for VSM may be classified from orders 2 to 7; among them the model of order 3 is popular. In VSM, voltage and current can be expressed by d and q axes as follows:

$$V_d = \omega(E_d + X_q I_q) + R_S I_d \quad (4.26)$$

$$V_q = \omega(E_q + X_d I_d) + R_S I_q \quad (4.27)$$

However, VSM-based nanogrid control operation is more complex than droop control and more effort will be required to implement VSM-based distributed nanogrid control system.

4.8 Selection of boost or buck converter

Selection of boost converter or buck converter as solar PV fed DC–DC converter for battery charger is normally decided by voltage levels. If solar PV array output voltage level is compatible with battery voltage level, buck converter may be used. If the voltage level of the battery is much higher than PV output voltage, boost converters are normally chosen. Usually, in small applications battery voltage levels are normally from 2 V, 12 V, 24 V, up to 48 V, and solar PV array output voltage is kept to be compatible with this voltage level.

4.9 Optimization for pulse width modulation (PWM)

Signal generation for the PVM is done in an optimized way. Different commonly used optimization techniques are as follows:

- *Current control charging mode*: Optimization takes place with a constant charging current.
- *Current control discharging mode*: Optimization takes place with constant discharging of current.
- *Voltage droop control (V–I) mode*: Optimization takes place by controlling voltage droop using *V–I* characteristics.
- *Constant I–V droop mode*: Optimization takes place considering constant droops.
- *Power flow control mode*: Optimization takes place by controlling power flow.

4.10 Grid power flow direction control

4.10.1 Power flow direction and control

The direction of power flow between NPG and the EG is provided in the gateway of the EG. Based on direction and its control power, flow between NPG and EG may be classified in the following way:

(a) Unidirectional power flow (UPF):
  (i) uncontrolled UPF (UUPF or U2PF);
  (ii) controlled UPF (CUPF).

(b) Bidirectional power flow (BPF):
   (i) uncontrolled BPF (UBPF);
   (ii) controlled BPF (CBPF).

Consider that an NPG system and an EG are connected through a gateway as shown in Figure 4.22. Two types of the direction of power flow have been shown in the Figure. Unidirectional mode of operation can be of two types:

(i) Type 1: from NPG to EG.
(ii) Type 2: from the EG to NPG.

In the bidirectional mode of operation, power flow is flexible and can flow in any direction based on power availability at the NPG and local load requirement.

Different combinations of control and directional features may be represented by digital logic "0" and "1".

Let,

Unidirectional power flow $= \overline{A} = 0$
Bidirectional power flow $= A = 1$
Uncontrolled $= \overline{B} = 0$
Controlled $= B = 1$

Possible digital combinations are presented in Table 4.2.

Based on the above digital combinations, four types of gateway control are possible as presented in Table 4.3.

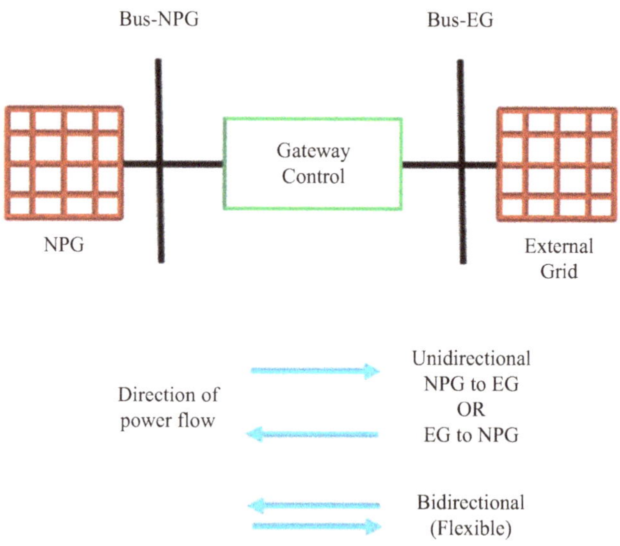

Figure 4.22 An NPG system and an EG connected through a gateway

Table 4.2 Possible grid power controls

	Uncontrolled ($\overline{B} = 0$)	**Controlled ($B = 1$)**
UPF ($\overline{A} = 0$)	00	01
BPF ($A = 1$)	10	11

Table 4.3 NPG gateway control

Gateway control	**Logic**	**Power flow**
$\overline{A}\overline{B}$	00	UUPF or U2PF
$\overline{A}B$	01	CUPF
$A\overline{B}$	10	UBPF
AB	11	CBPF

4.10.2 Factors of grid power flow control

Control of grid power flow depends on the following aspects:

- commercial relation between NPG and EG;
- NPG resource type;
- dynamic time-based characteristics or availability of NPG resource;
- local load;
- grid structure;
- guidelines of regulatory bodies.

4.10.3 Common practice

Most of the low-cost domestic utilities do not have gateway control that allows the uncontrolled bidirectional flow of power between NPG and the EG. In some specific applications, uncontrolled and controlled unidirectional power flow gateway is observed. However, in commercial relations, and to achieve optimum tariff, controlled bidirectional power flow is used in gateway control.

Let

P_{NPG} is the NPG generation capacity; P_{EG} is the power flow from NPG to EG;

P_L is the NPG load.

Based on gateway control schemes, different directional situations may arise for power flow between NPG and EG as presented in Table 4.4.

4.10.4 Control equipment

The direction of power flow between NPG and EG is controlled in the gateway employing the following equipment:

- directional relay;
- thyristor switch;

Table 4.4 Gateway control schemes based on different directional situations for power flow between NPG and EG

NPG-EG gateway control scheme	Condition	P_{EG}	Power flow between NPG and EG	
Controlled bi-directional power flow (CBPF)	$P_{NPG} = P_L$	$P_{EG} = 0$	Stop: no power flow	NPG can take power from EG or give power to EG (this option is most common)
	$P_{NPG} > P_L$	$P_{EG} =$ positive	NPG to EG	
	$P_{NPG} < P_L$	$P_{EG} =$ negative	EG to NPG	
Controlled uni-directional power flow (CBPF): Type 1	$P_{NPG} = P_L$	$P_{EG} = 0$	Stop: no power flow	NPG never takes power from EG but can give power to EG (this option is rarely chosen)
	$P_{NPG} > P_L$	$P_{EG} =$ positive	NPG to EG	
	$P_{NPG} < P_L$	$P_{EG} =$ negative	Stop: no power flow	
Controlled uni-directional power flow (CBPF): Type 2	$P_{NPG} = P_L$	$P_{EG} = 0$	Stop: no power flow	NPG can take power from EG but never gives power to EG (this option is chosen often)
	$P_{NPG} > P_L$	$P_{EG} =$ positive	Stop: no power flow	
	$P_{NPG} < P_L$	$P_{EG} =$ negative	EG to NPG	

- saturable current transformer;
- reactors;
- electromagnetic relay with normally open and normally closed contacts.

4.11 Clarke and Park transformations in NPG power flow control

Three-phase power flow control in NPG involves computation with three-phase voltages and currents. This computation deals with three voltages and three currents for a three-phase power flow control. The Clarke and Park transformations reduce these parameters into two voltages and two currents effectively reducing the computational task to a great extent. For this reason, these transforms are often used in internal power control in NPG systems as well as power flow control between NPG and EG. Currents and voltages are represented by space vectors in Clarke and Park planes for $\alpha - \beta$ reference frame and *d–q* reference frame, respectively.

Let, i_R, i_Y, and i_B represents currents of a three-phase system. If the system is balanced, then,

$$i_R + i_Y + i_B = 0 \tag{4.28}$$

The current space vector can be written as

$$\bar{i} = k\left(i_R + ai_Y + a^2 i_B\right) \tag{4.29}$$

where a is an operator used to represent angle in terms of the complex numbers and k is a constant of transformation.

The current space vector can be expressed in Clarke (α, β) plane as:

$$\bar{i} = (i_\alpha + i_\beta) \tag{4.30}$$

current components can be resented in terms of three-phase currents as follows:

$$\begin{bmatrix} i_\alpha \\ i_\beta \end{bmatrix} = k \times \begin{bmatrix} 1 & -\frac{1}{2} & -\frac{1}{2} \\ 0 & \frac{\sqrt{3}}{2} & -\frac{\sqrt{3}}{2} \end{bmatrix} \times \begin{bmatrix} i_R \\ i_Y \\ i_B \end{bmatrix} \tag{4.31}$$

The recommended value of k is

$$k = \frac{2}{3}$$

Therefore,

$$\begin{bmatrix} i_\alpha \\ i_\beta \end{bmatrix} = [\text{Clarke Matrix}] \times \begin{bmatrix} i_R \\ i_Y \\ i_B \end{bmatrix} \tag{4.32}$$

where,

$$[\text{Clarke Matrix}] = \frac{2}{3} \times \begin{bmatrix} 1 & -\frac{1}{2} & -\frac{1}{2} \\ 0 & \frac{\sqrt{3}}{2} & -\frac{\sqrt{3}}{2} \end{bmatrix} \tag{4.33}$$

The space vector components in the Clarke plane can be transformed into a space vector in the Park plane by Park transformation. The variables of the Clerke plane (α, β) can be converted to (d, q) by following way:

$$\begin{bmatrix} d \\ q \end{bmatrix} = \begin{bmatrix} \cos\theta & \sin\theta \\ -\sin\theta & \cos\theta \end{bmatrix} \times \begin{bmatrix} \alpha \\ \beta \end{bmatrix} \tag{4.34}$$

$$\text{or, } \begin{bmatrix} d \\ q \end{bmatrix} = [\text{Park Matrix}] \times \begin{bmatrix} \alpha \\ \beta \end{bmatrix}$$

where

$$[\text{Park Matrix}] = \begin{bmatrix} \cos\theta & \sin\theta \\ -\sin\theta & \cos\theta \end{bmatrix} \tag{4.36}$$

Similarly, the space vector components in the Park plane can be transformed into a space vector in the Clarke plane by inverse Park transformation. Therefore, (d, q) can be converted to (α, β) by following way:

$$\begin{bmatrix} \alpha \\ \beta \end{bmatrix} = \begin{bmatrix} \cos\theta & -\sin\theta \\ \sin\theta & \cos\theta \end{bmatrix} \times \begin{bmatrix} d \\ q \end{bmatrix} \tag{4.37}$$

or,

$$\begin{bmatrix} \alpha \\ \beta \end{bmatrix} = [\text{Inverse Park Matrix}] \times \begin{bmatrix} d \\ q \end{bmatrix} \tag{4.38}$$

where

$$[\text{Inverse Park Matrix}] = \begin{bmatrix} \cos\theta & -\sin\theta \\ \sin\theta & \cos\theta \end{bmatrix} \tag{4.39}$$

Thus, phase currents and voltages can be transformed in the Park plane using the Clarke matrix and Park matrix as shown in Figure 4.23. The angle θ is chosen 0°.

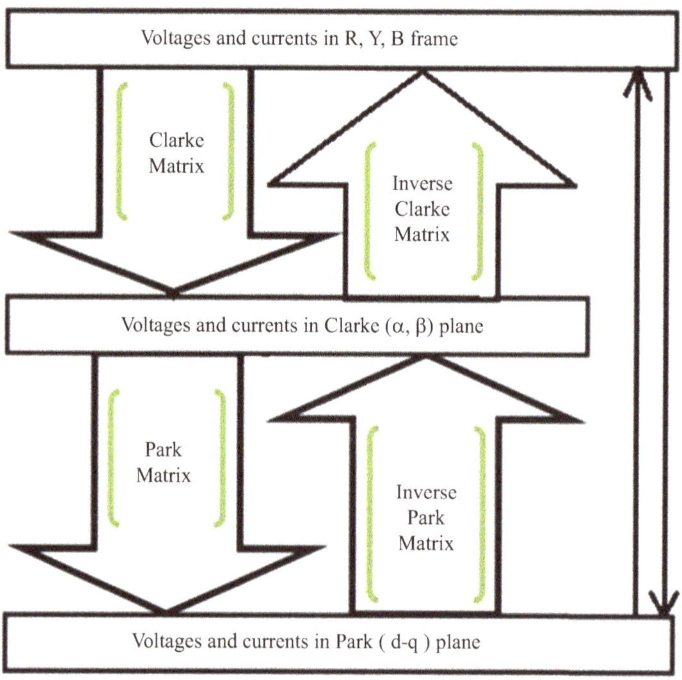

Figure 4.23 Currents and voltages transformation in Clarke and Park planes

Current in Clarke plane can be written as follows:

$$\begin{bmatrix} i_\alpha \\ i_\beta \end{bmatrix} = [\text{Clarke Matrix}] \times \begin{bmatrix} i_R \\ i_Y \\ i_B \end{bmatrix} \quad (4.40)$$

Currents in Park plane can be derived from current in Clarke plane as follows:

$$\begin{bmatrix} i_d \\ i_q \end{bmatrix} = [\text{Park Matrix}] \times \begin{bmatrix} i_\alpha \\ i_\beta \end{bmatrix} \quad (4.41)$$

Voltages in Clarke plane can be written as follows:

$$\begin{bmatrix} v_\alpha \\ v_\beta \end{bmatrix} = [\text{Clarke Matrix}] \times \begin{bmatrix} v_R \\ v_Y \\ v_B \end{bmatrix} \quad (4.42)$$

Voltages in the Park plane can be derived from voltages in the Clarke plane as follows:

$$\begin{bmatrix} v_d \\ v_q \end{bmatrix} = [\text{Park Matrix}] \times \begin{bmatrix} v_\alpha \\ v_\beta \end{bmatrix} \quad (4.43)$$

4.12 Solar-wind NPG model

A solar-wind-fed NPG model is shown in Figure 4.24. The battery unit is connected along with the solar PV unit and wind–turbine–generator unit; each of

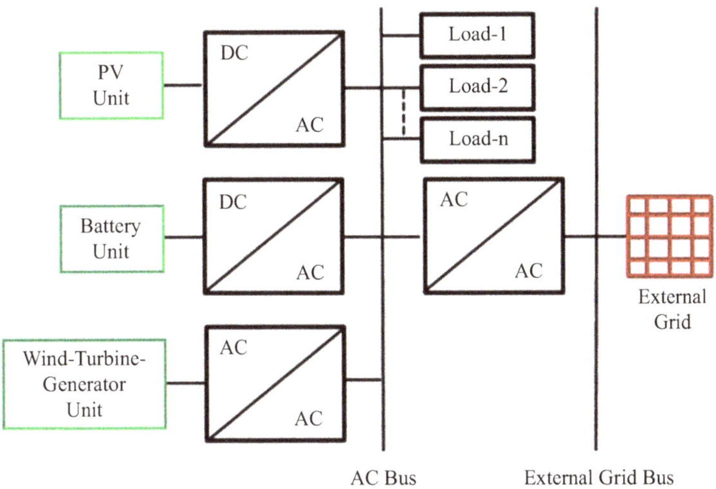

Figure 4.24 Solar-wind fed NPG model

them is connected with a separate converter unit. Outputs of all three converters are connected with an AC bus that feeds power to local loads. The AC bus is connected with an EG through a transformer that acts as AC to AC voltage converter. Wind turbines with high ratings are widely used in electric power generation. As wind turbines of small ratings (e.g., 400 W–20 kW) are also available, this model can be used to make nanogrids and picogrids as required. Often small wind turbine is easy to transport and installed from one place to another. Thus, the solar-wind picogrid model can be shifted, if required, from one place to another.

4.13 Case studies

4.13.1 Case study with a 20 kW nanogrid system

Case studies have been carried out on a 20 kW PV-connected nanogrid system as shown in Figure 4.25. Operations have been observed in the following conditions:

- different loads (no load to 150% overload),
- normal temperature,
- normal irradiance,
- variable string number, and
- different inverter configuration.

Different inverter configurations were tested that involves the following switching elements:

- Diodes.
- Thyristor.
- IGBT diodes.

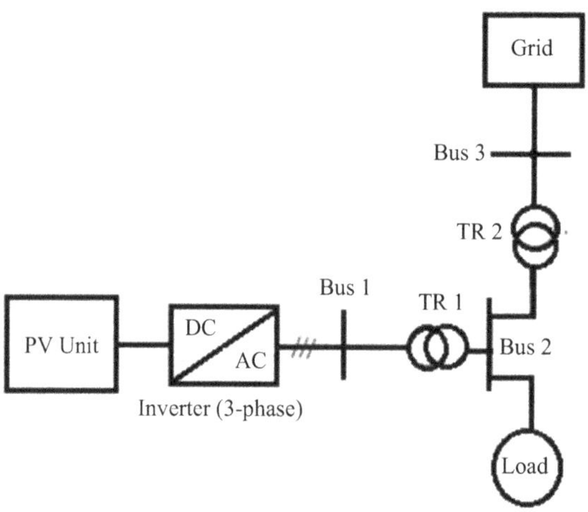

Figure 4.25 A 20 kW PV-fed nanogrid system with local load unit

- MOSFET diodes.
- GTO diodes.
- Ideal switches.

Switching operations are performed based on the logic of individual elements. Then the average model of voltage source converter (VSC) and switching function-based VSC is studied.

In both cases, total harmonics distortion (THD), DC, and sub-harmonics components are observed for starting transients. Two types of features were observed as follows:

Features type 1

- THD values are small.
- Magnitudes of DC components are too small.
- The amplitude of sub-harmonics components is large.

Features type 2

- THD values are large.
- Magnitude DC components are large.
- The amplitude of sub-harmonics components is small.

4.13.2 Case study with solar-PV fed DC–AC single-stage power converter connected with the utility grid

Initially, two similar types of PV units are connected in parallel to feed power to a DC to AC single-stage converter. Converter output is connected with the utility grid. The converter is bidirectional i.e., it can send power to the utility grid in the DC–AC mode and receive power from the utility grid in the AC–DC mode. The schematic diagram of the system is shown in Figure 4.26. Under the fixed level of illumination (1 000 W/m^2) and 25 °C temperature, the system is run with different numbers of cells. PV unit 1 and unit 2 generate power and send it to the utility grid. Output waveforms with the same number of series-connected cells (400) in each PV unit are shown in Figure 4.27.

Case 1: The number of active cells was increased equally in both PV units. The result is presented in Table 4.5. DC output of PV unit (i.e., DC input to the converter) is to increase slowly, converter efficiency remains above 96% with slightly reducing nature.

Case 2: The number of active cells is reduced equally in one of two PV units. The result is presented in Table 4.5. DC output of PV unit (i.e., DC input to the converter) is seen to reduce slowly, converter efficiency remains above 96% with slightly reducing nature. The voltage level is maintained by PV unit 1. However, DC output from PV reduced to a great extent when cells number went below a certain number (in this case, the number was 270).

Case 3: The number of active cells was reduced in one of two PV units. The result is presented in Table 4.6. DC output of PV unit (i.e., DC input to the converter) is seen to reduce slowly, converter efficiency remains above 96%

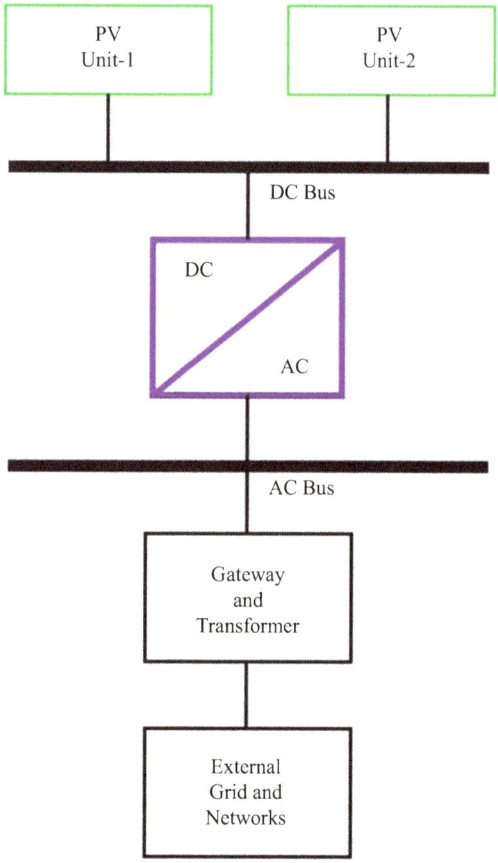

Figure 4.26 Schematic diagram of the case study with two PV units feeding power to EG

with slightly reducing nature. However, DC output from PV reduced to a great extent when cells number went below a certain number (in this case, the number was 270). During this situation, PV unit 2 has become a load to another PV unit 1. Power from PV unit 1 feeds power to PV unit 2 and as a result power flow to the utility grid.

Case 4: The number of active cells was reduced further in one of two PV units. The result is presented in Table 4.7. DC output of PV unit (i.e., DC input to the converter) is seen to reduce rapidly, converter efficiency also falls. DC power becomes negative when the cell's number went below a certain number (in this case, the number was 153). This indicates that total failure has occurred in PV unit 2 and generation by PV unit 1 is not sufficient to compensate for this loss. Thus, power will be absorbed from the utility grid. The situation should not be allowed and may be treated as a fault.

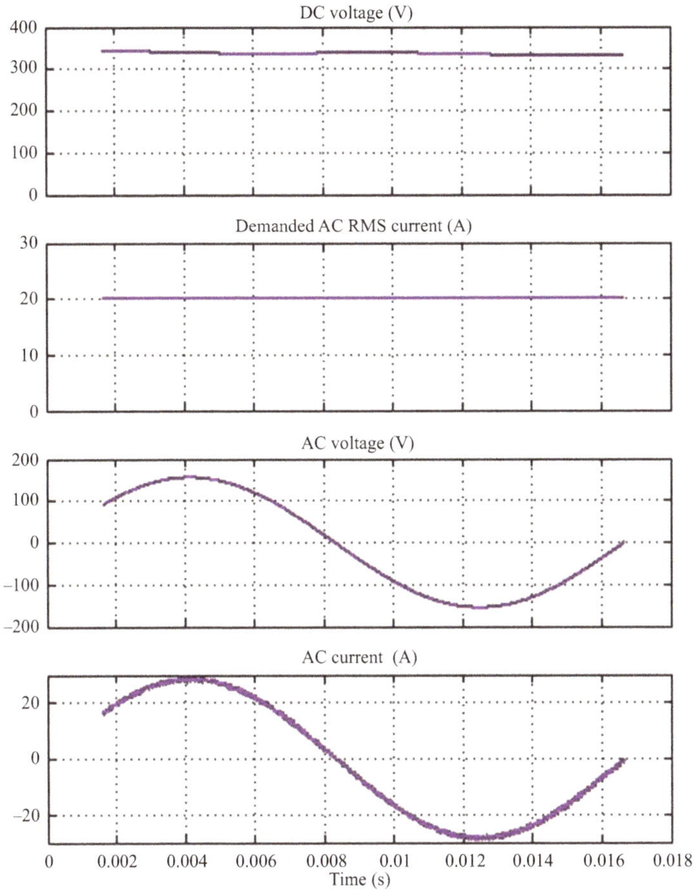

Figure 4.27 Single waveform with 400 cells connected in series in each panel

Table 4.5 Results of cases 1 and 2 with an equal number of cells in unit 1 and unit 2

PV unit 1		**PV unit 2**		**DC power input to the converter**	**AC output power of the converter**	**Efficiency**
Number of series cells	**Irradiance (W/m^2)**	**Number of series cells**	**Irradiance (W/m^2)**			
400	1 000	400	1 000	2 292		96.38
500	1 000	500	1 000	2 292		96.34
600	1 000	600	1 000	2 295		96.23
700	1 000	700	1 000	2 294		96.26
800	1 000	800	1 000	2 296		96.19

Table 4.6 Results of case 3 with a large reduction of generation unit 2

PV unit 1		PV unit 2		DC power input to the converter	AC output power of the converter	Efficiency
Number of series cells	Irradiance (W/m^2)	Number of series cells	Irradiance (W/m^2)			
400	1 000	400	1 000	2 292		96.42
400	1 000	300	1 000	2 291		96.36
400	1 000	270	1 000	1 468		14.14
400	1 000	268	1 000	1 420		4.65

Table 4.7 Results of case 4 with revers flow of power

PV unit 1		PV unit 2		DC power input to the converter
Number of series cells	Irradiance (W/m^2)	Number of series cells	Irradiance (W/m^2)	
400	1 000	400	1 000	2 292
400	1 000	300	1 000	2 291
400	1 000	200	1 000	861
400	1 000	160	1 000	114
400	1 000	155	1 000	18
400	1 000	154	1 000	1
400	1 000	153	1 000	−16
400	1 000	150	1 000	−68
400	1 000	100	1 000	−1155

4.13.3 Case study with step up flyback converter

A step-up flyback converter of 5 V/15 V DC–DC is considered in this case study as shown in Figure 4.28. It provides 5 V DC regulated supply to the connected load. Time-varying voltage is applied across a transformer primary that steps up the voltage. Transformer secondary voltage is rectified to DC.

MOSFET-based switching circuit for generation of time-varying voltage using 5 V DC is done by PWM device. The feedback controller controls PWM operation based on the feedback from the output of the converter. Three cases are considered to the performance of this converter.

Case 1: Flyback converter with fixed resistive load: The output waveforms and power are shown in Figure 4.29. A fixed resistive load is connected at the output of the flyback converter. Output voltage rises with a small overshoot and then becomes flat with regulation. During the rise time of output voltage, power shows high transient nature. Then, constant power is observed.

Case 2: Flyback converter with cyclic resistive load: The output waveforms and power are shown in Figure 4.30. A cyclic resistive load of 600 Hz frequency was applied. The voltage profile is not found flat. It increases and decreases at

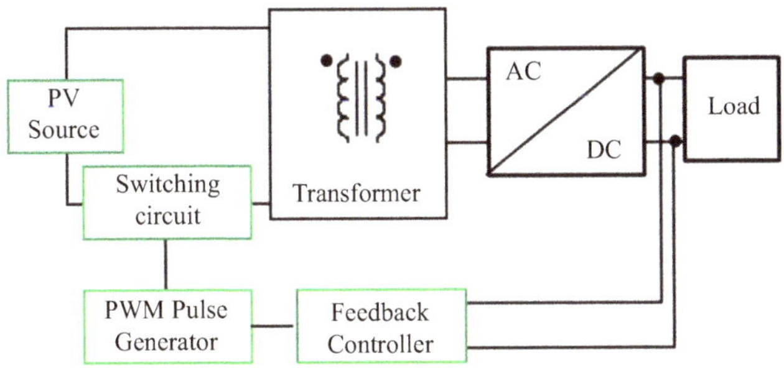

Figure 4.28 Step up flyback converter connected with load

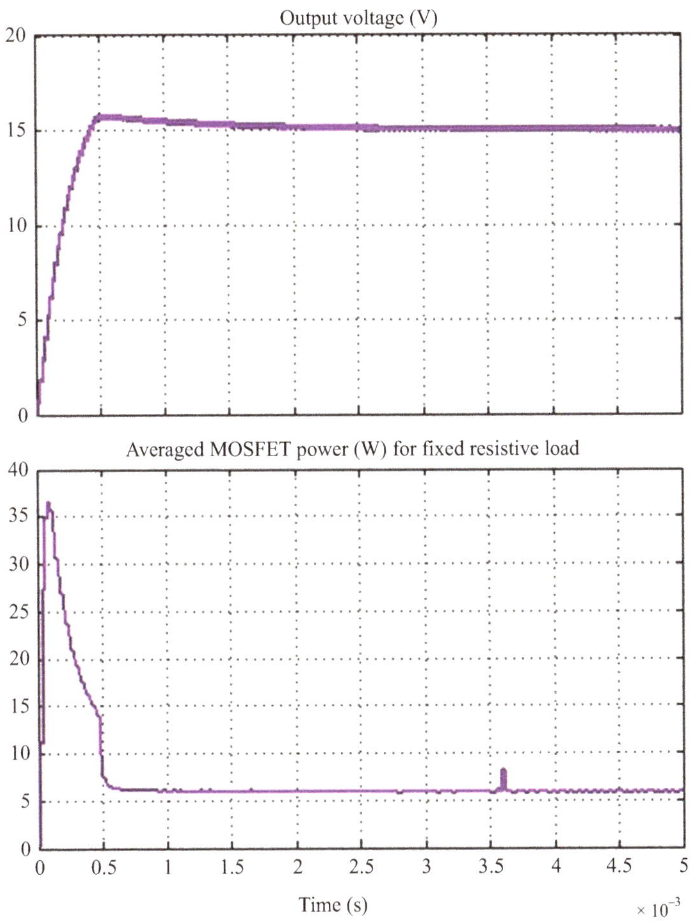

Figure 4.29 Output power and voltage at fixed resistive load

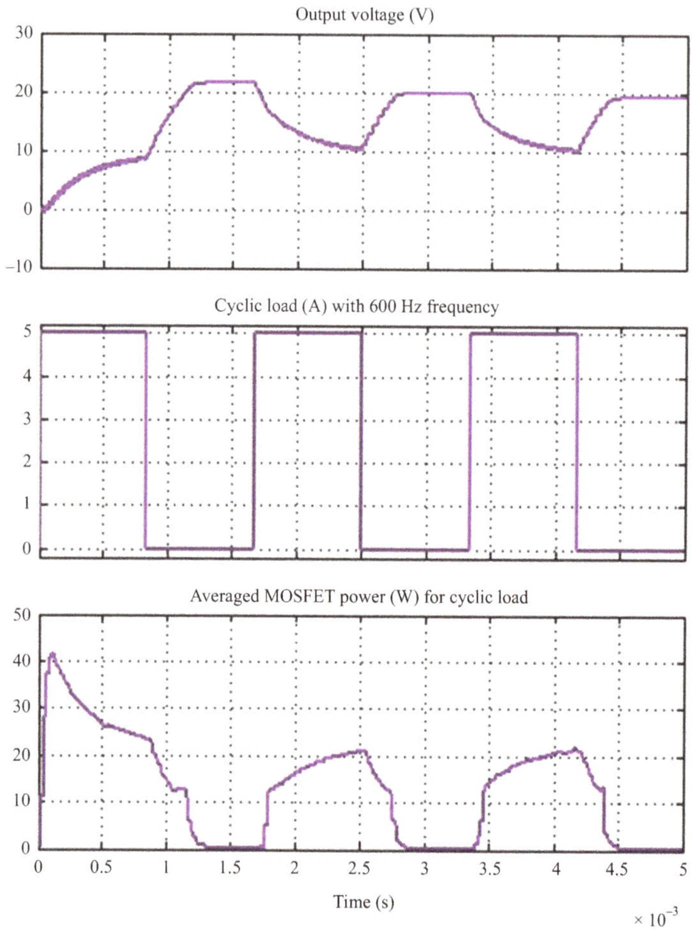

Figure 4.30 Output power and voltage at the cyclic load

cyclic frequency. However, the average voltage remains constant. Transient nature is observed for each cycle of the cyclic load in both voltage and power waveforms. There is no significant transient rise observed in the voltage waveform at starting. However, the transient rise is observed in the power profile at starting.

Case 3: Flyback converter with fixed and cyclic resistive load: The output waveforms and power are shown in Figure 4.31. Both fixed load of case 1 and a cyclic resistive load of 600 Hz frequency of case 2 are applied at the output of the flyback converter. The influence of cyclic load is prominent. The voltage profile is again not flat. It increases and then decreases at cyclic frequency. However, an average voltage remains constant. Transient nature is observed for each cycle of the cyclic load in both voltage and power waveforms that are slightly different from the previous case. Transient rise is not observed in voltage waveform at starting; However, it is observed in the power profile at starting.

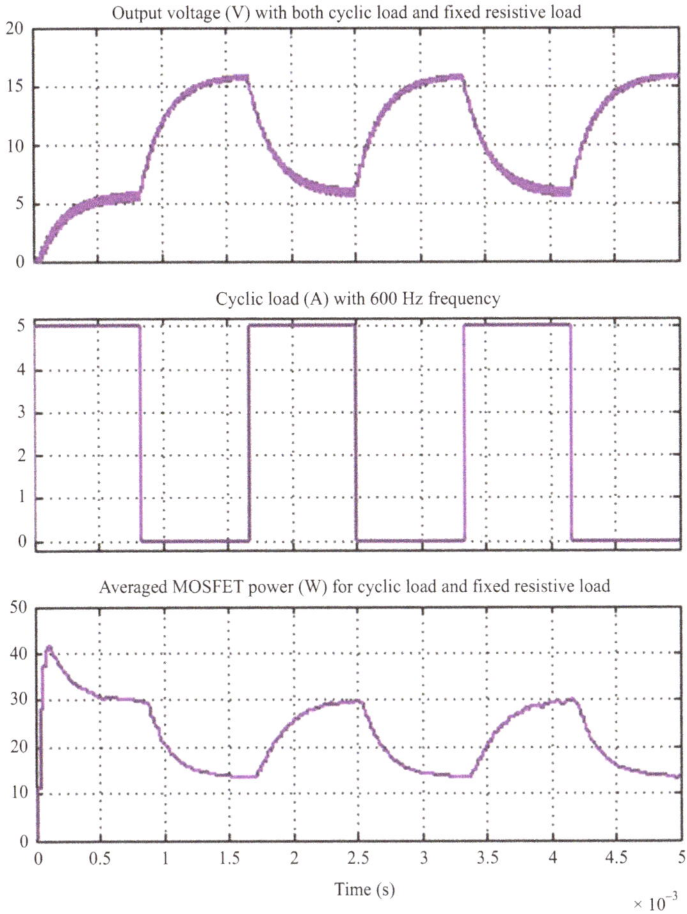

Figure 4.31 Output power and voltage at the fixed resistive load and cyclic load

4.13.4 Case study with DC–DC buck-boost IGBT/diode converter

DC–DC buck-boost IGBT-diode-made converter is connected with a fixed resistive load as shown in Figure 4.32. Output current and voltage of the converter are observed as presented in Figure 4.33. Small oscillations are observed in the current profile that may be minimized by adding filters.

4.13.5 Case study with DC–AC 2 and 4-pulse single-phase PWM inverter

DC–AC half-bridge and full-bridge inverters with PWM control schemes with 2 and 4 pulses are considered for the generation of single-phase supply. The inverter output is fed to a fixed R–L load.

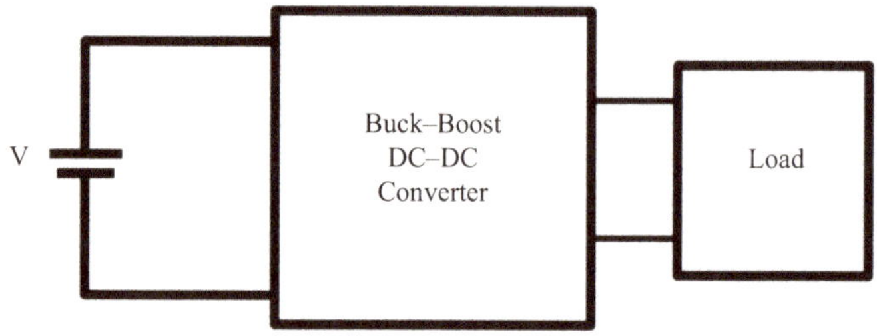

Figure 4.32 Buck-boost converter with resistive load

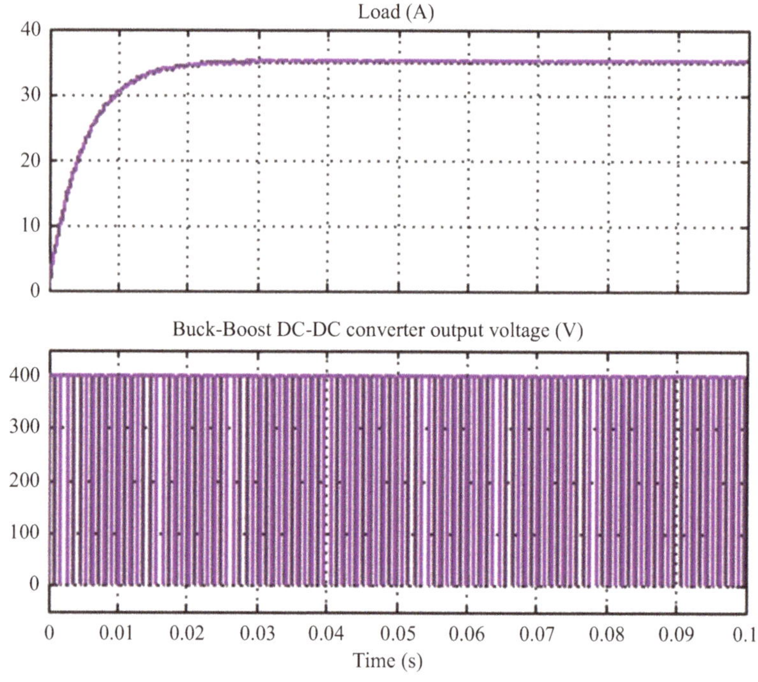

Figure 4.33 Output current and voltage of DC–DC buck-boost IGBT/diode converter

4.13.5.1 Half-bridge

DC–AC half-bridge inverter with PWM control schemes with 2-pulses is considered for the generation of single-phase supply as shown in Figure 4.34. Two 200 V DC sources are applied for the half-bridge inverter. Output current and voltage of the converter are observed as shown in Figure 4.35. The outputs of the FFT analysis of the current and voltage waveforms are shown in Figures 4.36 and 4.37, respectively.

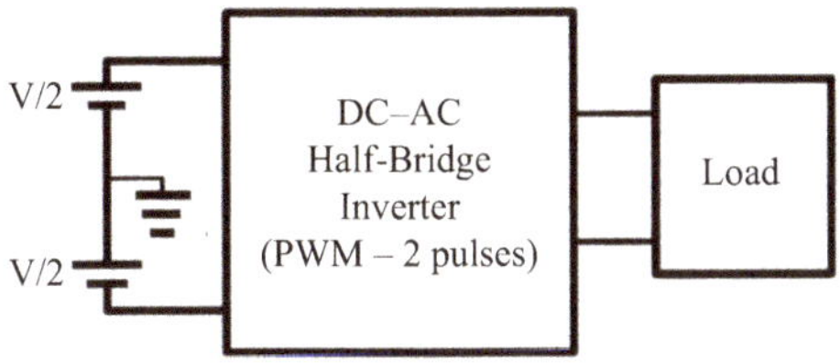

Figure 4.34 DC–AC half-bridge inverter with 2 pulses PWM control scheme

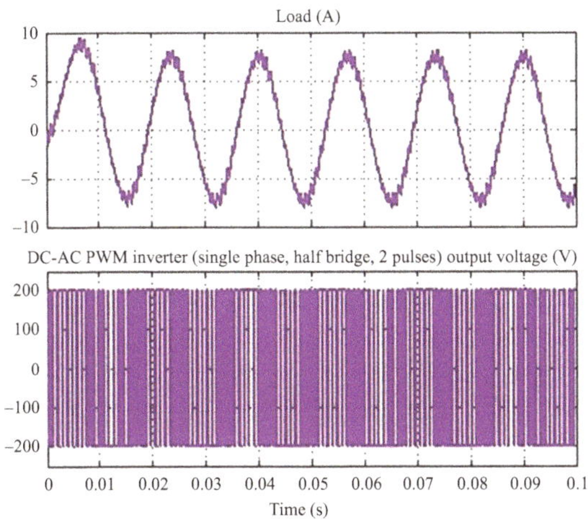

Figure 4.35 Output current and voltage of DC–AC single-phase half bridge IGBT/diode converter

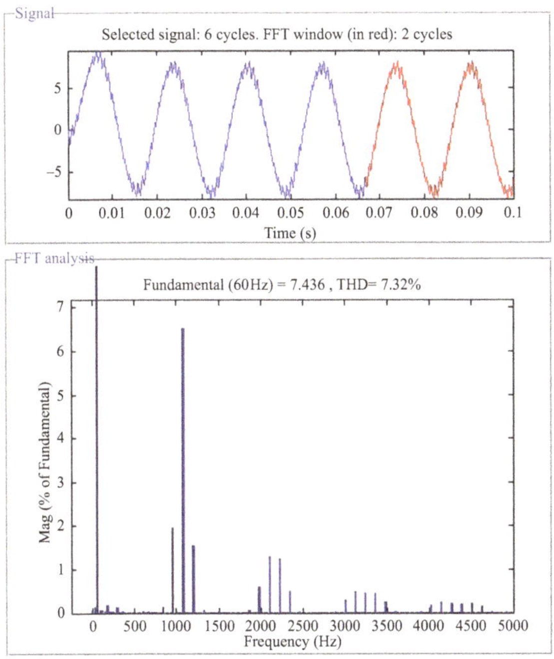

Figure 4.36 FFT analysis of output current of DC–AC single-phase half bridge IGBT/diode converter

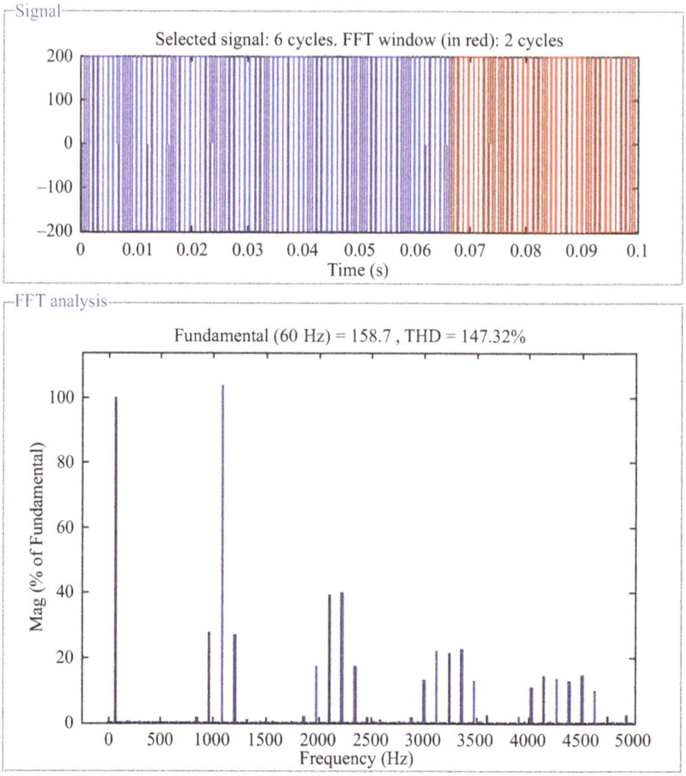

Figure 4.37 FFT analysis of output voltage of DC–AC single-phase half bridge IGBT/diode converter

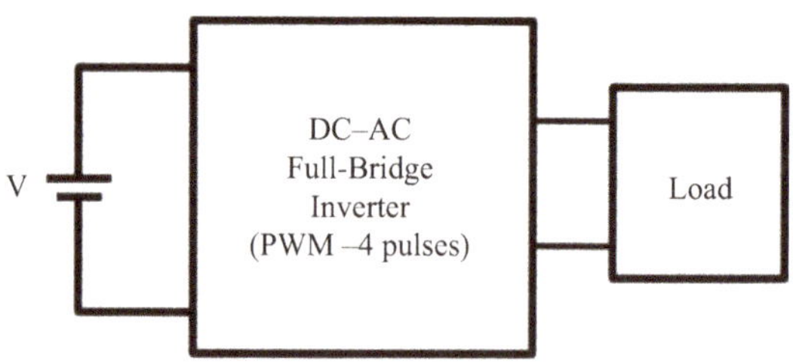

Figure 4.38 DC–AC full bridge inverter with 4 pulses PWM control scheme

4.13.5.2 Full bridge

DC–AC full-bridge inverters with PWM control schemes with 4-pulses are considered for the generation of single-phase supply as shown in Figure 4.38. The inverter output is fed to a fixed R–L load. A 400 V DC source is applied to the

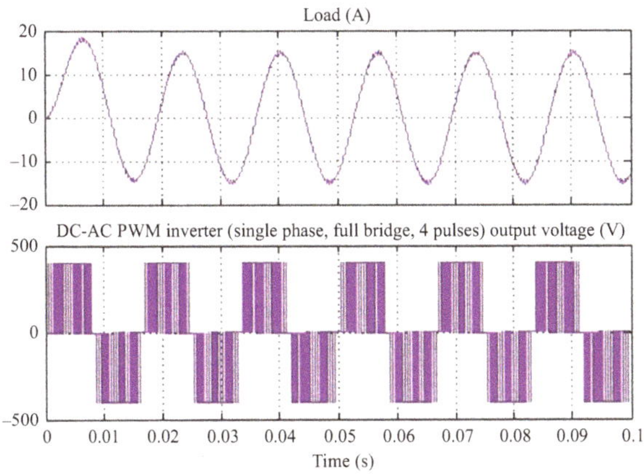

Figure 4.39 Output current and voltage of DC–AC single-phase full-bridge IGBT/diode converter

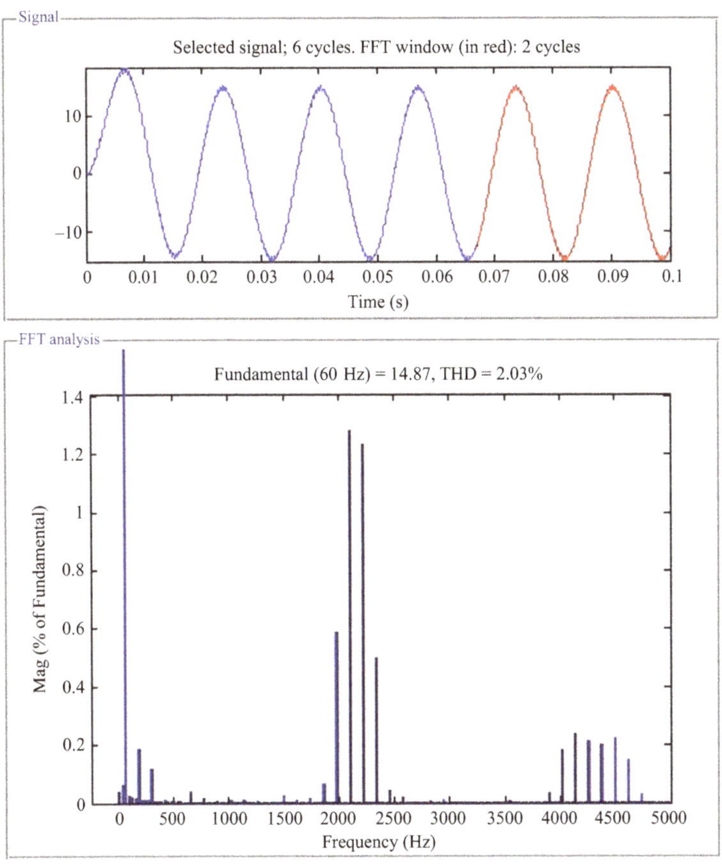

Figure 4.40 FFT analysis of output current of DC–AC single-phase full-bridge IGBT/diode converter

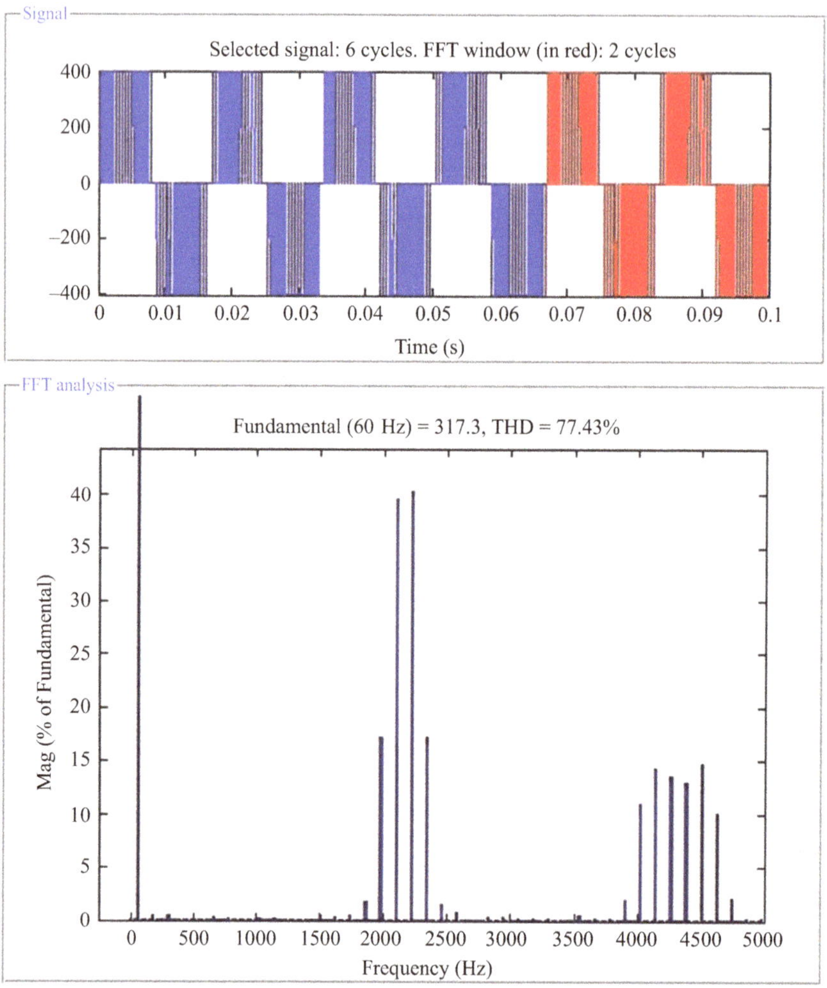

Figure 4.41 FFT analysis of output voltage of DC–AC single-phase full-bridge IGBT/diode converter

full-bridge inverter. Output current and voltage of the converter are observed as shown in Figure 4.39. The outputs of the FFT analysis of the current and voltage waveforms are shown in Figures 4.40 and 4.41, respectively.

DC component, fundamental component, and THD are observed at fixed R–L load with different inverter switching devices. The result is presented in Table 4.8. Very few variations are observed. Among different switching devices, inverters with IGBT and GTO show the same result, and MOSFET and switched VSC show the same result.

Table 4.8 Comparison of DC components, THD, and fundamental components for different inverter configurations

Converter switching devices		IGBT /diodes	MOSFET/ diodes	GTO /diodes	Switched VSC
Cycle considered for FFT		2	2	2	2
Fundamental frequency		60	60	60	60
Single-phase half-bridge two pulse PWM DC-AC converter	Fundamental current (A)	7.436	7.464	7.436	7.464
	THD % (current)	7.32	7.29	7.32	7.29
	Fundamental voltage (V)	158.7	159.3	158.7	159.3
	THD % (voltage)	147.32	146.75	147.32	146.75
Single-phase full-bridge four pulse PWM DC-AC converter	Fundamental current (A)	14.87	14.93	14.87	14.93
	THD % (current)	2.03	2.01	2.03	2.01
	Fundamental voltage (V)	317.3	318.5	317.3	318.6
	THD % (voltage)	77.43	77.12	77.43	77.12

4.14 Summary

The operation and performance analysis of nanogrid and picogrid systems have been covered in this chapter. Different operational aspects of solar PV-based NPG have been discussed. Closed-loop operation of solar PV-based grid has been described considering the variation of illumination level and temperature variations. Performance of PV changes with the variation of these parameters. Operation with a maximum power point tracking system has been discussed. Different useful control logics have been mentioned. Then few case studies have been presented. Results of waveform analysis on transient parts of NPG signals have been presented. Broad classification has been made based on transient behavior, harmonics distortion, etc. found from the result obtained by different converter configurations.

Further reading

[1] "IEEE DC Quick Charger for Use with Electric Vehicles Test Suite Specification – Version 1.0," in *IEEE DC Quick Charger for Use with Electric Vehicles Test Suite Specification – Version 1.0*, pp. 1–0, 17 September 2019.

[2] "IEEE Draft Standard Technical Specifications of a DC Quick and Bi-directional Charger for Use with Electric Vehicles," in *IEEE P2030.1. 1/D1*, February 2021, pp. 1–159, 9 February 2021.

[3] "IEEE Recommended Practice for Performance Testing of Electrical Energy Storage (EES) System in Electric Charging Stations in Combination with Photovoltaic (PV)," in *IEEE Std 2836-2021*, pp. 1–27, 8 June 2021, DOI: 10.1109/IEEESTD.2021.9447313.

[4] "IEEE Draft Standard for DC Microgrids for Rural and Remote Electricity Access Applications," in *IEEE P2030.10/D11*, February 2021, pp. 1–56, 10 March 2021.

[5] "IEEE Standard Conformance Test Procedures for Equipment Interconnecting Distributed Energy Resources with Electric Power Systems and Associated Interfaces," in *IEEE Std 1547.1-2020*, pp. 1–282, 21 May 2020, DOI: 10.1109/IEEESTD.2020.9097534.

[6] "IEEE Draft Recommended Practice for Personnel Qualifications for Installation and Maintenance of Stationary Batteries," in *IEEE P1657/D9*, July 2017, pp. 1–114, 1 January 2017.

[7] "IEEE Standard Glossary of Stationary Battery Terminology," in *IEEE Std 1881-2016*, pp. 1–42, 30 August 2016, DOI: 10.1109/IEEESTD.2016.7552407.

[8] "IEEE Guide for Technology of Unified Power Flow Controller Using Modular Multilevel Converter—Part 2: Terminology," in *IEEE Std 2745.2-2021*, pp. 1–25, 1 September 2021, DOI: 10.1109/IEEESTD.2021.9525319.

[9] "IEEE Recommended Practice for Monitoring Electric Power Quality," in *IEEE Std 1159-1995*, pp. 1–80, 30 November 1995, DOI: 10.1109/IEEESTD.1995.79050.

Chapter 5

Power quality issues in nanogrid and picogrid systems

This chapter deals with quality issues of electric power in nanogrid and picogrid systems. Different power quality (PQ) disturbances have been described. Their probable sources and effects have been discussed. Harmonics issues have been discussed in detail. Different useful parameters have been defined for quality analysis. Useful standards have been recommended and mathematical tools useful for PQ analysis have been presented.

5.1 Introduction

The quality of electric power concerning voltage and current waveforms in nano and pico grids (NPGs) is important for the performance, reliability, and satisfaction of consumers as well as the owner of the grid. In some cases, the performance of loads and sub-systems of the NPG itself is dependent on the quality of the power produced by the grid. Degradation of the quality of electric power produced by NPG may lead to a fall in reliability, sustainability, and consumer satisfaction. Different possible disturbances and their quality aspects in the NPG system have been discussed in this chapter.

The quality of electric power depends on both the voltage and current waveforms of the system. Magnitude and frequency are two important features of these waveforms. Any system is expected to operate with desired magnitude and frequency of currents and voltages. Thus, electric power quality (EPQ) in the AC system refers to providing near sinusoidal voltages at distribution buses and currents at rated magnitude and frequencies. This is also applicable for NPGs if they are completely AC by nature. In most cases, nanogrids and picogrids are found with both DC and AC parts in the systems. DC parts of these grids try to maintain direct "step" voltage and currents free from harmonics and other disturbances. Thus, combining the aspects of AC and DC systems, the quality term needs to be redefined for NPG.

EPQ in micro, nano, and pico grids refers to maintaining the step nature of bus voltages and currents, that is free from harmonics and other disturbances, in DC parts of the system and near sinusoidal voltages at distribution buses and currents at rated magnitude, and frequencies, in AC parts of the system.

NPG technology is advancing in multi-directions along with the change of power grid networks. NPG of different capacities and features are being introduced that sometimes make the system complex but more reliable. With the development of power electronics devices and communication technologies, uninterrupted power is demanded along with well-conditioned power for the utilities as well as the environment. Moreover, NPG systems are topologically located near the customer ends and therefore are responsible for customers' satisfaction. All these aspects demand the monitoring of the quality of electric power to be considered from the planning stage to the implementation and utilization of the NPG system.

5.2 Hexagonal approach for ensuring EPQ in NPG

A hexagonal approach can be followed for ensuring power quality (PQ) in NPG (Figure 5.1). Steps for ensuring EPQ in NPG systems involves the study of the following:

(i) fundamentals;
(ii) sources of disturbances;
(iii) effects of disturbances;
(iv) modeling, analysis, and learning;
(v) smart monitoring;
(vi) solution.

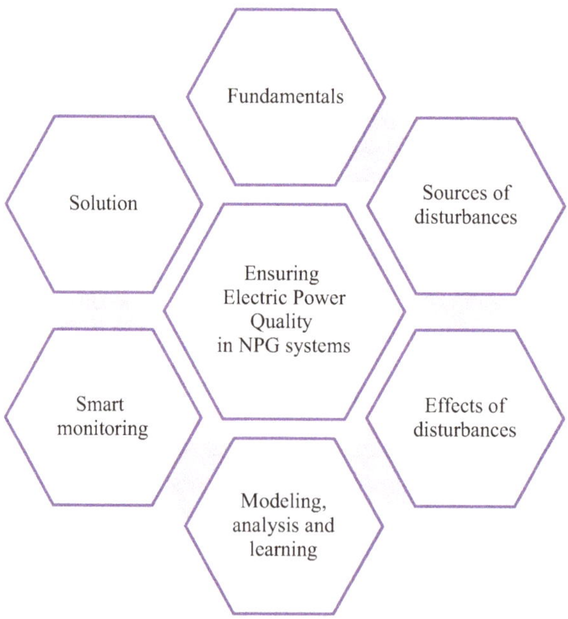

Figure 5.1 Hexagonal approach for NPG systems to ensure high EPQ

In the first steps, fundamentals about the NPG system and its interconnection with other systems should be clear. By the study of fundamentals, basic parameters and their nature of variations are identified that are the main cause of degradation of PQ in NPG based systems. All internal and external energy sources are to be brought under consideration. All incoming feeders and outgoing feeders along with connected utilities are to be considered for dealing with quality issues. Internal subsystems of NPG also play important roles in this regard. Internal subsystems cause disturbances and they are also most responsible for ensuring the quality of power to a high standard.

In the second step, possible sources of PQ-related disturbances are looked into. Sources of disturbance refer to the zones or particular locations or events causing unwanted variation of PQ parameters. Sources may exist both internal and external to the system. However, contributions of disturbances sources are not equal. Contribution varies from a normal situation to fault conditions. Conventional energy supply units and transformers are sources of disturbances due to their nonlinear nature. In addition, converter units, voltage, and power control units, etc. act as sources of PQ disturbances inside the NPG system. Currently taken by the utilities of different nature may contribute to distribution problems. The type of disturbances during faults may not match with the normal condition and they also vary in different load conditions.

In the third step, the effects of disturbances on different systems and subsystems are studied. PQ disturbances may result in damage of fully or some part of the system in a longer period as well as in a short duration. Sometimes, they may not damage the internal parts of the system but may cause dame to the external system connected with it. Often PQ disturbances may or may not cause damage, but can carry information of different types of fault and bigger problems. "Effects" of PQ degradation are reflected both on the system and consumer equipment. Therefore, prior modeling, system study, operational cases studies are required to get a clear picture of types of possible PQ disturbances and their effect on NPG systems.

In the fourth step, for detailed study, modeling and analysis are carried out, and then learning are made about possible PQ issues of the system. Best modeling practice should include the nonlinear nature of the systems. Sometimes, for simplicity, the linear approach is followed first and then gradually nonlinearities are inserted into the system. By 'modeling and analysis, the effort is given to configure the quality-related disturbance considering the probability of occurrence, sources, and effect. Mainly based on the mathematical models are used for this purpose; however, sometimes physical prototypes are developed for case studies. The prior modeling-based study helps to get an idea about probable features of PQ problems relevant to the system under consideration. However, the outcome may vary in real scenarios. The analysis includes planning-based analysis and operational analysis. Sometimes, it results in the large data set that may be stored or extracted features may be stored. Learning is the ultimate goal of modeling and analysis. For PQ-related issues of the NPG system, learning should be adaptive. Different, mathematical tools may be used for this purpose.

The fifth step involves instruments and measurements. Different relevant signals are captured with the help of sensors and data capture units. Signals causing PQ disturbances may be electrical as well as may be nonelectrical signals. Thus capturing different types of signals at different conditions needs the involvement of different types of transducers. For NPG, smart monitoring of disturbances is needed. Advanced monitoring technologies are applied for smart monitoring. Wireless monitoring may be required. Sometimes, monitoring may be done from the remote end. Application of Internet of Things and cloud-based communication and data handling techniques are seen to be used for PQ monitoring of both single NPG systems and distributed multiple NPG systems.

The last step deals with the solution to PQ problems. In reality, sustained PQ-related disturbance-free system conditions are difficult to achieve. However, attempts are taken to minimize disturbances as much as possible to keep them with tolerable limits. Tolerant limits of quality disturbances are often fixed by regulatory bodied and guided by international standards. For medical purposes and sophisticated control applications, the demand for good quality power is high. In an open and deregulated market, the best solution is essential to provide the utmost customer satisfaction.

The above-mentioned steps are for the generalized structure of NPG applications. It may vary for specific applications. Some steps may be omitted or some more steps may be added to solve the PQ issues.

5.3 Major sources of EPQ disturbances in NPG systems

Major sources of PQ-related disturbances may be divided into the following categories:

(i) Energy resources that may be internal as well as external to the NPG system.
(ii) Converter units responsible for the integration of grids, sub-parts, and utilities, etc.
(iii) Actual loads, equipment, and components.
(iv) Distribution systems and other subsystems.
(v) Connected external network.

5.4 Classification of major type PQ disturbances

Different types of PQ disturbances are observed which may belong to a known category or maybe of a completely new type. One event may consist of multiple types of disturbances. Definitions of power disturbances along with relevant guidelines are given by international bodies like International Electro-Technical Commission (IEC), IEEE standard, etc. Commonly faced quality-related disturbances faced in NPG and other systems are presented in Table 5.1.

Table 5.1 Major power quality-related disturbances in NPG-based system

Disturbance	Significance in a short
Interruption	Voltage magnitude is zero
Outage	Power interruption for not exceeding 60 s duration due to fault or mal-tripping of switchgear/system
Under voltage	Voltage magnitude is below its nominal value
Over voltage	Voltage magnitude is above its nominal value
Voltage/current unbalance	Deviation in the magnitude of voltage/current of any one or two of the three-phases
Harmonics	Non-sinusoidal waveforms
Transients	The sudden rise of the signal
Flicker	A visual effect of frequency variation of voltage in a system
Ringing waves	A transient condition that decays gradually
Voltage sag	A reduction in root-mean-square (RMS) voltage over a range of 0.1–0.9 pu for a duration greater than 10 ms but less than 1 s
Voltage swell	An increase in RMS voltage over a range of 1.1–1.8 pu for a duration greater than 10 ms but less than 1 s

5.5 Harmonics

NPG systems are expected to operate with voltage and current waveforms of particular system frequency keeping zero frequency in DC parts of the system. The system frequency of AC parts is referred to as fundamental frequency. However, practical systems, both voltage and current signals, consist of waveforms of frequencies other than fundamental. Waveforms having frequencies other than the fundamental following which the system is expected to operate are known as harmonics. These frequencies may be integer multiple of fundamental or may not be integer multiple of the fundamental frequency.

5.5.1 Fundamental wave

An alternating waveform having a frequency at which the system's alternating signals are expected to remain operative is known as a fundamental waveform. For an AC power system, fundamental voltage and current waveforms should be pure sine waves of the same frequency. If any system consists of both AC and DC parts, then at all AC parts of the system, all voltage and current waveforms should be the sine of the same frequency. The frequency of fundamental waveforms is known as the fundamental frequency. It is normally designed as either 50 Hz or 60 Hz. The fundamental frequency is also referred to as first-order frequency and denoted by suffix "1" in the study of the harmonics. Mathematically a fundamental voltage waveform may be expressed as follows:

$$v_1 = V_1 \sin 2\pi f_1 t \tag{5.1}$$

where v_1 is the fundamental voltage of the system; V_1 is the amplitude of the fundamental component of voltage; f_1 is the fundamental frequency.

5.5.2 Harmonic components

Harmonics refers to the waveforms with frequencies other than the fundamental frequency. The frequency of the harmonic component is known as harmonic frequency. Thus, the harmonic component may be expressed as

$$v_n = V_n \sin 2\pi f_n\, t \ (f_n \neq f_1) \tag{5.2}$$

where v_n is a harmonic component of voltage; V_n is the amplitude of the harmonic component of voltage; f_n is the harmonic frequency; n is called the order of the harmonics $(n \neq 1)$.

5.5.3 Classification of harmonic components

Based on the magnitude of n harmonics are classified into two main categories:

(a) Integer harmonics.
(b) Fractional harmonics.

5.5.3.1 Integer harmonics

When the harmonic frequency is integer multiple of the fundamental frequency, the harmonic component is called integer harmonic. Thus, for integer harmonic

$$v_n = V_n \sin 2\pi f_n\, t$$

where n is integer i.e., 2, 3, 4, ...

Based on the type of integer, integer harmonics are divided into two categories as follows:

(a) Odd harmonics.
(b) Even harmonics.

Odd harmonics

Integer harmonics with an odd integer multiple of the fundamental frequency are known as odd harmonics. Mathematically,

$$v_n = V_n \sin 2\pi f_n\, t$$

where n is an odd integer i.e., 3, 5, 7, ...

Even harmonics

Integer harmonics with even integer multiple of the fundamental frequency are known as even harmonics. Mathematically,

$$v_n = V_n \sin 2\pi f_n\, t$$

where n is even integer i.e., 2, 4, 6, ...

5.5.3.2 Fractional harmonics

When the harmonic frequency is non-integer or fractional multiple of the fundamental frequency, the harmonic component is called fractional harmonic. Thus, for fractional harmonic

$$v_n = V_n \sin 2\pi f_n t$$

where n is not integer e.g., 0.2, 1.3, 2.4, etc.

Based on the type of n, fractional harmonics are divided into two categories as follows:

(a) Sub-harmonics.
(b) Inter-harmonics.

Sub-harmonics
Fractional harmonics with less than fundamental frequency are known as sub-harmonics. Mathematically,

$$v_n = V_n \sin 2\pi f_n t$$

where $n < 1$.

Inter-harmonics
Fractional harmonics with greater than but non-integer multiple of the fundamental frequency are known as subharmonics. Mathematically,

$$v_n = V_n \sin 2\pi f_n t \tag{5.3}$$

where $n > 1$ and n is not integer.

5.5.4 Non-sinusoidal waveform

Non-sinusoidal wave is the resultant of all harmonic components. Mathematical,

$$v = \sum_{n=0}^{N} V_n \sin 2\pi f_n t \tag{5.4}$$

where

$$f_n = nf$$

n may or may be an integer; N is the highest value or order of n; $n = 0$ refers to step function with zero frequency.

Waveforms may be divided into two main categories as follows:

- Stationary wave.
- Nonstationary wave.

5.5.4.1 Stationary wave

For stationary waves, amplitudes of individual frequency components remain constant throughout the event, Mathematically,

$$v = \sum_{n=0}^{N} V_n \sin 2\pi f_n t \tag{5.5}$$

where V_n is constant i.e., $V_n \neq f(t)$ for the whole period of the event under consideration.

5.5.4.2 Nonstationary waves

For non-stationary waves, the amplitude of each harmonic component may change with time. Thus, Harmonics' amplitude is dependent on time.

A non-sinusoidal wave is expressed as

$$v = \sum_{n=0}^{N} V_n \sin 2\pi f_n t \tag{5.6}$$

where V_n is also variable i.e., $V_n = f(t)$ for the whole period of the event under consideration.

5.5.4.3 Harmonics with different phase angle

Phase angles of the individual harmonic component may not be the same as the fundamental waveform. Let, β_n be the phase angle of harmonics component with frequency f_n, Then,

$$v = \sum_{n=0}^{N} V_n \sin (2\pi f_n t - \beta_n) \tag{5.7}$$

5.5.5 The lowest order of harmonic

In a practical system, the lowest frequency component varies with system conditions. Ideally zero frequency i.e., direct "step" component is the lowest. However, in practice, the frequency component that shows greater than or at least equal to 3% of the amplitude of the fundamental component is considered as the lowest frequency component. Corresponding multiplication factor (n) is considered the order lowest harmonic component present in the system.

5.5.6 Useful harmonics parameters

Let voltage and current waveforms of systems follows:

$$\begin{aligned} v &= \sum_{n=0}^{N} V_n \sin (2\pi f_n t - \beta_n) \\ i &= \sum_{n=0}^{N} I_n \sin (2\pi f_n t - \alpha_n) \end{aligned} \tag{5.8}$$

Various parameters about harmonic components are presented in Table 5.2.

5.5.7 Harmonics-related PQ indices

The performance of NPG largely depends on the performance of different types of converters. Converters are major sources of harmonics. Thus, the quality of output

Table 5.2 Useful harmonics parameters

Parameter	Mathematical expression
The average value of the *n*th component	$i_{nav} = \left(\frac{2}{\pi}\right)\left(\frac{I_n}{n}\right)$
Average value	$i_{av} = \sum_{n=1,2,3,\ldots\ldots} \left(\frac{2}{\pi}\right)\left(\frac{I_n}{n}\right)$
RMS value	$i_{rms} = \sqrt{average\ of\ i^2} = \sqrt{\sum_{n=1,2,3,\ldots} \frac{I_n^2}{n}}$
Form factor	$Form\ Factor = \frac{RMS\ Value}{Average\ \ Value} = \frac{\sqrt{\sum_{n=1,\ 2,\ 3,\ldots} \frac{I_n^2}{n}}}{\sum_{n=1,\ 2,\ 3,\ldots} \left(\frac{2}{\pi}\right)\left(\frac{I_n}{n}\right)}$
Active power of harmonics of order *n*	$P_n = \frac{1}{2\pi}\int_0^{2\pi} i_n v_n d(2\pi ft) = \frac{I_n}{\sqrt{2}}\frac{V_n}{\sqrt{2}}\cos(\alpha_n - \beta_n)$ $= i_{n\ rms} v_{n\ rms} \cos\varphi_n$ where $\varphi_n = (\alpha_n - \beta_n)$ = phase difference between harmonic the component of voltage and current waveforms of order n
Total active power	$P = \sum_n P_n = \sum_n i_{n\ rms} v_{n\ rms} \cos\varphi_n = i_{1\ rms} v_{1\ rms} \cos\varphi_1 + i_{2\ rms} v_{2\ rms} \cos\varphi_2 + i_{3\ rms} v_{3\ rms} \cos\varphi_3 + \ldots\ldots\ldots\ldots\ldots\ldots$
Power factor	$Power\ Factor = \frac{Active\ Power}{Apparant\ Power} = \frac{\sum_n i_{n-rms} v_{n-rms} \cos\varphi_n}{\sum_n i_{rms} v_{rms}}$ $where, i_{rms} = \sqrt{average\ of\ i^2} = \sqrt{\sum_{n=1,2,3,\ldots} \frac{I_n^2}{n}}$ $v_{rms} = \sqrt{average\ of\ v^2} = \sqrt{\sum_{n=1,2,3,\ldots} \frac{V_n^2}{n}}$

powers given by the converters influences the performance of the NPG system and the power delivered by it to the utilities.

Many parameters have been introduced to define the quality of signals and power [1]. Different useful PQ indices related to harmonics are presented in Table 5.3.

5.5.8 Sources of harmonics

Important sources of harmonics in NPG-based systems are different types of converter units connected with it. However, Harmonics are contributed by other connected devices like generators, rotating machines, transformers, integrated grid buses, etc. nature of harmonics varies with the source and operational conditions. Converters and other power electronics devices-based control units contribute most of the fractional harmonics to the systems. Harmonics found in NPG-based systems depend on system conditions: normal operation and faults.

Table 5.3 Harmonics-related power quality indices

Power quality indices	**Definition**
Harmonic factor for *n*th harmonic	$HF_n = \frac{V_n}{V_1}$
Total harmonic distortion in voltage waveform (THD_V)	$THD_V = \frac{\sqrt{\sum_{n=2}^{n=\infty} (V_n)^2}}{V_1}$
Total harmonic distortion in current waveform (THD_I)	$THD_I = \frac{\sqrt{\sum_{n=2}^{n=\infty} (I_n)^2}}{I_n}$
Distortion for *n*th harmonic	$HF_n = \frac{V_n}{V_1 n^2}$
Distortion factor (*DF*)	$DF = \frac{\sqrt{\sum_{n=2}^{n=\infty} (V_n/n)^2}}{V_1}$
Active power distortion factor (*PDF*)	$PDF = \frac{P_H}{P_1} \times 100\%$ $PDF = \frac{\sqrt{\sum_{n=2}^{n=\infty} (P_n/n)^2}}{P_1} \times 100\%$
Reactive power distortion factor (*QDF*)	$QDF = \frac{Q_H}{Q_1} \times 100\%$ $QDF = \frac{\sqrt{\sum_{n=2}^{n=\infty} (Q_n/n)^2}}{Q_1} \times 100\%$
Apparent power distortion factor (*ADF*)	$ADF = \frac{A_H}{A_1} \times 100\ \%$ $ADF = \frac{\sqrt{\sum_{n=2}^{n=\infty} (A_n/n)^2}}{A_1} \times 100\%$
Flicker factor (*FF*)	$FF = \Delta V / \lvert V \rvert$
Crest factor (*CF*)	$CF = V_{peak} / V_{RMS}$
PQIndex(*PQI*)	$PQI = \left[\left\lvert \frac{\mathrm{V} - 100\%}{V_{CBEMA}(t) - 100\%} \right\rvert \right] .100\%$

As fault can take place almost all parts should be brought under consideration. The nature of harmonics during fault may vary from normal conditions. Some NPG components, that may not contribute harmonics significantly at normal conditions can influence the production of harmonics in other parts of the system. Harmonics nature during fault depends on location, time, and type of fault. Harmonics sources include all units of the NPG system. In addition to that, the lines, system buses, and grid integrating buses have a great influence on harmonics during faults. Thus, major sources of harmonics in the NPG system may be listed as follows:

(a) converters and controllers;
(b) generators and other rotating machines;
(c) transformer;
(d) controllers;
(e) nonlinear loads;
(f) arcing devices;
(g) lines, system buses, and grid integrating buses.

5.5.9 Transformers

Transformers' magnetic properties to transformer bring nonlinearity into its characteristics. Nonlinearity causes the generation of harmonics. However, the nature of the harmonics varies with system conditions. Also, natures of harmonics are different in power transformer, distribution transformer, and instrument transformer i.e., current transformer and potential transformer. Some important influencing factors on harmonics in the transformer are given as follows:

(a) *Normal excitation*: Nonlinearity exists during normal excitation because of the B–H characteristics of the magnetic materials. Along with other harmonics, it produces triplet harmonics. The effect may be reduced by the Delta connection of the poly-phase transformer.
(b) *Symmetrical over excitation*: Nature of symmetry of B–H characteristics of the magnetic property of the transformer changes slightly during symmetrical overexcitation. It tends to reduce odd symmetry in the characteristics. Patterns of harmonics change during overexcitation. Normally, this condition produces odd harmonics (like 5, 7, 11, 13, 17, 19....etc.) i.e., harmonics of orders $6k \pm 1$, where k is an integer.
(c) *Inrush current harmonics*: During re-energization of the transformer after switching off, bring additional flux due to the existence of residual flux. Thus, during starting it produces current inrush with even harmonics dominated by second-order harmonics.
(d) *DC magnetization*: DC magnetization changes core flux characteristics. Thus, it causes harmonics in the transformer.
(e) *Saturation*: Effect saturation of magnetic flux, produce new patterns of harmonics. This condition may appear during overload or fault and have a great influence on the performance of the transformer, especially, the current transformer used for measurement, control, and protection units.

5.5.10 Rotating machine

Rotating machines are also an important source of harmonics. Magnetic nonlinearity, air gap, and nonlinear flux distribution due to structural features are the main reason for harmonics generation. Rotating machines are found in the form of generators as an energy resource and the form of the motor as loads in the NPG system. NPG may be integrated with other grids having lots of rotating machines (both generators and motors) in their network. Harmonics effects of other networks may influence connected NPG systems. The main factors influencing harmonics generations in rotating machines are as follows:

(i) magnetic nonlinearities of the core material;
(ii) nonuniform flux distribution in the air gap;
(iii) slot structures for slot harmonics;
(iv) design parameters like pitch factor and distribution factors;
(v) rotor saliency;
(vi) crawling;
(vii) cogging;
(viii) rotor misalignment;
(ix) mass unbalance;
(x) bar breakage;
(xi) fractal error;
(xii) unsymmetrical fault;
(xiii) inter-turn short circuit.

5.5.11 Arcing devices

Arcing devices are highly nonlinear. Arc ignition contributes lots of harmonics. Major arcing devices are of three types:

(i) electric arc furnace;
(ii) discharge type lighting;
(iii) arc welders.

5.5.12 Converters and controllers

In NPG-based systems, converters and controller units are the main sources of harmonics. These units are made of semiconductor-based devices. Harmonics contributed by controllers and converters are dominated by fractional harmonics [2]. Both in AC and DC parts of the systems, they appear during usual transient conditions, load change, faults, etc.

Converter units include:

- Inverters
- Rectifiers
- Choppers

Controller units mainly include:

- Phase controller
- AC/DC voltage regulators
- Reactive power controller

5.5.13 Effects of harmonics

Common effects of harmonics on various parts of NGP-based systems are given in Table 5.4.

5.5.14 Harmonics reductions

In power system applications, harmonics is a big challenge. In most cases, the frequencies of harmonic components and their amplitudes are variable and unpredictable. Thus modeling, case studies and adaptive learning help us to get an idea about harmonic possibilities and probabilities. Based on learning, different attempts are taken to reduce harmonics in the system. The common methods used for harmonics reduction are as follows:

(a) filtering,
(b) harmonic injection,
(c) adaptive flexible control for harmonic reduction, and
(d) proper setting of design parameters.

5.5.14.1 Filtering

Filtering is an old technique for harmonics reduction and quality improvement of power signals. By filtering, unwanted harmonics are blocked or directed to other paths. There are many types of filtering methods and circuits that have been introduced. For power frequency harmonics, normally main filter units are made of passive elements, whereas for electronics or very small power applications active elements are used to make filters. Broadly, filters are divided into four categories:

- low-pass filter,
- high-pass filter,
- bandpass filter, and
- bandstop filter.

Filters are characterized by their characteristics frequency. This frequency helps to decide allowable frequency components. Logically, low-pass filter and high-pass filter are basic filters, the other two categories of filters i.e., bandpass filter and bandstop filter can be formed by suitable combinations of low-pass and high-pass filters. Some common configurations of passive filters connected at the output end of inverters are as follows:

- C filter.
- L–C filter.
- C–L–C filter.

Table 5.4 Effects of harmonics on various parts of NGP-based system

SN	Effects on	Type of effects
1	Rotating machines	(a) Increase of harmonic losses (b) Increase of harmonic torque (c) Change of speed-torque characteristics (d) Cogging and crawling of induction motor (e) Increase of capacitive current, like a high capacitive current through stray capacitance (f) Increase of voltage stress on insulation
2	Transformer	(a) Increase of core loss (b) Increase of copper loss due to harmonic currents (however, it is negligible in large transformers) (c) Increase in voltage stress (d) Increase of core vibration (e) Increase of saturation problem
3	Transmission lines	(a) Skin effect and proximity effect (b) Loss (c) Voltage drop (d) Dielectric stress (e) Corona
4	Measuring instruments	(a) Change of sign and magnitude of error (b) Harmonic torque (c) Change of sensing features (d) Mal operation
5	Power system protection	Malfunctioning Change of characteristics
6	Capacitor banks	(a) Dielectric loss in capacitors (b) Thermal stress increasing the stress on the capacitor bank and the system connected to the bank (c) Resonance causing an increase of losses and overheating of capacitors, and often leading to their destruction (d) Reactive power causing an increase of change of harmonic contents sometimes increases reactive power over permissible manufacturer tolerances (e) Power factor correction capacitors causing malfunction of the operation of power factor correction capacitors
7	Consumer equipment	(a) Imperfect or wrong operation (b) Overheating (c) Change of characteristics (d) Shortening of life span (e) Decrease of efficiency
8	Cables	(a) Skin effect (b) Proximity effect (c) Dielectric stress
9	Communication circuits	(a) Noise (b) Interference
10	Others	(a) Poor damping (b) Resonance • Series • Parallel • Composite

5.5.14.2 Harmonic injection

Harmonics injection is an advanced technique for harmonics reduction. Here, a frequency component is added or injected to the system that has a polarity opposite to the frequency component to be eliminated. For a satisfactory outcome of the method, one should have prior knowledge about the harmonics components to be eliminated.

5.5.14.3 Adaptive flexible control methods for harmonic reduction

As most of the converters operate based on pulse width modulation, by proper control, harmonics can be mitigated. Let us name all these techniques having adaptive features incorporated with control units of converters as adaptive flexible control for harmonics reductions or mitigations (AFCHR or AFCHM). Some used techniques are given below.

Multiple pulse width modulations

Adaptive flexible control for harmonic reduction: By using multiple pulses at each half cycle of the output signals, harmonics are reduced. This method is also called uniform pulse width modulation. The output frequency (normally it is the reference frequency, (f_r) and carrier frequency (f_c)) decides the width of the pulses. The number of pulses (i) is determined by

$$i = \frac{f_c}{2f_r} \tag{5.9}$$

Sinusoidal pulse width modulation

This method helps to reduce low-order harmonic components. Thus, it results in a reduction of harmonic distortion. Here the width of the pulse is set proportional to the amplitude of the sine wave.

Modified sinusoidal pulse width modulation

The previous method is followed in this method. In addition to that, the amplitude of fundamental components is raised concerning the amplitudes of harmonic components. Thus, it helps in the reduction of harmonics distortion factors.

Phase displacement control

Output signals of different inverter units are mixed after adding variable phase delay. Phase delays are adjusted in such a way that the resultant signal has less amount of harmonics. Displacement angle is an important controlling parameter. To reduce nth order harmonics, initially, displacement angle is assumed as follows:

$$\alpha = \left(\frac{360}{n}\right)^{\circ} \tag{5.10}$$

Then gradually displacement angle is adjusted as required.

5.5.14.4 Design parameters

Design parameter control should be considered from the beginning i.e., from the time of design. By proper choice of design parameters, harmonic generations can be minimized. This method should be adopted as best practice for the design of electrical machines.

5.6 Transients

Transient concerning NPG applications refers to voltage and current components that occurred during the transition from one steady-state to another steady state. Transients are normally associated with a change in the steady-state due to sudden big load change, switching action, faults, etc. These phenomena occur for a short duration in a power system resulting in a very high rise in voltage and current. They may come into the system generating in the system itself or propagating from another external system. Transients are treated as major PQ problems that can occur in both AC and DC parts of the system.

5.6.1 Classifications of transients

The classification of transients is shown in Figure 5.2. Transients commonly observed in the power system may broadly be classified into two main categories as follows:

(a) unidirectional impulsive transients and
(b) oscillatory transients.

Unidirectional impulsive transients: Unidirectional impulsive transients may occur due to lightning, switching effect of inductive or capacitive circuits, etc.

Oscillatory transients: These types of transients show oscillations. Based on the frequency, oscillatory transients are divided into two categories as follows:

(i) *Power frequency oscillatory transients*: frequency of oscillation matches with the frequency of the system, e.g., the effect of big load change or fault in a synchronous generator.

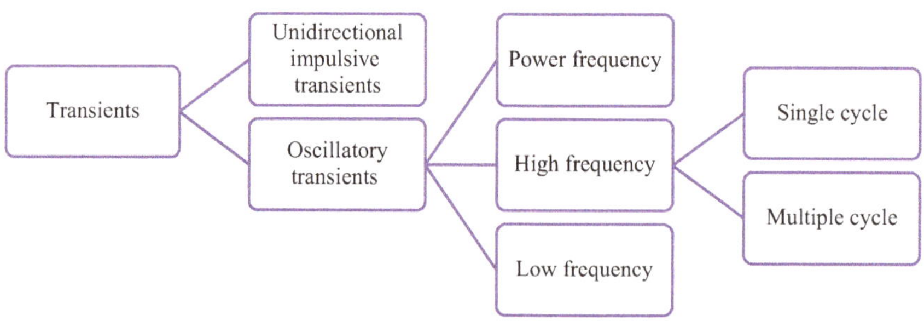

Figure 5.2 Classifications of transients

(ii) *High-frequency oscillatory transients*: frequencies of oscillations are much higher than that of fundamental or system frequency. High-frequency transients may be of two categories: single cycle and multiple cycles.
(iii) *Low-frequency oscillatory transients*: frequencies of oscillations are lower than that of fundamental or system frequency.

5.6.2 Sources of transients

Major causes of transients are summarized in Table 5.5.

5.6.3 Effects

Transients cause degradation of the quality of the electric power. It hampers the operation and performance of different subsystems and utilizes. The peak value of transients may damage to fully or some part of the system. It may cause the malfunction of relay and switchgear units. High-frequency transients may change the properties of magnetic materials as desired. Thus, minimization of transients is expected. Lots of fault-related information may be extracted from the transients of the system.

5.7 Sag

Sag refers to a short-duration reduction of RMS voltage. Reduction of voltage in between 0.1 and 0.9 pu for the duration in between 0.5 cycles and 1 min is normally considered as sag. It is also known as voltage dip. Common causes and effects of sag are presented in Table 5.6.

5.8 Swell

Swell refers to the short-duration phenomenon of an increase of RMS voltage. An increase of voltage magnitude in between 1.1 and 1.8 pu for a duration ranging

Table 5.5 Classification of transients

Transients	Major causes
Impulsive transients	• Lighting • Switching effect or state change of inductive or capacitive circuit
Oscillatory transients	• Capacitor energizing • Re-strike during capacitor de-energizing • Line or cable energizing • Current chopping • Multiple re-strikes • Repetitive switching actions • Faults • Switching operations

Table 5.6 Common causes and effects of sag

Sag	Factors and results
Causes	• Energization of heavy loads • Starting of large induction motors • Single line-to-ground faults • Line-line and symmetrical fault • Load transferring from one power source to another
Effects	• Voltage stability problem • malfunctions of low-voltage devices • malfunctions of uninterruptible power supply • malfunction of measuring and control units fed by the power • Interfacing with communication signals

Table 5.7 Common causes and effects of swell

Swell	Factors and results
Causes	• Sudden switching off of a large load • Energizing a capacitor bank • Voltage increases at healthy phase during unsymmetrical ground fault at another phase • Momentary overvoltage due to power frequency surge or transients
Effects	• Voltage stability problem • Malfunctions of low-voltage devices • Malfunctions of uninterruptible power supply • Malfunction of measuring and control units • Interfacing with communication signals

from 0.5 cycles to 1 min is normally considered as swell. Swell is the opposite of sag and occurs less in numbers as compared to sags. Common causes and effects of swell are presented in Table 5.7.

5.9 Interruption

Interruption refers to a reduction of the supply voltage (or load current) to less than 0.1 pu. Based on the duration of the event of reduction, interruption is classified into many groups.

According to European Standard EN-50160:

1. *Short interruption*: Interruption that is up to 3 min.
2. *Long interruption*: Interruption that is longer than 3 min.

According to IEEE Standard-1250:

1. *Instantaneous interruption*: Interruption that is between 0.5 and 30 cycles.
2. *Momentary interruption*: Interruption that is between 30 cycles and 2 s.
3. *Temporary interruption*: Interruption that is between 2 s and 2 min.
4. *Sustained interruption*: Interruption that is longer than 2 min.

Causes:
Common causes of interruption are as follows:

- sudden failures of equipment,
- improper control operations,
- sudden blown of fuse, and
- sudden breaker opening.

5.10 Sustained interruption

Sustained interruption refers to voltage drops to zero that does not return or recover automatically. It is the most severe of all types of interruptions. As per IEC, the duration of sustained interruption is more than 3 min and as per IEEE definition, the duration is more than 2 min. The occurring frequency and period of sustained interruptions is one of the deciding factors of power reliability.

Common causes of sustained interruptions are

- fault occurrence,
- improper relay coordination,
- the incorrect intervention of relay, and
- planned interruption in some part of the system.

5.11 Under-voltage

The under-voltage refers to a reduction of RMS voltage to 0.8–0.9 pu for more than 1 min. The period for under-voltage is greater than sag. The main causes of under-voltage are

- overload,
- less supply capability,
- fault, etc.

Under-voltage results in many problems like

- voltage instability,
- high motor current,
- high reactive power demand, etc.

5.12 Overvoltage

The overvoltage refers to the increase of RMS voltage to 1.1–1.2 pu for more than 1 min. The timespan of overvoltage is greater than swell. Common causes of overvoltage in power systems are as follows:

- insulation fault,
- ferro resonance,
- open faults,
- rise of voltage for fault at another phase,
- lightning overvoltages, and
- switching overvoltages.

Overvoltage results in the following:

- overstress on insulation,
- problems of voltage instability,
- demand for reactive power, etc.

5.13 DC offset

DC offset refers to the appearance of DC and/or voltage components in an AC system. By Fourier series expansion of wave, it may be analyzed. Common causes of DC offset in power systems are as follows:

- operation of rectifiers,
- operation of electronic switching devices, and
- disturbances and faults.

Common effects of DC offset are as follows:

- half-cycle saturation of transformer core,
- increase of even harmonics,
- additional heating effect, and
- electrolytic erosion of grounding electrodes, connectors, etc.

5.14 Electric noise

Electric noise refers to unwanted electrical signals in the power system, Normally, unwanted signals of lower than 200 kHz superimposed are considered as severe noise. However, magnitudes of noise signals are very small as compared to power signals. The main causes of electric noise are

- faulty connections,
- arc furnaces,
- electrical furnaces,
- power electronic devices,

- control circuits,
- welding equipment,
- loads with solid-state rectifiers,
- improper grounding,
- turning off capacitor banks,
- adjustable-speed drives,
- corona, and
- interference with communication circuits.

Though the magnitude of the noise is very small, it can create lots of problems in the operation of electronic devices, controllers, etc. Noise may be mitigated by filters, line conditioners, shielding, etc.

5.15 Voltage fluctuation

Voltage fluctuation refers to variations of voltage magnitude that do not normally exceed specified voltage ranges. As per ANSI C84.1-1982, it is from 0.9 to 1.1 pu. In the time domain, voltage fluctuations may be classified as follows:

- step-voltage fluctuations of either regular or irregular fashion,
- cyclic voltage fluctuation, and
- random voltage fluctuation.

Common causes of voltage fluctuation are as follows:

- pulsed output,
- resistance welders,
- start-up of drives,
- arc furnaces,
- drives operation,
- rapidly changing loads, and
- rolling mills.

Normally, voltage fluctuations of less than 10% do not result in severe problems. Some common effects of voltage fluctuations are as follows:

- degradation of the equipment performance,
- instability, and
- improper reactive power compensation.

5.16 Flicker

Flicker is "continuous and rapid variations in the load current magnitude which causes voltage variations." Flicker is often observed in illumination applications, arc furnace operation, etc.

5.17 Power frequency variations

Power frequency variation is a change of system fundamental frequency from its specified nominal value (e.g., 50 or 60 Hz). Common causes of power frequency variation are as follows:

- difference between active power demand and power generation;
- sudden load change beyond a control limit;
- improper coordination between load characteristics and control circuit for automatic load frequency control (ALFC).

5.18 Useful tools for signal assessment

Harmonics and transients are most common among PQ-related issues. The frequency components and their magnitudes are normally variable in real cases and the signals are non-stationary. Lots of research have been carried out and are still going on to study the nature of various signals found in power systems. The analytical approach found for signal analysis may broadly be categorized into three types:

(a) time-domain analysis,
(b) frequency domain analysis, and
(c) time–frequency domain analysis.

In practice, different approaches are found. The analog method is a very old concept for implementation. Last few decades, digital technology has advanced to a great extent. In some cases, combinations of analog and digital instrumentations are found. Thus, ways of implementing signal assessing mathematical tools may be of following categories:

- analog-based implementation,
- digital-based implementation, and
- analog–digital implement.

Along with the signal analysis, different optimization, communication techniques, and big data assessing tools have been introduced. Those are being utilized to extract features, store data, and communicate among different parts of interconnected systems. In the following subsections, few commonly used mathematical approaches have been discussed.

5.18.1 Fourier series

5.18.1.1 Small history

In the era of Napoleon, Baron Jean-Baptiste-Joseph Fourier (March 21, 1768, May 16, 1830), professor of mathematics, was born in a poor family in Auxerre. From the early stage of his profession, he used to think that any arbitrarily defined function could be represented by a single analytical expression. A similar type of idea was given much before by Leonhard Euler (1707–1783). He published a formula:

$$\frac{1}{2}x = \sin x - \frac{1}{2}\sin 2x + \frac{1}{3}\sin 3x + \ldots \tag{5.11}$$

Fourier applied this in solving real heat transfer problems through solid material. He noticed some limitations in Euler's expression. He wrote, "The equation is no longer true when the value of x is between π and 2π. However, the second side of the equation is still a convergent series but the sum is not equal to $x/2$. Euler, who knew the equation and gave it without comment," (Quotation from J Herivel, Joseph Fourier, the Man and the physicist, Clarendon Press, Oxford, 1975, p. 319).

Then Fourier explained the theory of heat conduction and introduced expression for temperature distribution in the solid medium in 1807. Initially, it faced lots of criticism with doubts by other scientists like Laplace, Lagrange, Biot, and Poisson. But slowly his theory was accepted. He was awarded for this purpose in the field of mathematics in 1811.

5.18.1.2 Fourier series expansion

Fourier expands single-valued continuous function in the time domain by DC component, and series of integer harmonics in 1822 [3]. The expansion series is known as the Fourier series. Let $f(t)$ be a continuous periodic function in the time domain. Then by Fourier series expansion, it can be written as follows:

$$f(t) = a_0 + \sum_{n=1,2\ 3,\ldots\ldots} a_n \cos n\omega t + \sum_{n=1,2\ 3,\ldots\ldots} b_n \sin n\omega t \tag{5.12}$$

where a_0 is the DC component present in the function; n is the order of harmonics.

Coefficients a_0, a_n, and b_n are

$$\begin{aligned} a_0 &= \frac{1}{T}\int_{-\frac{T}{2}}^{\frac{T}{2}} f(t)dt \\ a_n &= \frac{2}{T}\int_{-\frac{T}{2}}^{\frac{T}{2}} f(t)\cos\ n\omega t\ dt \\ b_n &= \frac{2}{T}\int_{-\frac{T}{2}}^{\frac{T}{2}} f(t)\sin\ n\omega t\ dt \end{aligned} \tag{5.13}$$

Odd symmetry

For functions with odd symmetry,

$$\begin{aligned} a_n &= 0 \\ b_n &= \frac{4}{T}\int_{0}^{\frac{T}{2}} f(t)\sin\ n\omega t\ dt \end{aligned} \tag{5.14}$$

Even symmetry
For a function with even symmetry

$$b_n = 0.$$

$$a_0 = \frac{2}{T}\int_0^{\frac{T}{2}} f(t)dt \tag{5.15}$$

$$a_n = \frac{4}{T}\int_0^{\frac{T}{2}} f(t)\cos n\omega t\, dt$$

$$a_n = \frac{4}{T}\int_0^{\frac{T}{2}} f(t)\cos n\omega t\, dt \tag{5.16}$$

Single-valued continuous periodic function $f(t)$ in the time domain can be written as follows:

$$f(t) = \sum_{n=1,2,3..}^{\infty} c_n e^{jn\omega t} \tag{5.17}$$

where

$$c_n = \frac{1}{2}(a_n - jb_n) = \frac{1}{\pi}\int_{-\pi}^{\pi} f(\omega t)e^{jn\omega t}d(\omega t) \tag{5.18}$$

$$c_0 = a_0 = \frac{1}{2\pi}\int_{-\pi}^{\pi} f(\omega t)\, d(\omega t) \tag{5.19}$$

5.18.1.3 Fourier transform

Fourier transform expresses a time-domain periodic single-valued and continuous function $f(t)$ into frequency domain function $F(f)$ [2], [3].

By Fourier transform,

$$F(f) = \int_{-\infty}^{+\infty} f(t)\, e^{-j2\pi fT} dt \tag{5.20}$$

When

(a) $f(t)$ is well-defined single-valued and continuous function in time interval T.
(b) There is no discontinuity in time T if discontinuity exists, then it will be finite.
(c) The number of maxima in T is finite.

5.18.1.4 Limitations

The transform will not give a satisfactory result if the above conditions are not satisfied. In reality and discrete approaches, problems with non-stationary signals cannot be directly assessed by Fourier transform.

5.18.2 Discrete Fourier transform (DFT)

If the signal is found in the form of series of independent numerical values with time-discrete intervals, the frequency domain spectrum can be written by DFT as follows:

$$F(f_k) = \frac{1}{N}\sum_{n=0}^{N-1} f(t_n)e^{j2\pi kn/N} \tag{5.21}$$

where N is the total samples number in one period.

DFT finds wide application for the analysis of discrete signals as most of the data capturing techniques is discrete.

Limitation

The main limitation of DFT is that it requires a large execution time.

5.18.3 Fast Fourier transform (FFT)

FFT can also be used to analyze discrete functions. It is much faster than DFT [4]. Matrix multiplication required in FFT is made more simple and hence it requires a lesser time of execution. FFT is a popular mathematical tool for analyzing power signals.

5.18.4 Hartley transform and discrete Hartley transform (DHT)

Hartley transform also transforms time-domain function into frequency domain [1], [4]. For a time-domain function $f(t)$, the frequency spectrum can be given as follows:

$$H(\omega) = \frac{1}{2\pi}\int_{-\infty}^{\infty} f(t)(\cos \omega t + \sin \omega t)dt \tag{5.22}$$

Hartley transform-based spectrum is related to Fourier spectrum in the following way:

$$H(\omega) = Re[F(f)] - Im[F(f)] \tag{5.23}$$

The time-based function can be recovered by inverse Hartley transform as follows:

$$f(t) = \frac{1}{2\pi}\int_{-\infty}^{\infty} H(\omega)(\cos \omega t + \sin \omega t)d\omega \tag{5.24}$$

Like DFT, there is discrete Hartley transform (DHT) that can be written as follows:

$$H(f_k) = \frac{1}{N}\sum_{n=0}^{N-1} f(t_n)\cos 2\pi kn/N \tag{5.25}$$

The factor $\frac{1}{N}$ of DHT is similar to the DFT.

DHT is real and symmetrical. Complex DHT, CAS–CAS transform (CCT) and discrete sine and cosine transform (DSCT) are other advanced mathematical tools based on the Hartley transform.

5.18.5 Wavelet transform (WT)

WT can analyze a function both in the time–frequency domain [5]. In WT, another function, called mother functions $g(t)$, plays an important role. It can be altered for analysis purposes. The function $g(t)$ is made of a decaying function and an oscillatory function. Mathematically, it may be written as follows:

$$g(t) = e^{-\alpha t^2} e^{j\omega t} \tag{5.26}$$

They are many types of mother wavelets. Daubechies wavelet is a common type of mother wavelet. Other functions can be derived from these functions.

Let us consider a derived function from mother wavelets as follows:

$$g'(a, b, t) = \frac{1}{\sqrt{a}} g\left(\frac{t-b}{a}\right) \tag{5.27}$$

Then by WT, a function $f(t)$ can be expressed as follows:

$$WT(a, b) = \frac{1}{\sqrt{a}} \int_{-\infty}^{\infty} f(t) g\left(\frac{t-b}{a}\right) dt \tag{5.28}$$

It has the following advantages.

- WT can analyze non-stationary signals.
- WT execution is fast.
- WT can decompose a signal in the frequency domain.
- Frequency resolution in WT can be adjusted. It helps to analyze transients.
- Both low-interval high-frequency components and long-interval low-frequency components can be analyzed by WT.
- Localized impulses and oscillations can be analyzed by WT.

In common practice, de-noising techniques are followed before applying WTs. Prior knowledge on effective de-noising will be helpful for specific applications.

Two types of WT are used as follows:

- Continuous WT (CWT).
- Discrete WT (DWT).

5.18.5.1 CWT

CWT deals with continuous signals. High computation effort is required to perform CWT. It is slower than DWT. CWT represents a signal in the time–frequency domain; the representation is highly localized concerning the "time" variable. Thus CWT refers to a convolution of a sequence of input data with a function set made by the mother wavelet. The output is normally a real-valued function. However, when the mother wavelet is complex, CWT results in complex value function.

5.18.5.2 Discrete WT

It deals with discrete signals. DWT needs less effort for computation and hence it is faster than CWT. In practice, DWT is often used for fault diagnosis [3,6]. DWT is also applied to preprocess data before the application of principal component analysis. The number of decompositions is normally known as the level of decomposition. The level of decompositions is normally set by signals' size. Coefficients i.e., approximate and detail can be determined by the following approach:

$$A_{j+1}(q) = \sum_{n=-\infty}^{\infty} h(n-2q)a_j(n) \tag{5.29}$$

$$D_{j+1}(q) = \sum_{n=-\infty}^{\infty} g(n-2q)d_j(n) \tag{5.30}$$

where j is the wavelet decomposition level; a_j denotes the approximate coefficients of jth level; d_j denotes detail coefficients of jth level; h and g denote filters.

5.18.6 Principal component analysis (PCA)

PCA is a computational approach to extract the principal significant component from signals. It is very much useful in feature extraction. PCA refers to a multivariate method of big data signal processing extracting features of similarities and differences in the data. It also helps to reduce the data sets' dimension by projecting original data in a set of orthogonal components carrying the main information of the original signal [5,7].

Normally four steps are followed before applying PCA. Thus, steps can be written as follows:

- Step 1: determine the covariance matrix of the data set and, subsequently.
- Step 2: determine the matrix of eigenvectors of the covariance matrix.
- Step 3: perform SVD.
- Step 4: perform PCA to rebuild data set and interpret.

Step 1:
Consider a matrix X having k number of elements as follows:

$$X = \{X_1, X_2, X_3, \ldots X_k\}$$

Let

$$\overline{X} = \text{average of } X = \text{average of } \{X_1, X_2, X_3, \ldots X_k\}$$

The covariance of X can be written as follows:

$$COV_X = \frac{1}{k-1}\sum_{k=1}^{K}(X_1 - X) - (X_1 - X)^T \tag{5.31}$$

T refers to an operation of transposition of a matrix.

Step 2:

An orthonormal matrix U is to be formed by eigenvectors of covariance of X (COV_X) as elements. The first few (e.g., 3) columns of U carry mostly of around from 80% to 90% of the information of the initial set of data; these columns are normally empirical orthogonal functions.

Step 3:

Performing single-value decomposition (SVD), covariance can be modified as

$$COV_X = USU^T$$

where S is the diagonal motrix for SVD [7].

Step 4:

Then PCA is performed to rebuild the initial data set, facilitate interpretation along with data reduction.

5.18.7 K-means, Silhouette index, and precision

All these parameters are used to extract features in signal processing and can be used in the machine learning process. Definitions of these parameters are given below:

- *K-means*: It refers to clustering data from a given data set. Cluster effectiveness can be judged by this parameter.
- *Silhouette index*: It refers to the quality of the data clustering presented from -1 to 1. It indicates a comparison between how close one sample is to its neighbors and how far it is from the other clusters.
- *Precision value*: It refers to how often an algorithm is successful in classifying the results. The precision value indicates the performance of a control process, quality, and effectiveness of an algorithm used for monitoring purpose.

5.18.8 Machine learning and artificial intelligence

Machine learning is applied to extract features from a large data set. Various types of machine learning tools are used to process, monitor, and extract features from the signals. Neural network-based artificial intelligence is often applied for feature extraction, decision making, and optimization. Along with it, fuzzy-based techniques are also applied. Different genetic algorithms have been introduced in this field. These techniques are also helpful in minimizing PQ disturbances, control units, etc. Machine learning-based supervision of NPG systems improves control operation and the reliability of the system.

5.19 Standards and guidelines

Poor quality-related many standards and guidelines have been introduced by international technical bodies like IEEE, ANSI, IEC, etc. Some useful references of guidelines are presented in Table 5.8. The guidelines may be followed for quality assurance of NPG systems.

Table 5.8 Guidelines by IEEE, ANSI, and IEC

Standards	Issues
IEEE-4	High-voltage testing
IEEE-100	Electrical and electronic engineering—standard dictionary
IEEE-120	Electrical measurements in power circuits
IEEE-141	Power distribution for industrial plants with the effect of voltage disturbances—recommended practice
IEEE-142	Grounding in industrial and commercial power systems
IEEE-213	Measurement of emissions from television and FM broadcast receivers in the range of 300 kHz–25 MHz to power lines
IEEE-241	Power systems in commercial buildings—recommended practice
IEEE-281	Power system communication equipment—standard service conditions
IEEE-299	Electromagnetic shielding enclosures
IEEE-367	Power station ground potential rise and induced voltage from a power fault
IEEE-376	Measurement of impulse strength and impulse bandwidth
IEEE-430	Measurement of radio noise from overhead power lines and substations
IEEE-446	Emergency and standby systems for industrial and commercial applications
IEEE-449	Ferro resonance voltage regulators standards
IEEE-465	Surge protective devices—test specifications
IEEE-472	Event recorders
IEEE-473	Electromagnetic site survey practice (10 kHz–10 GHz)
IEEE-493	The design practice of reliable industrial and commercial power systems
IEEE-519	Harmonic control and reactive compensation of static power converters
IEEE-539	Corona and field effects of overhead power lines
IEEE-859	Reporting and analyzing outage occurrences and outage states of in transmission
IEEE-944	Uninterruptible power supplies for generating stations
IEEE-998	Direct lightning strike shielding of substations
IEEE-1048	Protective grounding of power lines.
IEEE-1057	Digitizing waveform recorders and their standards
IEEE-Pl100	Powering and grounding sensitive electronic equipment in commercial and industrial power systems.
IEEE-1159	Monitoring EPQ. Categories of power system electromagnetic phenomena
IEEE-1250	Service guides for equipment sensitive to momentary voltage disturbances
IEEE-1346	Evaluation practice for electric power system compatibility with electronics process equipment
IEEE-P-1453	Flicker
IEEE/ANSI-18	Hunt power capacitors standards
IEEE/ANSI-C37	Surge withstand capability (SWC) tests
IEEE/ANSI-C50	Harmonics and noise in synchronous machines
IEEE/ANSI-C57.110	Establishing transformer capability
IEEE/ANSI-C57.117	Failure data reporting for power transformers and shunt reactors in utility systems
IEEE/ANSI-C62.45	Surge voltage in low-voltage AC power circuits, lightning arresters applications

(Continues)

Table 5.8 (*Continued*)

Standards	Issues
IEEE/ANSI-C62.48	Interactions between power system disturbances and surge protective devices
ANSI-C84.1	Electric power systems and equipment voltage ratings (60 Hz)
ANSI-70	National electric code
ANSI-368	Telephone influence factor
ANSI-377	Spurious radio frequency emission from mobile communication equipment
IEC-38	Standard voltages
IEC-816	Measurement guide for short-duration transients on low-voltage power and signal lines
IEC-868	Flicker meter
IEC-1000	EMC
EN-50160	Voltage characteristics of electricity supplied by public distribution systems.
EC/EN-60868	Flicker meter implementation
IEC-61000	EMC

5.20 Summary

Different quality issues of electric power that may appear in grid systems have been presented in this chapter. Commonly faced PQ disturbances have been discussed along with reasons for their appearance and probable effects. Some quality disturbances have been further classified into different categories. NPG systems face lots of harmonics-related PQ issues. Mostly fractional harmonics are found and signals are seen to be nonstationary. Parameters that are useful for estimating the quality of power and indicating the presence of disturbances have been presented followed by useful mathematical tools and standards.

References

[1] S. Chattopadhyay, M. Mitra, and S. Sengupta, *Electric Power Quality*, 1st ed., Springer, Netherlands, 2011.

[2] H. Samet, T. Ghanbari, E. Kazemzadeh, and E. Moshksar, "Full-selective contribution of sources in DC nanogrids using a technique based on buck/boost converter," *IET Energy Systems Integration*, vol. 1, no. 1, 2019, pp. 14–22.

[3] T. S. Abdelgayed, W. G. Morsi, and T. S. Sidhu, "Fault detection and classification based on co-training of semi-supervised machine learning," *IEEE Transactions on Industrial Electronics*, vol. 65, no. 2, pp. 1595–1605, 2018.

[4] S. Karmakar S. Chattopadhyay, M, Mitra, and S, Sengupta, *Induction Motor Fault Diagnosis*, 1st ed., Springer, Singapore, 2016.

[5] Z. Wang, G. Tian, M. Meo, and F. Ciampa, "Image processing based quantitative damage evaluation in composites with long pulse thermography," *NDT & E International*, vol. 99, pp. 93–104, 2018.

[6] M. Zhao, M. Kang, B. Tang, and M. Pecht, "Multiple wavelet coefficients fusion in deep residual networks for fault diagnosis," *IEEE Transactions on Industrial Electronics*, vol. 66, no. 6, pp. 4696–4706, 2019.

[7] W. P. Winfree, K. E. Cramer, J. N. Zalameda, P. A. Howell, and E. R. Burke, "Principal component analysis of thermographic data," in *Thermosense: Thermal Infrared Applications XXXVII*, vol. 9485, International Society for Optics and Photonics, 2015, p. 94850S.

Further reading

[1] "IEEE Draft Standard Technical Specifications of a DC Quick and Bidirectional Charger for Use with Electric Vehicles," in *IEEE P2030.1.1/D1*, February 2021, pp. 1–159, 9 February 2021.

[2] "IEEE Standard for Utility Industry Metering Communication Protocol Application Layer (End Device Data Tables)," in *IEEE Std 1377-2012 (Revision of IEEE Std 1377-1997)*, pp. 1–576, 10 August 2012, DOI: 10.1109/IEEESTD.2012.6264065.

[3] "IEEE Guide for Safety in AC Substation Grounding," in *IEEE Std 80-2013 (Revision of IEEE Std 80-2000/ Incorporates IEEE Std 80-2013/Cor 1-2015)*, pp. 1–226, 15 May 2015, DOI: 10.1109/IEEESTD.2015.7109078.

[4] "IEEE Draft Recommended Practice for Personnel Qualifications for Installation and Maintenance of Stationary Batteries," in *IEEE P1657/D9*, July 2017, pp. 1–114, 1 January 2017.

[5] "IEEE Standard Glossary of Stationary Battery Terminology," in *IEEE Std 1881-2016*, pp. 1–42, 30 August 2016, DOI: 10.1109/IEEESTD.2016.7552407.

[6] "IEEE Recommended Practice for Monitoring Electric Power Quality," in *IEEE Std 1159-1995*, pp. 1–80, 30 November 1995, DOI: 10.1109/IEEESTD.1995.79050.

[7] "IEEE Recommended Practice for Power Quality Data Interchange Format (PQDIF)," in *IEEE Std 1159.3-2019 (Revision of IEEE Std 1159.3-2003)*, pp. 1–185, 1 May 2019, DOI: 10.1109/IEEESTD.2019.8697192.

[8] "IEEE Recommended Practice for Monitoring Electric Power Quality," in *IEEE Std 1159-2019 (Revision of IEEE Std 1159-2009)*, pp. 1–98, 13 August 2019, DOI: 10.1109/IEEESTD.2019.8796486.

[9] "IEEE Recommended Practice for Monitoring Electric Power Quality," in *IEEE Std 1159-2009 (Revision of IEEE Std 1159-1995)*, pp. 1–94, 26 June 2009, DOI: 10.1109/IEEESTD.2009.5154067.

Chapter 6

Faults and protection in nanogrids and picogrids

This chapter deals with faults and protection in nanogrid and picogrid systems. Faults commonly faced in these grids have been discussed along with their causes and effects. Schemes used for protection against different faults have been presented. Internal faults, as well as external faults affecting the system, have been considered. Probable faults in each section of the grids have been presented.

6.1 Introduction

The operation of nano and pico grids (NPGs) systems should be safe and well protected against different types of fault that may appear in the system. To design protection schemes for NPG, one should have adequate knowledge about the variety of sources, loads, and the internal topology of the NPG system. A fault may occur inside the system and can cause considerable damage to the system and connected loads and other utilities. On the other hand, faults may occur outside the NPG system and the fault signals may propagate to the NPG system causing lots of damage. Some faults may not produce significant damage but can hamper the system performance, efficiency, reliability, and quality of electric power produced by the system. Thus, these types of faults are also important and must be paid attention for early diagnosis and protection. Like other systems, NPG systems also need adequate safety measures for operators and users. Different types of safety measures, faults and their protection have been discussed in the following sections.

6.2 Fault zones

For reliable and smooth operation of the NPG, adequate information and knowledge should be gathered on possible types of faults that may occur in different parts of the system. Based on the knowledge and experience, prior care must be taken to minimize the damage that may take place because of the occurrence of faults. Each sub-zone of the NPG system is prone to the occurrence of various types of faults. The nature and the severity of different types of fault are different.

Internal faults can occur at any part of the NPG systems. We can divide the NPG system into different zones for a better understanding of faults and then plan for zone-wise protection schemes. The protection scheme should cover all zones and care should be taken so that no part remains unprotected. Overlapping of protective zones will be there for good design of protection schemes for NPG. For providing overall protection, NPG systems can be divided into different zones as follows:

- resource area,
- storage unit,
- power electronics converters and controllers,
- protective elements,
- link bus and load bus,
- gateway,
- load distribution sections,
- feeders and other lines, and
- transformers.

6.3 Faults in photovoltaic resources

6.3.1 Common faults

Different types of sources are used in the NPG system. Here, some faults in the solar photovoltaic (PV) systems will be discussed. Some common faults in the solar PV systems are

- partial shading,
- undesired level of irradiance,
- string fault, and
- hot spot or temperature rise.

6.3.2 Partial shading

Partial shading is a common problem of PV modules. It refers to covering a part of PV area by dust or other meaning resulting in less energy generation. Monitoring and cleaning after a certain interval are the common solutions to this problem. However, many advanced methods have been introduced to monitor it remotely.

6.3.3 Irradiance

Irradiance decides the amount of power generation by PV module. For satisfactory power generation, it should be within the limit. Different schemes are available to monitor irradiance. Maximum power point tracking based solar PV controller works closed-loop operation based on internal system parameters as well as external parameters like temperature and irradiance. Insufficient irradiance may result in no or less production of electric energy from solar PV units.

6.3.4 String fault

Presently, solar PV-based nanogrids and picogrids are widely used in the distributed form as compared to other resources of non-conventional types for electrical energy generation. In NPG applications, solar PV-based generations are connected with local grids. The PV modules are generally interconnected with each other in various design combinations like series and parallel, and thus PV modules form PV arrays. Each series combination of PV modules is referred to as PV string. These modules may also be connected in parallel combinations. Thus, integration of PV modules may be of any of combinations like series or parallel or both.

String fault refers to a fault in any one or many strings that do not contribute to power generation. The faults in strings of solar photovoltaic (SPV) arrays may occur due to many reasons. Two common causes of string fault are long outdoor operations and extreme weather conditions. This fault may degrade output, performance and sometimes may lead to fire hazards.

6.3.4.1 Diagnosis practice

String fault may be detected by spot verification and at the time of routine maintenance. However, by only visual monitoring of solar string, the faulty string may be overlooked. String fault affects the performance of the converter connected with them. Therefore, if current signals of converter outputs are monitored, string fault can be detected. For this, prior system information and learning are necessary. String fault affects the harmonics nature of converter output. If fast Fourier transform (FFT), discrete wavelet transform (DWT), or other mathematical tools are applied, the change of features may be extracted. Then by choosing the best parameter, string fault may be detected.

6.3.4.2 Algorithm

A simplified algorithm for string fault detection may be made as follows:

(a) Capture current from the inverter output.
(b) Perform FFT, wavelet decomposition, and mathematical tools on the signals.
(c) Extract system-specific features of different statistic parameters at normal and string fault conditions.
(d) Perform several case studies and self or adaptive learning.
(e) Apply specific features for string fault detection.

6.3.4.3 Comparison of different approaches for string fault monitoring

A comparison of different approaches for string fault monitoring is presented in Table 6.1.

6.3.4.4 Case study

A model of 20 kW nanogrid system was used in the case study. The output of the PV unit was converted into three-phase AC by inverter as shown in Figure 6.1. PV unit is made of several strings connected in parallel. Each PV string consists of several PV panels connected in series. The case study considers 72 strings. The maximum output capacity of each string was 300 W. The open-circuit voltage was 60 V. The short-circuit current string was 5.6 A, the temperature co-efficient was –0.24%/°C, the diode ideality factor was 0.94, the shunt resistance was 460 ohms and the series resistance was 0.55 ohms.

Table 6.1 Comparison of different approaches for string fault monitoring in solar PV-based NPG system

String fault monitoring approach	Remarks
Outdoor visual monitoring	• It is not possible daily • A fault may be overlooked
Testing of PV string during maintenance	• It is not possible daily • It is a time-consuming process
Converter output current signal monitoring	• Continuous monitoring is possible • Needs analysis in frequency or time–frequency domain • System information and learning are needed for accurate fault isolation

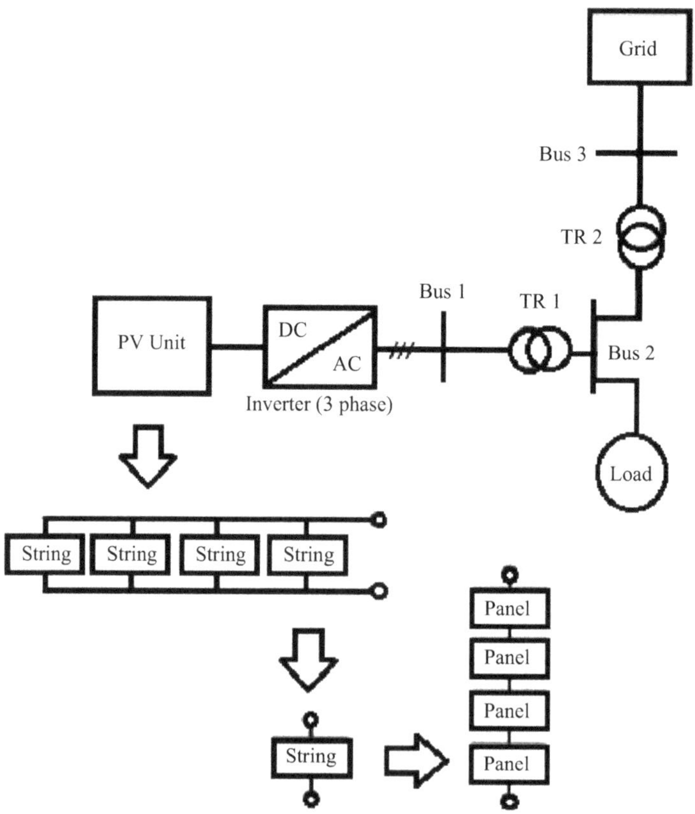

Figure 6.1 Single line representation of grid-connected nanogrid network

The link bus was connected to the local load and utility grid through a transformer. A voltage source converter-based inverter with three bridge arms was used. It was designed to provide 50 Hz, 220–260 V phase voltage. A star-delta transformer step-up was connected with the utility grid.

String fault detection was detected by monitoring output currents of inverter connected with PV arrays suffering from string fault. Analysis of output inverter currents of the nanogrid model was carried out for both at normal and different string fault conditions.

The FFT-based analysis was carried out on output inverter currents for the determination of sub-harmonic components, total harmonic distortion (THD), and DC components.

Furthermore, wavelet decomposition was performed on the inverter output currents. Then, skewness and kurtosis values of approximate and detail coefficient at different DWT levels under normal and different string fault conditions were determined. The inverter output currents have been monitored.

Variation of THD and DC components for different percentages of string faults (S %) are shown in Figures 6.2 and 6.3, respectively. The THD variation is almost linear in the first part and decreasing with the increase in the percentage of string faults from normal and up to a certain percentage (13% for this case study) of string faults, and then it is increasing with the increase of the percentage of fault.

The variation of the DC component is almost linear and increasing with the increase in the percentage of string faults from normal and up to a certain percentage (15%, for this case study) of string fault and then it becomes almost constant.

Converter output currents are nonstationary, therefore, DWT will be more suitable than FFT for specific feature extraction. Different parameters were determined from the wavelet decomposition-based coefficient. Few of them are found capable of detecting string faults accurately and can measure the percentage of string faults. After learning from several cases, the best parameter(s) may be selected for string fault monitoring.

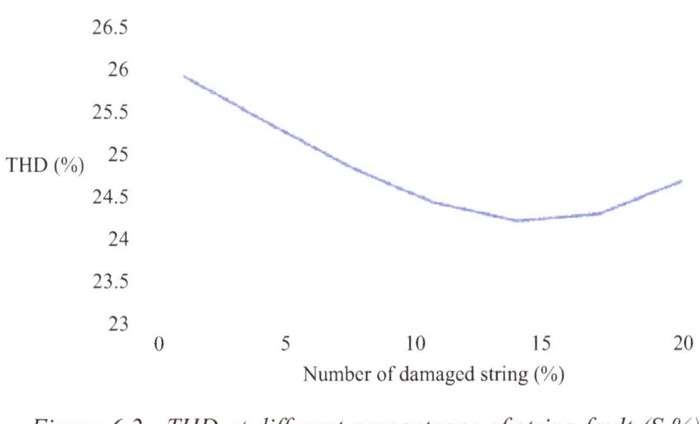

Figure 6.2 THD at different percentages of string fault (S %)

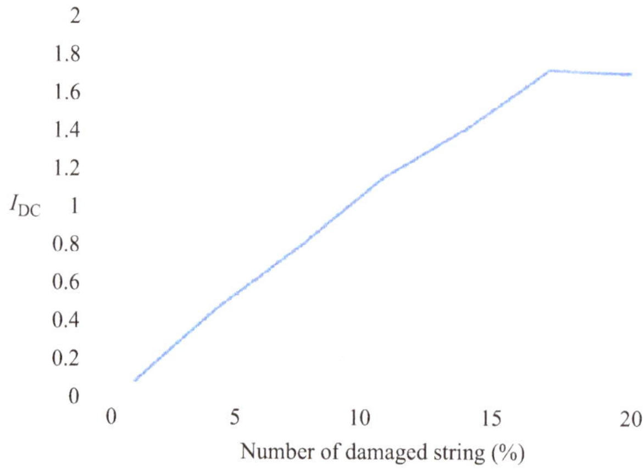

Figure 6.3 DC component at different percentages of string fault (S %)

6.3.5 Hot spot or temperature rise

Hot spot refers to the abnormal rise of temperature at certain parts of the solar PV module. This may happen due to the damage to PV material by any cause. Gradually, the area of hot spot spreads in surrounding areas and may result in total damage module. Therefore, early detection of the hot spot is necessary and it helps to isolate that faulty area or panel from the main string connections.

6.3.5.1 Case study findings

A case study was carried out on a 12 kW nanogrid system fed solar PV connected with 48 battery units. Inverter output was monitored and the effects of temperature rise on the inverter output were observed. It shows that the temperature rise has a considerable effect on converter output waveforms. The percentage value of THD is seen to increase and DC components at AC output waveform are seen to decrease with the increase of temperature as shown in Figures 6.4 and 6.5, respectively.

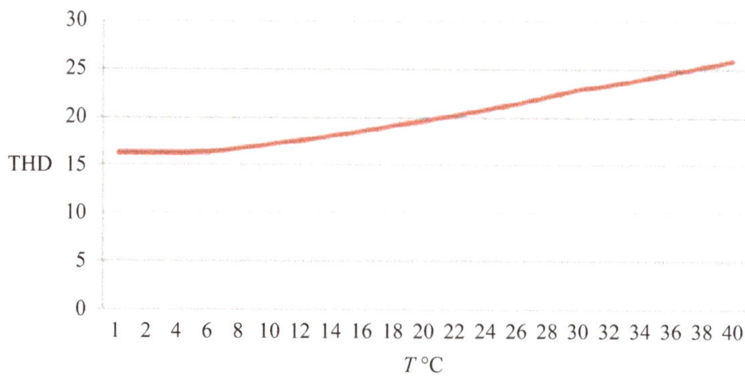

Figure 6.4 Temperature versus THD

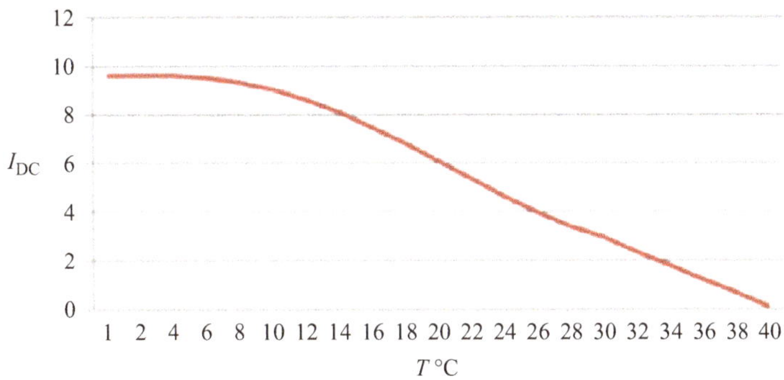

Figure 6.5 Temperature versus DC component curve

6.3.5.2 Detection

Temperature inspection of the pannel, routine and regular testing of pannels may be done for hot spot detection. Also, various advanced techniques have been introduced based on thermal monitoring and converter output waveform analysis that may be followed. FFT, DWT, and other tools may be used to study the effect of a hot spot on the system output.

An algorithm that may be developed based on optimization are as follows:

(i) capture inverter output current waveforms,
(ii) perform FFT,
(iii) determine THD, DC component,
(iv) perform DWT,
(v) determine different parameters,
(vi) observe the variation of those parameters with the increase of temperature and perform learning, and
(vii) apply learnings for fault detection.

6.4 Faults in the battery storage unit

Different types of batteries are used in the NPG system. Some common faults in battery units are

- dielectric failure,
- change of state of charge (SOC) characteristics,
- decrease of ampere–hour capacity,
- leakage, and
- wrong connection.

With aging dielectric properties of battery materials, SOC characteristics and leakage property change. Testing may be carried out during maintenance and inspection time. For maintenance-free batteries, replacement after a certain interval is the solution to these problems.

Wrong connection with charging circuits and loads may lead to rapid discharge and flashes. After installation, care is to be taken to check regularly. Following common steps are taken for battery protection:

- testing on a regular interval,
- condition monitoring,
- polarity checking,
- inspection on leakage,
- inspection on whether there is any hidden load or hidden short circuit path,
- use of diode with converter circuit to avoid the reverse flow of current,
- use of reactor with converter circuit to limit current, and
- proper converter protection provides indirect protection for the battery.

6.5 Faults in power electronics converters and controllers

Different types of power electronics converters and controllers are used in NPG systems. Some common faults in converter and controller units are

- total conversion failure,
- partial conversion failure,
- element burn out,
- short circuit,
- failure of the free-wheeling diode,
- failure in the feedback control path,
- absence of proper reference signals, and
- the sudden absence of reference signal.

Total conversion failure: It may happen during the following events:

- supply failure,
- open circuit in the supply line,
- breaker operation due to fault,
- blown of fuse due to fault, and
- failure of control logic.

Partial conversion failure: It may occur during the following events:

- failure of switching elements,
- failure in free-wheeling diodes,
- failure of the heat sink, and
- improper or failure in control logic.

Element burn out: It occurs due to

- increase of current,
- failure of the heat sink,
- failure in the cooling system, and
- the flow of short circuit current.

Short circuit: Flow if short circuit current through converters may take place due to internal as well as external reasons. It may occur due to the following main reasons:

- bus short circuit,
- short circuit in the load,
- the short circuit inside the converter is due to the burning of some elements, the creation of shorted path, etc.

Failure of free-wheeling diode: It occurs mainly due to the flow of excess current beyond its capacity.

Failure in feedback control path: It may occur due to sensor failure and open or short circuits in the feedback path.

Absence of proper reference signals: It is the consequent effect of previous failure i.e., failure in feedback path that does not provide feedback signals.

The sudden absence of reference signal: It may occur due to the following main reasons:

- improper selection of reference signal or data,
- unavailability of the reference signal or data, and
- memory loss containing data for reference signals.

6.6 Faults in basic protective elements

6.6.1 Common fault

Different types of basic protective elements are used in the NPG system. Three common faults that may occur in basic protective elements units are

(a) failure in the grounding system,
(b) failure in the current limiting reactor, and
(c) "snubber circuit" or "snubber" fault.

Grounding materials, reactors, and snubber circuits are three important parts of protective devices for NPG systems. Occasionally, faults may appear in these elements causing another fault to the system or larger damage to the system. There, monitoring after a certain interval should be carried out to avoid such incidents.

6.6.2 Failure in the grounding system

Ungrounded system and improper grounding cause difficulties in fault detection. Failure in grounding makes the system unsafe that may affect operating persons and create damage in the system during fault. Both neutral and equipment groundings are important for safety and fault detection. Faults may occur in the line for both types of groundings. "Grounding practice" as per the guidelines of international standards should be followed.

6.6.3 Failure in the current limiting reactors

Reactors are used with many converter circuits to limit current. Therefore, the failure of reactors will allow the unlimited flow of current through the converter

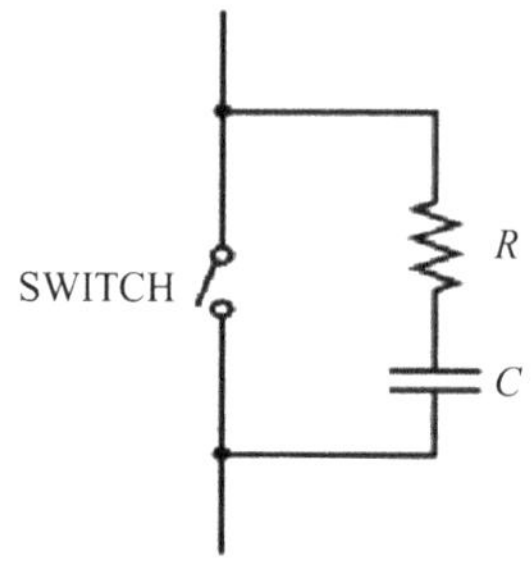

Figure 6.6 Inverter switch RC snubber circuit used for brushless DC (BLDC) motor

resulting in damage. This may lead to a total shutdown of the system. Therefore, the conditions of the reactors should be monitored regularly. Overcurrent relay settings should keep in coordination with reactors' limit values.

6.6.4 Snubber fault

The snubber fault refers to faults that occur in the snubber circuit. The snubber circuit plays an important role in the converter circuit. Different types of snubber circuits are used in converter applications. Among different types of snubbers, RC snubber (shown in Figure 6.6) is a widely used converter to feed power to inductive loads such as electric motors, etc. It uses a small resistor in series with a small capacitor. This limits the rate of rising in voltage (dv/dt) across the thyristor to a value that prevents the inaccurate turn-on of the thyristor. "Snubber resistance fault" may occur due to aging, temperature, and other factors. The snubber itself plays a key role in the minimization of electromagnetic levels and mitigation of voltage spikes on the switches. Thus, it gives protection against high current and voltage. However, for continuous operation and multiple events of abnormal situations, the snubber circuit itself may be damaged. Snubber resistance fault refers to damage of snubber resistor. It initiates from a gradual decrease of the effective resistance value of the snubber resistor. At extreme conditions, it becomes zero and the snubber path gets short-circuited.

Typical design formulas for selecting capacitor and resistor of snubber circuits are

$$R_{snub} = 2.\delta\sqrt{\frac{L}{C}}$$

$$C_{snub} = L\left(\frac{I_r}{K.V_s}\right)^2$$

where R_{snub} is the snubber resistance; C_{snub} is the snubber capacitance; δ is the optimum damping factor; K is the optimum current factor; I_{r} is the recovery current of the semiconductor switch; V_{s} is the source voltage; L is the circuit inductance.

Detection

If the inverter switch snubber circuit resistance fault (ISSCRF) occurs, it should be detected to prevent the system from further expensive damage. Short circuit in snubber circuit is easy to detect and may be done by comparator circuit or by overcurrent relay. But the gradual decrease of snubber resistance is difficult to detect. Ignorance of this may cause larger damage and greater repairing time. Therefore, continuous monitoring may be followed to detect a small and gradual decrease of the resistance value of the snubber resistor. The value of the snubber resistor beyond the 5% tolerance range may be considered as a set value for fault discrimination. It may be noted that the selection of the fault level depends on designers' choices and the system specifications. Snubber fault starting can be detected by monitoring current waveforms and performing waveform analysis in the time–frequency domain. FFT and DWT-based current signature analysis may be followed for continuous monitoring and fault detection.

6.7 Faults in link bus and load bus

6.7.1 Classification

Both AC-type and DC-type buses are found in the NPG system. Based on the utility, different types of link buses and load buses are used in the NPG system. Two common faults in link bus and load bus are

- short circuit and
- open circuit.

Different types of faults that can occur in the bus are shown in the chart in Figure 6.7. Among different types of faults, symmetrical faults are comparatively less in number experienced during operations.

6.7.2 Detection from station

Bus faults can be detected in the station by proper selections of relays. Earth relays are common for earth fault detections. For the line-to-line short circuit, a negative sequence relay is used. Thus, combinations of various types of relays define the art of protections for bus faults.

6.7.3 Detection from the remote end

With the advancement of communication technology and smart sensors, grid parameters may be monitored from the remote end. Therefore, it enables remote conditions and fault monitoring. However, for severe faults, time delay inserted in remote monitoring is a big challenge to be handled. Remote fault detection should be followed by action for alarm generation or circuit breaker operation.

6.7.4 Case study on load bus fault detection in nanogrid system from the remote end

Bus faults have a considerable effect on the converters. Thus, monitoring low-voltage inverter output, different types of bus faults can be discriminated against situations in normal. Case study on nanogrid system connected with load bus and utility grid bus

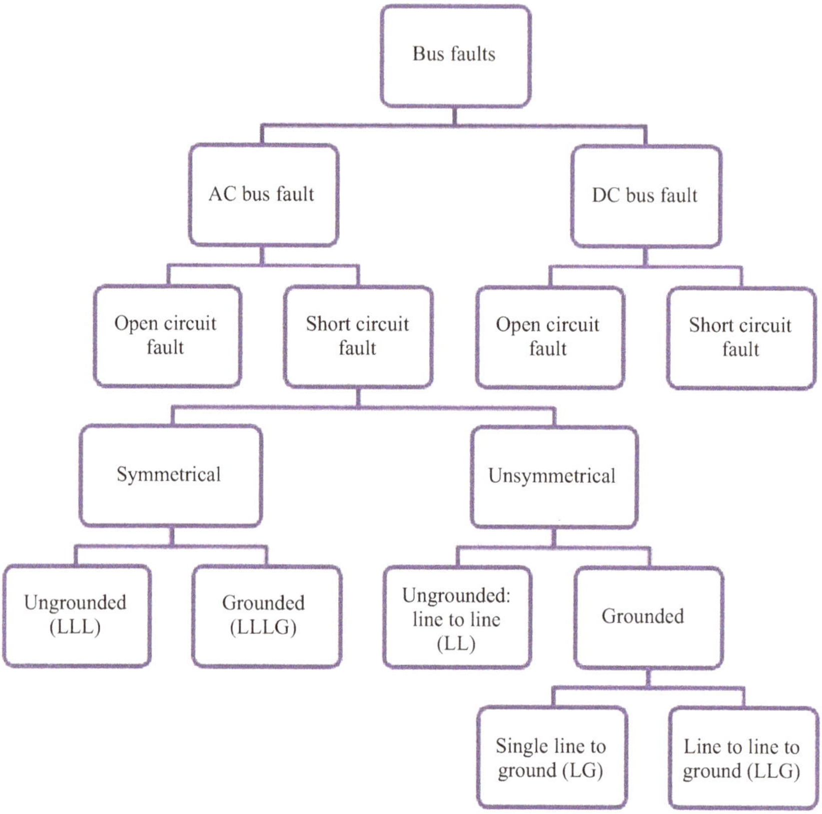

Figure 6.7 Different types of bus faults

has shown that FFT and DWT-based analysis of statistical parameter of nanogrid low-voltage converter output waveform can discriminate different faults occurred at high-voltage bus located at the remote end.

Voltage source converter-based 20 kW nanogrid system was considered as shown in Figure 6.8. Ungrounded symmetrical fault (LLL), grounded symmetrical fault (LLLG), ungrounded double line faults (LL), grounded double line faults (LLG), and a single line to ground faults (LG) have been considered at bus five and current signals connected with bus one have been analyzed for fault assessment.

FFT and DWT are applied on inverter output. THD, DC component, Kurtosis, and Skewness at DWT level – 1 were seen to change significantly for different types of faults. Observations are presented in Table 6.2. Parameters obtained from FFT and DWT-based signal analyses were seen to vary from normal. By FFT, LG and LLL faults can be discriminated clearly, but other faults are difficult to discriminate, and the faulty phase cannot be identified clearly. By DWT-based approach, all types of faults can be discriminated clearly, the faulty phase can be

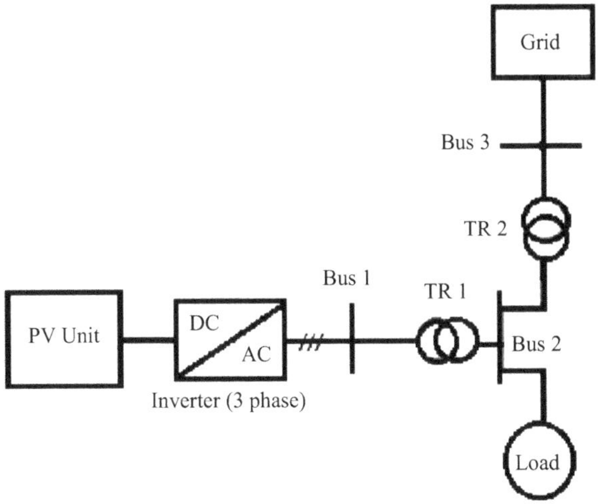

Figure 6.8 Layout of 20 kW nanogrid system used in the case study

Table 6.2 Observations from the case study

Observation made at	Faults location	Fault types	Current signature analysis by	Observation
Bus 1	Bus 2 and bus 3	LG, LL, LLG, LLL, and LLLG	FFT based THD DC component, sub-, and inter-harmonics	Seen to vary from normal LG and LLL faults can be discriminated Other faults are difficult to discriminate The faulty phase can not be identified
			DWT-based Skewness and Kurtosis of wavelet coefficients	Seen to vary from normal All types of faults can be discriminated The faulty phase can be identified clearly In a noisy environment, Kurtosis is more effective

identified clearly. In a noisy environment, Kurtosis is found more effective. For continuous monitoring with large data, a machine learning approach may be applied for fault discriminations.

6.7.5 Case study on grid bus and load bus fault in a solar-wind hybrid system

Solar-wind-fed NPG systems are blooming in the present scenario of pollution and an acute dearth of fossil fuels. Concerning this, this case study has been pursued on

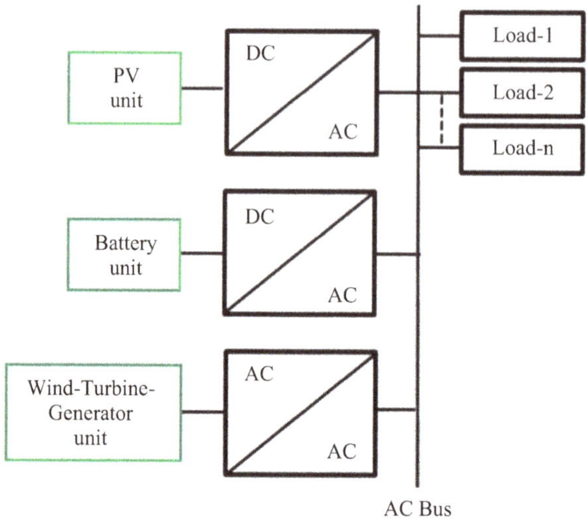

Figure 6.9 Solar-wind-battery hybrid system

the solar-wind-battery fed hybrid system as shown in Figure 6.9. Solar PV unit has the capacity of 5 kW, wind-turbine-generator has the capacity of 10 kW and 3-phases balanced loads were connected.

A study has been made to discriminate normal healthy conditions and various unsymmetrical faults in this hybrid nanogrid system. Following types of faults along with normal conditions have been considered in load bus:

(a) line to ground,
(b) line to line, and
(c) line to line to line.

Current signals are monitored from the converter output end. Stockwell transform and multistep DWT-based decompositions are performed. Then, output coefficients are judged by different statistical parameters like, mean, median, RMS, Skewness, Kurtosis, etc. The analysis has been done considering the following resource combinations:

(a) Only solar PV resource of 5 kW is available.
(b) Only wind-turbine-based generation of 10 kW is available.
(c) Both solar PV and wind-turbine-based generations 15 kW are available.

Findings: Faults were considered at load bus connected with full load i.e., 5, 10, and 15 kW, respectively, for the above three resource combinations. Fault detections were done at converter output ends. The findings of the case studies are as follows:

(a) DWT-based statistical parameters are capable of extracting features for fault discrimination.

(b) Results of fault detections are accurate in all three resource combinations.
(c) Feature extraction reduces data size and therefore reduces the execution time. Cloud-based Internet of Things enabled remote monitoring is possible.

6.8 Faults in feeders and other connecting lines

Different types of feeders and other connecting lines are used in the NPG system. Feeders and other connecting lines are also probable zones for fault occurrence. Common faults in feeders and other lines are

- overload,
- short circuit, and
- open circuit.

Overload refers to an increase of load beyond line capacity that may burn if it is ignored for long or may reduce the lifespan of the line. The short and open circuit faults may be classified in the same way done for bus faults as shown in Figure 6.7. These faults can be detected by conventional approach with different types of overcurrent, earth fault, and voltage relays synchronized with breakers as well as by advanced waveform analysis-based approach using different signal processing tools.

6.9 Faults in transformer

Transformer plays an important role in the power system. Both power transformer and distribution transformer demand a very reliable protection scheme to achieve better performance of the network. Most of the transformers are oil-immersed. This helps in reducing the temperature of the transformer. Fault found in the operation of the transformer may be external or internal. But both of them have some adverse effect on the transformer which may spread out and damage the transformer and another device which are connected with the transformer. Some faults may result in the rise of temperature and have other thermal effects. So immediate fault detection must be done and necessary isolation of the transformer is very much important.

6.9.1 Classification of faults

Faults that are faced by transformers in power systems are broadly classified as

(a) external faults and
(b) internal faults.

External faults: Faults which, occurred outside the transformer and affects it are known as external faults. Thus, these faults occur externally and may damage the transformer. Therefore, transformers need protection against external faults. One such example of an external fault is a short circuit fault that has occurred on the secondary side. The fault is external, but the fault current flows through the

Table 6.3 Different types of faults appeared in the transformer

Fault type	Faults
External faults in the primary side	(a) *Overvoltage*: This may come from the supply. It may damage insulation at the first stage and then other parts (b) *Under voltage*: It is not dangerous for the transformer under the operation, but is undesirable concerning the supply side and loads connected to it
External faults in the secondary side	(a) *Overload*: sudden overloading of the secondary side is not very dangerous, but continuous overloading increases the temperature which may lead to damage (b) *Short circuit*: this is a very serious fault. When this fault occurs, the fault current flows through the transformer. The temperature rises rapidly and may burn the winding. So immediate detection of fault and isolation of transformer is necessary (c) *Under load*: when under load occurs there is no possibility of damage to the transformer. However, it is not desirable to run the transformer at low efficiency. So protection is needed to generate an alarm signal
Internal faults	(a) Core balance leakage fault (b) Earth fault (c) Inter turn short circuit fault (d) Failure of solid insulating material (e) Failure of insulating oil

transformer. This raises the temperature. At high temperature in the transformer, windings may burn out. External faults can be classified as follows:

(i) external faults in the primary side and
(ii) external fault in the secondary side.

Internal faults: These faults occur in internal parts of the system.

Different types of faults observed in transformers are presented in Table 6.3.

6.9.2 Choice of protection scheme

Protection schemes for a transformer are designed considering the following factors:

(a) type of transformer,
(b) connection of transformer windings,
(c) ratings of transformer,
(d) application of transformer,
(e) types of faults, and
(f) importance of service

Different protection schemes and relays used for transformer protection are presented in Table 6.4.

Table 6.4 Different protection schemes for transformers

Protection scheme	Purpose
Buchholtz relay	Overload
Fuse	Overload
Differential protection	Internal faults. Two types: (a) Current balance differential protection (b) Voltage balance differential protection
Biased differential protection	It is used for the detection of an internal fault. It can cope with the change of CT ratio
Harmonic restraining relay	Offer reliable operation in presence of inrush current
Core balance leakage protection	Protection against leakage or earth fault
Frame leakage protection	The relay operates if there is a flow of current through the frame leakage path which should be normally insulated from the ground
Condition monitoring	Protection against failure of oil and solid insulating material

6.10 Faults in the battery unit

Battery input is provided by charger or charge control. On the other hand, battery output provides power to DC load or inverter units. In NPG applications, common battery-related faults are listed below:

(a) overcharging current,
(b) overcharging voltage,
(c) battery reverse polarity,
(d) SPV reverse polarity, and
(e) severe current flow from the battery to the solar PV array.

6.11 Grounding practice

Grounding is important for the safety and detection of faulty reliably. Grounding practice includes the following:

(a) Equipment grounding:
    (i) each resource device of the NPG system must be grounded;
    (ii) all controllers, converters must be grounded.

(b) Neutral grounding:
    In AC parts of the system, neutral of three-phase connection of transformers or generators must be grounded.
(c) *Standard practice*: The practice of solid grounding, resistance grounding, and reactance grounding should be followed as per the guidelines international standards and guidelines given by regulatory bodies.

6.12 Lightning protection

6.12.1 Lightning

Lightning can damage directly or indirectly propagating through the line used for connecting the external grid with NPG. Adequate precaution and safety measures should be taken to avoid damages due to lightning.

Lightning refers to unidirectional voltages with a very high peak. Severe damage can occur in the NPG system due to both direct and indirect lightning strokes. It can damage insulators, lines, and other parts of the NPG systems.

Lightning takes place due to the electrophysical properties of the cloud in three following steps:

(a) charge accumulation,
(b) streamer formation, and
(c) lightning discharge.

Streamers of lightning may be very fast streamers or relatively slow streamers and may be directed downward or upwards or in other directions. Lightning discharge can take place in the following three major routes:

(a) intracloud (IC);
(b) cloud to cloud (CC);
(c) cloud to ground (CG).

Cloud to ground lightning can discharge in two ways:

(i) negative discharge i.e., from negatively charged cloud to the positive ground surface;
(ii) positive discharge i.e., positively charged cloud to the negative ground surface.

6.12.2 Features

Some features of lightning are given below:

- *Discharge property*: Lightning discharge involves ionization of air, plasma formation, heat generation, etc.
- *Waveform characteristics*: With time, lightning is unidirectional rising (front) sharply with a very short time reaching peak value (crest) and then decaying (known as tail) for time. Front ranges up to 10 μs, crest reaches up to 450 kA, and tail ranges up to 10–100 μs and takes up to 1 000 μs to die out.
- *Return strokes*: Streamer propagating toward the cloud with opposite polarity of mainstreamer is referred to as return strokes of lightning discharge.
- *Multiple strokes*: Lightning strokes followed by return strokes and other several forward strokes are known as multiple strokes.
- *Frequency and intensity*: The lighting discharge waveform is unidirectional. The frequency and the intensity of lightning strokes are normally used to express the number of lightning strokes that occurred in a small particular

period (say up to 0.0005 s) and their peak or crest value. Air pollution and global warming have increased the frequency of occurrence.

6.12.3 Damages in NPG system due to lightning

Lightning has many adverse effects and can damage an unprotected NPG system:

- It can create a flow of high current and high voltage.
- High current can burn equipment.
- High voltage can damage insulating materials.
- It can damage the mechanical support structure.

6.12.4 Protection against lightning

NPG system should be protected against lightning. Following protective measures are taken against lightning:

(a) earthing,
(b) shielding, and
(c) lightning arrestors (LA).

Different types of LA or surge arrestors are used for the protection against lightning are as follows:

- rod gap type LA,
- horn gap type LA,
- sphere gap type LA,
- multiple gaps type LA,
- impulse type LA,
- valve type LA,
- expulsion type LA,
- auto valve type LA,
- thyrite LA, and
- metal oxide LA.

6.13 Common protection schemes practiced in NPG

While setting up picogrid or nanogrids, some protections schemes are very common and they should not be ignored. These are listed below:

(i) overload protection,
(ii) under-voltage protection,
(iii) short circuit protection,
(iv) overcharging current protection,
(v) overcharging voltage protection,
(vi) battery reverse polarity protection,
(vii) SPV reverse polarity protection, and
(viii) reverse current flow from the battery to solar PV array protection.

6.14 Safety measures and routine tests practiced in NPG

For safe and reliable operation, NPG systems require some safety measures and routine tests. Some of these are listed below:

(a) use of cables and wire specifications as per standard guidelines,
(b) proper earthing system,
(c) maintaining proper insulation levels for cables and equipment,
(d) no inflammable items near the grids,
(e) proper selection of fuse wires,
(f) proper relay setting,
(g) earth resistance tests,
(h) checking ground voltage,
(i) insulation testing, and
(j) battery tests.

6.15 Summary

This chapter has described different types of faults that may commonly appear in NPG-based systems. Possible faults exist in each internal unit of the NPG. Therefore, no zone or area, or unit should be kept unprotected. Commonly used approaches for the detection of that fault have been presented in the chapter. Few case studies have also been presented. Some advanced signal processing tools have been highlighted for fault diagnosis. Along with the installation of different protection schemes, common safety measures and regular inspection should be followed to ensure the reliable operation of NPG.

Further reading

[1] "IEEE DC Quick Charger for Use with Electric Vehicles Test Suite Specification – Version 1.0," in *IEEE DC Quick Charger for Use with Electric Vehicles Test Suite Specification – Version 1.0*, pp. 1-0, 17 September 2019.

[2] "IEEE Draft Standard Technical Specifications of a DC Quick and Bi-directional Charger for Use with Electric Vehicles," in *IEEE P2030.1.1/D1*, February 2021, pp. 1–159, 9 February 2021.

[3] "IEEE Recommended Practice for Performance Testing of Electrical Energy Storage (EES) System in Electric Charging Stations in Combination with Photovoltaic (PV)," in *IEEE Std 2836-2021*, pp. 1–27, 8 June 2021, DOI: 10.1109/IEEESTD.2021.9447313.

[4] Continental Automated Buildings Association, "Impact of Smart Grid on Connected Homes Landmark Research Study," in *Impact of Smart Grid on Connected Homes*, pp. 1–106, 17 May 2013, DOI: 10.1109/IEEESTD.2013.6786299.

[5] "IEEE Standard for Utility Industry Metering Communication Protocol Application Layer (End Device Data Tables)," in *IEEE Std 1377-2012 (Revision of IEEE Std 1377-1997)*, pp. 1–576, 10 August 2012, DOI: 10.1109/IEEESTD.2012.6264065.
[6] "IEEE Draft Standard for Synchrophasor Measurements for Power Systems," in *IEEE PC37.118.1/D4.2*, May 2011, pp. 1–56, 23 June 2011.
[7] "IEEE/ANSI Draft Recommended Practice for Electromagnetic Compatibility Limits and Test Levels," in ANSI PC63.12/D12g, May 2015, pp. 1–49, 1 January 2015.
[8] "IEEE Draft Standard for DC Microgrids for Rural and Remote Electricity Access Applications," in *IEEE P2030.10/D11*, February 2021, pp. 1–56, 10 March 2021.
[9] "IEEE Approved Draft Standard for Test Methods for Surge Protectors and Protective Circuits Used in Information and Communications Technology (ICT) circuits, including Smart Grid Data Circuits," in *IEEE PC62.36/D38*, February 2016, pp. 1–146, 1 January 2016.
[10] "IEEE Standard Conformance Test Procedures for Equipment Interconnecting Distributed Energy Resources with Electric Power Systems and Associated Interfaces," in *IEEE Std 1547.1-2020*, pp. 1–282, 21 May 2020, DOI: 10.1109/IEEESTD.2020.9097534.
[11] "IEEE Guide for Safety in AC Substation Grounding," in *IEEE Std 80-2013 (Revision of IEEE Std 80-2000/ Incorporates IEEE Std 80-2013/Cor 1-2015)*, pp. 1–226, 15 May 2015, DOI: 10.1109/IEEESTD.2015.7109078.
[12] "IEEE Draft Recommended Practice for Personnel Qualifications for Installation and Maintenance of Stationary Batteries," in *IEEE P1657/D9*, July 2017, pp. 1–114, 1 January 2017.
[13] "IEEE Standard Glossary of Stationary Battery Terminology," in *IEEE Std 1881-2016*, pp. 1–42, 30 August 2016, DOI: 10.1109/IEEESTD. 2016.7552407.
[14] "IEEE Guide for Technology of Unified Power Flow Controller Using Modular Multilevel Converter—Part 2: Terminology," in *IEEE Std 2745.2-2021*, pp. 1–25, 1 September 2021, DOI: 10.1109/IEEESTD.2021.9525319.
[15] "IEEE Recommended Practice for Monitoring Electric Power Quality," in *IEEE Std 1159-1995*, pp. 1–80, 30 November 1995, DOI: 10.1109/ IEEESTD.1995.79050.
[16] "IEEE Recommended Practice for Monitoring Electric Power Quality," in *IEEE Std 1159-2019 (Revision of IEEE Std 1159-2009)*, pp. 1–98, 13 August 2019, DOI: 10.1109/IEEESTD.2019.8796486.
[17] S. Karmakar, S. Chattopadhyay, M. Mitra, and S. Sengupta, *Induction Motor Fault Diagnosis*, 1st ed., Springer, Singapore, 2016.
[18] S. Chattopadhyay and A. Das, *Overhead Electric Power Lines – Theory and Practice*, 1st ed., IET, Stevenage (UK), 2021.

Chapter 7
Utilization of nanogrids and picogrids

This chapter deals with the utilization of nanogrids and picogrids. Different utilizations have been classified based on various grid aspects. Tariff and load considerations have been discussed. Examples of many grid utilization in various sectors have been provided. Some case studies have been presented.

7.1 Introduction

Reliability and continuity are the two most important requirements of a power supply system. With this aim, grids are being converted or upgraded into smart grids. It should handle variable, increasing, and extreme conditions of the demand efficiently. Along with this, it aims at minimizing carbon emission, integrating renewable sources, and reduction of the tariff. It allows distributed generations (DGs) and opens the door of both way communications between suppliers and customers. The main advantageous features of the smart nano and pico grid (NPG) are as follows:

- use of more digital devices,
- use of smart meters,
- information sharing with customer,
- better demand response,
- better load management,
- more power safety and security,
- better self-restoration capacity, and
- remote monitoring and remote control.

Additional main components of the smart grid are as follows:

- smart metering;
- monitoring and digital control;
- interconnected transmission facility;
- advanced communication, Internet of Things (IoT) with remote monitoring and control opportunity;
- virtual storage facility;
- flexible distributed load management.

7.2 Classification of NPG utilization

Different types of NPG utilizations are found. Different types of grid NPG utilizations are shown in Figure 7.1. NPG utilizations may be classified in various ways.

Based on the load unit number or load area, NPG utilization may be divided into two categories as follows:

- Individual or self-utilization.
- Utilization in a small area.

Based on the shape of output voltage and current waveforms, NPG utilization may be divided into two categories as follows:

- AC.
- DC.

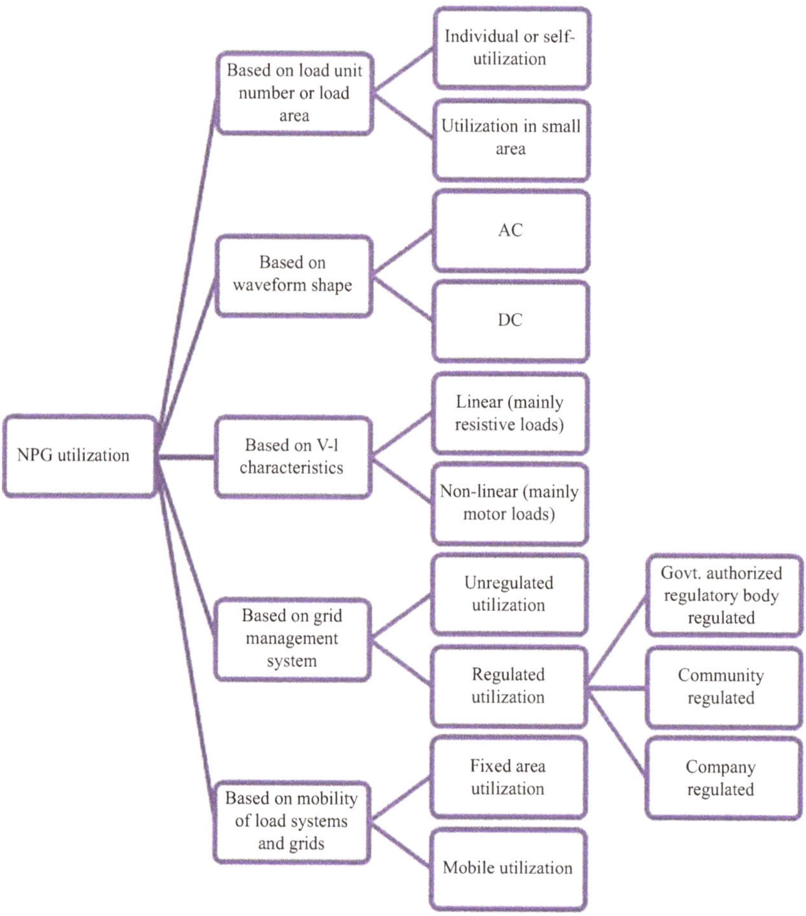

Figure 7.1 Different types of grid NPG utilizations

Based on load voltage–current characteristics, NPG utilization may be divided into two categories as follows:

- Linear (mainly resistive loads).
- Nonlinear (mainly motor loads).

Based on the grid management system, NPG utilization may be divided into two categories as follows:

- Unregulated utilization.
- Regulated utilization.

Based on the type of regulatory body, NPG utilization may be divided into three categories as follows:

- Government-authorized body-regulated utilization.
- Community-regulated utilization.
- Company-regulated utilization.

Based on the mobility of grid and load systems, NPG utilization may be divided into two categories as follows:

- Fixed area utilization.
- Portable or mobile utilization.

7.3 NPG tariff schemes

NPG-based energy trading involved different types of tariffs in different countries and communities. In some regions, it is regulated by the government, in other regions, it is governed by the community management system or private companies. Different types of tariff schemes used for NPG systems are shown in Figure 7.2. Tariffs followed for energy trading through NPG may broadly be classified into two categories:

- Flat rate tariff.
- Dynamic tariff.

Flat rate tariff: This type of tariff is simple in computations and sets the prices based on units of power consumption. Its nature is proportional to the price of per unit energy consumption.

Dynamic tariff: It is the variable type and prices depend on various factors of trading, supplies, and demand aspects. Some commonly used examples of the dynamic tariff are given below:

- Real-time pricing: it considers real-time generation cost.
- Time of use: it considers the time and duration of energy consumption.
- Variable peak point: it considers the maximum power demand that has been experienced by the system.
- Critical peak pricing: it considers an increase in price due to the event of high demand.
- Critical price rebate: it considers rebate of the price at the events of low demand and high resource availability.

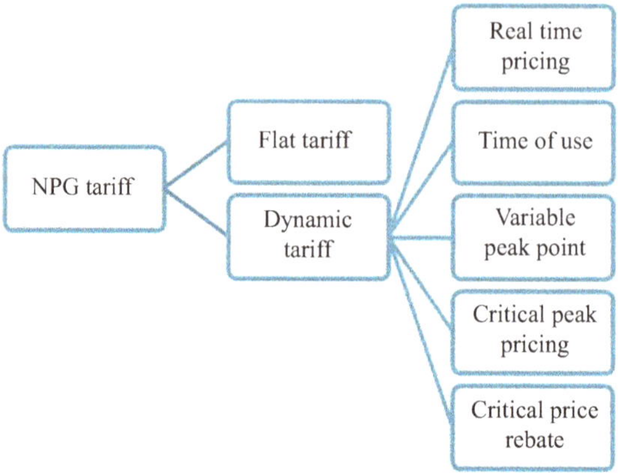

Figure 7.2 Different types of tariff schemes used for NPG systems

7.4 Picogrids utilization

The utilization of picogrids is increasing rapidly. Comparatively off-grid picogrid systems share a major part of utilization. Picogrids are preferred when the load unit is small in size. Solar photovoltaic (PV)-fed picogrid systems can be utilized for meeting energy demands on a small-scale need basis. Few examples are given below:

- picogrid in street lighting;
- picogrids in small domestic applications;
- picogrid for small motor loads;
- picogrid for rural domestic application;
- picogrid for rural primary schools or libraries;
- picogrids in agricultural applications;
- picogrids in small rural markets;
- picogrids for small religious temples or community centers.

7.4.1 Picogrid in street lighting

If street lighting with several supports or light posts covering the full length of the roads is considered, micro or nanogrid is chosen. But, for the same street, if each or few street lighting arrangements are considered separately installed in individual support/towers/poles, then picogrid offers the optimum solution. Picogrids for street lighting may be operated in standalone isolated mode or with the on-grid condition. Picogrid fed each street lighting system gets separate and independent control facility. Both solar-fed picogrid model and solar-wind hybrid picogrid model (where the wind is available) are used in street lighting applications. Solar-fed street lights are used in various applications like road lighting, construction site lighting, garden lighting, farm area lighting, remote area lighting, security lighting, etc. Sometimes, a camera is fitted to the lighting system so that it can serve dual purposes of lighting and security monitoring. Mobile apps-based control is sometimes used to control lighting and security features.

7.4.2 Picogrids in small domestic applications

Domestic applications where each apartment or building requires small energy, picogrids can be utilized. It will minimize the cost of utilization, cost of the battery, and be easy to monitor. The combination of picogrids and LED lighting systems together reduces energy consumption to a great extent. This combination also helps in cost reduction and carbon emission reduction at large.

7.4.3 Picogrid for small motor loads

In domestic applications, many motor loads are used for a small period for the regular interval. For example, water pumps, washing machines, motor pumps in gardening applications, etc. operate single or two times a day. This type of individual unit can be run by picogrids that will ultimately save the electric bill of the house.

7.4.4 Picogrid for rural domestic application

Energy consumption per family or house in rural areas is less than that of urban areas. These energy demands can be met by picogrids. Interconnected picogrids' networks can be designed for load sharing. Local community-based management systems may be formed to look after the grid network and maintenance. The remote villages or islands, interconnected picogrids are a good and sustainable solution to provide electricity. In some countries, bank loans and/or government subsidies are available to set up such picogrid networking in village areas.

7.4.5 Picogrid for rural primary schools or libraries

Rural primary schools or libraries normally have few rooms and demand is comparatively less. In these cases, electricity can be provided by picogrids. For academic institutions, government subsidies are available in many countries. Also, as these institutions normally run in the daytime, it is not mandatory to install battery units. However, they can be included and power can be delivered to the utility grid at night time.

7.4.6 Picogrids in agricultural applications

In agricultural applications, small pumping units can be powered by solar PV-based picogrids. Other agricultural appliances like drier, trillers, etc. can be run by electric power. Lighting arrangements can be made at agricultural houses by picogrids mode. The main advantage of this is that investment is one time and it will provide power for several years at almost no running cost.

7.4.7 Picogrids in small rural markets

Rural markets normally operate in the daytime. Operation of the rural market at night is not possible in many regions of the world because of either nonavailability of power or high price of electricity that cannot be afforded by very small businessmen of the poor economy. The time of marketing can be extended many hours at night by ensuring power by installing solar-fed picogrids. Initial installation costs may be collected by co-operative or society concepts or by monthly installment systems from the sellers. In some places, it is observed that market place is not fixed and varies with days in a week. As picogrid structure has the capacity of easy portability, they can be moved whenever the market place moves from one place to another.

7.4.8 Picogrids for small religious temples or community halls

Temples and community halls play important role in society. They are often used as venues for cultural activities. Many small religious temples or community halls are used daytime or at night for few hours. This short-duration energy demand can be met by solar PV-fed picogrid systems. Loads in this type of application are normally small (less than 2.5 kW). For electric power requirements at night, battery units need to be installed. Expenditure and management can be handled by forming small groups, society of communities formation.

7.4.9 Picogrids fed mobile charging station

Picogrids can be utilized to make a small power station feed power to mobiles phones or laptops for charging. By solar PV arrangements, power can be made available where there is no conventional electricity. The power station can even be integrated with conventional grid supply. Using a battery storage unit, power can be made available for 24 h. Based on charging outlet facilities, picogrid-fed mobile charging stations can be made in three categories:

(a) charging station with only DC outlet facility;
(b) charging station with only AC outlet facility;
(c) charging station with both DC and AC outlet facility.

Picogrid-fed mobile charging station is a good solution for providing charging facilities in remote areas.

7.5 Domestic utilization of NPG

In domestic applications, choices of nanogrids or picogrids depend on power demands. Total loads in the domestic area are calculated by the number of rooms, number of buildings, etc. Based on the load demand, the grid type is selected. Often a cluster of picogrids and nanogrids forming microgrid is chosen to provide electricity. Major domestic loads include the following equipment:

- lights,
- fans,
- TV,

- refrigerator,
- pump,
- washing machines,
- iron,
- room heater, and
- air conditioning machine.

Domestic loads are broadly classified into two categories:

- light loads and
- motor loads.

In most cases, motor loads (fans, pumps, etc.) need AC supply whereas light loads may need AC supply of DC supply. If solar PV and battery-fed grids are used, then DC–DC and DC–AC conversions are needed to supply AC power and only DC–DC conversion is needed to supply DC power. Thus AC grid, the efficiency of grid conversion is around 60–75% whereas for the DC grid, as less number of conversions is needed, efficiency is slightly higher around 80–90% or even more.

7.6 Efficiency in NPG utilization

7.6.1 Efficiency

Grid efficiency may be defined by the ratio of grid output to grid input. Let P_D is the load demand; P_L is the grid loss; P_{IN} is the grid input.

Power input to the grid from solar PV or battery will be

$$P_{IN} = P_L + P_D$$

Therefore, grid efficiency may be written as follows:

$$\eta_G = \frac{P_O}{P_{IN}} = \frac{P_D}{P_L + P_D}$$

7.6.2 Number of grid conversions and grid efficiency

Grid efficiency is very much dependent on grid conversion. With the increase in the number of conversions, grid loss increases. Thus, grid efficiency decreases with the increase in the number of grid conversions.

If the loads require DC supply, then for solar PV and battery-fed NPG systems, the DC grid needs fewer conversions than AC grids. In these cases, the DC grid offers greater efficiency.

7.7 Utilization for EV

Electric vehicles (EV) are going to be one of the main consumers of nanogrids and picogrids. In the last few decades, electric mobility (E-mobility) has become an alternate and sustainable solution to cope with the increased level of carbon emission and other pollutions along with rapid and alarming climate change throughout the world. EVs are driven by electricity and equipped with a battery of different types. EV batteries required regular charging which creates the demand for electric power. This demand for EV batteries can be met by conventional grids. But the green and economic solution are to use an alternate source of energy resources for this purpose. EV batteries require a large number of charging stations with charging points in all countries in the upcoming days. To meet this, rapidly increasing demand, economic, environment-friendly, and sustainable solution is expected to utilize nanogrids and picogrids for feeding power to charge EV batteries. More detail about E-vehicles with be discussed in Chapter 8 and NPG integration with charging stations will be discussed in Chapters 9 and 10.

7.8 Portable or mobile picogrid utilization

In many applications, it is observed that the loads are not located at a fixed place. They are mobile based on the requirement. For example, boats moving on the river, mobile vans, mobile pumps set for agricultural purposes, etc. Solar PV and/or battery-fed picogrids are normally chosen to provide power to these loads. Grids in these applications are small in size and normally run in off-grid mode. However, when, movement is stopped and loads are fixed in one place, picogrids can be clusters to improve capacity and also run on grid conditions. Clustering of picogrids may be required for larger mobile grid utilization. For example, power demands in small village fairs or book fairs, or other festivals. In these cases, the load requirement is comparatively high. Also, demand locations are not fixed and move from time to time in different places. In those cases, the clustering of many picogrids is chosen to meet the power demand.

7.9 Some examples

7.9.1 Example of NPG fed indoor mini agricultural pump

This example shows the utility of NPG for meeting water demand for agricultural purposes. In most of the small to medium range applications, 1–5 HP underground submersible pumps of mini size are used. If there is no other load or very little extra load for lighting, the NPG system is designed to have approximately 30% higher capacity than the calculated power demand. A mini-water pumping system with a 5 HP motor is shown in Figure 7.3. For this, solar PV-fed nanogrid capacity is set as 6.5 kW. As agricultural water need is not needed continuously of daily, the motor does not run all time. Also, there is no mandatory need to run it at night. Thus, it runs after intervals of several hours or days at day-time. These operational features eliminate the need for battery storage for the power setup. It may or may not be connected with the utility grid. The major advantages of this scheme are a large reduction of running cost and dependency of power supply from conventional grid improving overall agriculture-based rural economy.

Figure 7.3 Mini water pumping system with 5 HP motor

7.9.2 Example of an outdoor 12 kW nanogrid for deep tubewell pumping system for agricultural needs

This example shows the utility of nanogrid to provide water for agricultural purposes. An outdoor 12 kW nanogrid is shown in Figure 7.4. It feeds power to a 5-HP underground submersible pump. It operates during the daytime and does not have any battery storage unit. An array of three strings connected in series has been used for electric power generation from solar energy; each string has four photovoltaic panels connected in series. It needs very low maintenance due to the absence of battery units. The area under the PV panels may be used as sheds that may further be utilized for various needs.

7.9.3 Example of NPG fed mini fisheries

This example shows the utility of NPG for meeting water demand and air mixing for fisheries. In these cases, the production of fish needs the replacement of water after a certain interval and regular air mixing for balancing maintaining oxygen level in water and creation of small waves that are good for the growth of fish. Thus, 1–5 HP underground submersible pumps of mini size may be used for watering and small motors of DC type of capacity less than 1 HP are used for creating airflow and waves in the water. A solar PV-fed nanogrid run application for fisheries is shown in Figure 7.5. For this, solar PV-fed picogrid capacity is set as

Figure 7.4 Outdoor 12 kW nanogrid for deep tubewell pumping system for agricultural needs

Figure 7.5 A solar PV-fed picogrid run indoor application for fisheries

2.5 kW. The system operates only in the daytime and it eliminates the need for battery storage for the pico-grid setup. It may or may not be connected with the utility grid. The main advantages of this scheme are zero electricity bill and reduction of and dependency of power supply from the conventional grid. This can be implemented indoor as well as outdoor applications and thus can help to boost up rural economy.

7.9.4 Example of NPG fed agricultural farmhouse

This example shows nanogrid installed in a farmhouse. A 3 kW nanogrid fed farmhouse is shown in Figure 7.6. It can feed power to a small 2 HP pump, drier, lights, etc. The farmhouse can be used both in the daytime and at night (use at night requires extra battery storage). It can be installed in the remotely located farmhouse where conventional grid supply is not or rarely available. This can also be used to fulfill the rural need and boost up the rural economy.

Figure 7.6 Nanogrid-fed farmhouse

Figure 7.7 Energy independent shop

7.9.5 Example of NPG-fed shop

One energy-independent shop is shown in Figure 7.7. A pico or nano grid from 2 kW to 6 kW capacity is sufficient to run a small shop in urban, rural as well as at very remote locations where conventional grid supply may or may not be available. It must have battery storage to run the shop both day and night time.

7.9.6 Example of NPG-fed petrol pump

The use of pico or nano grid in petrol pumps is becoming popular. A 5.7 kW solar PV-fed grid is shown in Figure 7.8. It has its battery storage unit and petrol pump is expected to run both at day and night time. It carries loads of fuel pumps, water pumps, air compressors, lights, fans, and meters. It can be operated in connection with conventional grid supply. This type of arrangement enables to plan and set up petrol pumps at remote locations where conventional grid supply is not reliable and available all time or not available at all. It helps greatly in reducing the electricity bills.

Figure 7.8 NPG-fed petrol pump

7.9.7 Example of NPG-fed street lighting

Street lights can be operated by a solar-fed picogrid system very efficiently. It is equipped with individual battery storage for lighting arrangement mounted on each pole and may have a centralized storage system. For street lights at roadside isolated street lighting mounts on the separate pole are preferred; it does not require an interconnection between poles. Sometimes, for lighting arrangements in a small area like a playground, swimming pool, parks, exhibition centers, etc. interconnected picogrids are designed with centralized battery storage units.

7.9.8 Mobile or portable picogrid in fishing boats for fisherman

Mobile or portable picogrid is becoming very popular for fishermen as they roam over water where there is no conventional electric power supply. In the early days, they used to burn kerosene, they used battery-fed emergency lights. Now solar-fed picogrid equipped with a battery and LED light offers the best solution to them. This can help them in lighting as well as operate a radio, charge mobile phones, etc. This mobile picogrid can be utilized when they come back home. Countries like Sri Lanka, where a large number of people are dependent on fishing for their economy, are utilizing mobile picogrid helps to boost up the economy and living standard.

7.9.9 Picogrid in robotics applications

In most cases, batteries are used to feed power to controllers, actuators, etc. in robotic systems. They need regular charging. Mobile robots and flying robots draw high currents from the batteries and they need frequent charging. It becomes a real problem in remote and outdoor applications. Picogrid instead of nanogrid will be more suitable to solve this problem. Solar PV and battery-integrated picogrids improve the state of charge of the batteries. The picogrids used for robotic applications may be categorized into two groups as follows:

(a) picogrids installed in the field area of robots' application;
(b) picogrids mounted in the robot structure.

7.10 Case studies

7.10.1 Case study with DC load and AC load in domestic application

Consider a picogrid system delivering energy to a small domestic application. Various loads are presented in Table 7.1.

7.10.1.1 Full AC load

Therefore, the sum of maximum demands or loads

$$P_O = P_D = 2\,650 \text{ Watt}$$

Let us assume conversion efficiency is (considering one-step DC–DC and one-step DC–AC types of conversions) is 70%. Therefore,

$$\eta_G = 0.7$$

total input

$$P_{IN} = \frac{P_O}{\eta_G} = \frac{2\,650}{0.7} = 3\,785.714 \text{ W}$$

Loss will be

$$P_L = P_O - P_{IN} = 1\,135.714 \text{ W}$$

7.10.1.2 Fully DC load

For this case, DC–AC conversions are not needed. This will reduce conversion loss. Let us assume conversion efficiency is (considering one step DC–DC type of conversion) is 90%. Therefore,

$$\eta_G = 0.9$$

total input

$$P_{IN} = \frac{P_O}{\eta_G} = \frac{2\,650}{0.9} = 2\,944.444 \text{ W}$$

Table 7.1 Loads in the case study

Serial no.	Items	Load
1	Lights	1 000
2	Fans	400
3	Pump	750
4	TV	100
5	Refrigerator	200
6	Others	200

Loss will be

$$P_L = P_O - P_{IN} = 294.444 \text{ W}$$

7.10.1.3 Partial DC and AC loads

For DC load, DC–AC conversions are not needed, where conversion efficiency is (considering one step DC–DC type of conversion) is 90%.

For AC load, the conversion efficiency is (considering one step DC–DC and one step DC–AC types of conversions) is 70%

DC output is $P_{O\_DC} = 1\,000\text{W}$

Total input for feeding DC load is

$$P_{IN\_1} = \frac{P_{O\_DC}}{\eta_G} = \frac{1\,000}{0.9} = 1\,111.111 \text{ W}$$

Loss will be

$$P_{L1} = P_{O\_DC} - P_{IN} = 111.111 \text{ W}$$

AC output is $P_{O\_AC} = 1\,650\text{W}$

Total input for feeding AC load

$$P_{IN\_2} = \frac{P_{O\_AC}}{\eta_G} = \frac{1\,650}{0.7} = 2\,357.143 \text{ W}$$

Loss will be

$$P_{L2} = P_{O\_AC} - P_{IN} = 707.1429 \text{ W}$$

Total input for feed power to both DC and AC load will be

$$P_{IN} = P_{IN\_1} + P_{IN\_2} = 1\,111.111 + 2\,357.143 = 3\,468.254 \text{ W}$$

Total loss for feed power to both DC and AC load will be

$$P_L = P_{L1} + P_{L2} = 111.111 + 707.1439 = 818.254 \text{ W}$$

Therefore, overall efficiency may be written as follows:

$$\eta_G = \frac{P_O}{P_{IN}} = \frac{2\,650}{3\,468.254} = 0.76407 = 76.407\%$$

The results have been summarized and presented in Table 7.2.

7.10.2 Case study on loss and ampere–hour requirement in-home application

Domestic utilization of NPG has been considered in this case study. Six different types of AC loads have been considered like lights, fans, pumps, TV, refrigerator, etc. The Watt–hour requirements without considering NPG conversion loss have

Table 7.2 Loss calculation for DC- and AC-type loads

Load type		Total load	Conversion efficiency (%)	Input power (W)	Loss (W)	Loss per input
Full AC load		2 650	70	3 785.714	1 135.714	30G
Full DC load (assumption)		2 650	90	2 944.444	294.444	10%
Partial DC and AC load	For AC load	1 650	70	2 357.143	707.143	30%
	For DC load	1 000	90	1 111.111	111.111	10%
	Overall effect	2 650	76.407	3 468.254	818.254	23.59%

been calculated based on the load and average working hours as presented in Table 7.3.

Then, the converter efficiency of the NPG system has been considered 65% (i.e., conversion loss of 35%). Accordingly, input powers and Watt–hour required to feed the above six categories loads are determined as presented in Table 7.4.

Based on Watt–hour requirements with and without considering NPG loss, ampere–hour capacities have been determined for batteries of 24 V and 48 V ratings as presented in Table 7.5. This shows that conversion loss in NPG increases

Table 7.3 Loads, average working hours, and Watt–hour requirement without considering NPG conversion loss for different home utilities

Serial no.	Items	Type	Load	Hour (average)	Watt–hour (without considering loss)
1	Lights	AC	1 000	12	12 000
2	Fans	AC	400	16	6 400
3	Pump	AC	750	0.5	375
4	TV	AC	100	12	1 200
5	Refrigerator	AC	200	24	4 800
6	Others	AC	200	4	800
Total			2 650		25 575

Table 7.4 Input power and Watt–hour requirement considering NPG conversion loss for different home utilities

Serial no.	Converter efficiency of NPG (%)	Conversion loss NPG (%)	Power input (W)	Hour (average)	Watt–hour (considering loss)
1	65	35	1 350	12	16 200
2	65	35	540	16	8 640
3	65	35	1 012.5	0.5	506.25
4	65	35	135	12	1 620
5	65	35	270	24	6 480
6	65	35	270	4	1 080
Total			3 577.5		34 526.25

Table 7.5 Ampere–hour capacities of battery needed to meet the load

NPG conversion loss	Watt–hour	Battery output voltage (V)	Ampere–hour
With no loss	25 575	24	1 065.625
With loss	34 526.25	24	1 438.594
With no loss	25 575	48	532.8125
With loss	34 526.25	48	719.2969

the ampere–hour requirement for battery units. Obviously, in practice, NPG systems have certain conversion losses that can be calculated from the specification of efficiency; ampere–hour requirement for battery should be adopted considering conversion losses.

7.10.3 Case study centralized NPG

A centralized nanogrid of 12.7 kW is shown in Figure 7.9. Solar power is generated at two locations. Solar power is used to charge battery units. Solar powers produced at two locations and battery storage power are inverted by a single inverter system. The inverter output is feed power to local loads and is also connected (in bidirectional mode) with the conventional grid.

7.10.4 Case study with LED-driven lighting system

Carbon emission and cost of electricity are reduced by generating electric power from solar PV-based resources. These can further be reduced by using LED replacing conventional type bulbs. Reduction of energy consumption from the conventional grid means a reduction of the emission of greenhouse gas (carbon dioxide, etc.). In comparison between conventional bulbs and LED bulbs, we can assume:

(a) A 60 W light bulb (conventional type) serves the same purpose that can be served by a 10 W LED bulb; both offer the same luminance.
(b) The average life span of a 60 W bulb is 1 year and the average life span of an LED bulb is 15 years (operating 12 h per day)
(c) Price per 60 W bulb 1$ and price per 10 W LEV bulb 8$.
(d) 1 kWh energy costs 0.288$ (say for example).
(e) 1 kWh energy production in the conventional method results in CO_2-emission of around 0.559 kg.

Figure 7.9 A 12.7 kW centralized nanogrid system domestic application

Table 7.6 Year energy consumption

Bulb type	Utility per day (hour)	Utility per year (hour)	Energy consumption per year (kWh)
Light bulb	12	4 380	262.8
Light bulb	12	4 380	43.8

Table 7.7 Cost and carbon emission

Year	Light bulb			LED			Net carbon reduction
	Energy consumption (kWh)	Carbon emission (kg)	Cost ($)	Energy consumption (kWh)	Carbon emission (kg)	Cost ($)	
1	262.8	146.90	76.68	43.8	24.48	20.61	83.33%
15	3 942	2 203.57	1 150.29	657	367.26	197.21	

Based on the above assumptions, yearly energy consumptions by two types of bulbs are calculated as presented in Table 7.6. Considering the longer average life span of LED bulbs and their approximate prices, expenditure and carbon emission are calculated as presented in Table 7.7. The case result shows a great reduction in prices and greenhouse gas emissions. The use of LED in domestics, corporates, and street lights results in a net carbon reduction of 83.3% that is a great amount for 15 years duration. Therefore, this proves that energy utility with solar-fed NPG together with LED made lighting system will offer less pollution and economic solution. Different international organizations are taking a leading role to use environmentally friendly LEDs along with solar power.

7.10.5 Case study on the aging effect

A case study on the aging effect was carried out on a lithium-ion battery. Lithium-ion ($LiFePO_4$) battery of 12 V-rated voltage and 40 Ah-rated capacity was considered. The battery undergoes regular charging and charging modes and the maximum ampere–hour (Ah) capacity was monitored for 3.5 years. The case study shows that the maximum capacity of the battery decreases with aging in approximate linear nature; though the value remains above the rated value. Maximum capacity variation with aging is shown in Figure 7.10:

$$\text{Initial maximum capacity} = Ah_{\max 0} = 43.17 \text{ Ah}$$

$$\text{Maximum capacity after 3.5 years} = Ah_{\max t} = 42.31 \text{ Ah}$$

$$\therefore \text{Decrease of maximum capacity after 3.5 years}$$

$$= Ah_{\max t} - Ah_{\max t} = 43.17 - 42.31 \text{ Ah} = 0.86 \text{ Ah}$$

$$\therefore \text{Average decrease of maximum capacity}$$

$$= \frac{Ah_{\max_t} - Ah_{\max_t}}{t} = \frac{0.86}{3.5} \text{Ah} = 0.2425 \text{ Ah/year}$$

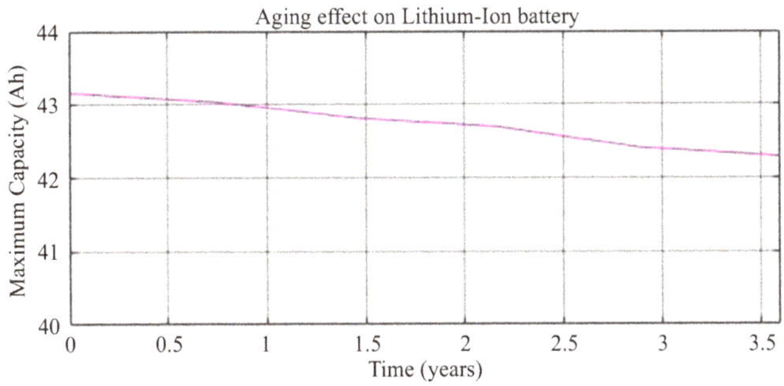

Figure 7.10 Aging effect on lithium-ion ($LiFePO_4$) battery

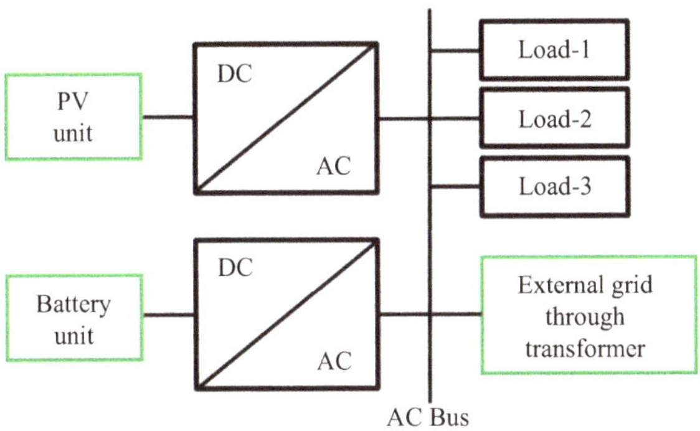

Figure 7.11 A 6 000 W nano grid system connected with three load units and external grid

Therefore, the average decrease of maximum capacity with respect to

initial maximum capacity after 3.5 years

$$= \frac{Ah_{\max_t} - Ah_{\max_t}}{Ah_{\max t} \times t} \times 100\% = 0.56\% \text{ per year}$$

However, this decrease is very small concerning the initial maximum capacity or even rated capacity in a normal life span of the battery. But beyond the normal life span of a battery, it drastically reduces.

7.10.6 Case study on 6 000 W nanogrid system

A 6 000 W nanogrid system was considered under this case study as shown in Figure 7.11. In the grid, solar PV feeds power three following units:

- battery unit;
- local domestic loads: it consists of three house loads;
- external grid: 6.6 kV external grid is connected through a transformer.

PV output power, loads, external grid power, battery power, and state of charge (SoC) were monitored for 24 h. The result with a battery of 1 200 Ah capacity is shown in Figure 7.12. At the bottom, scaling time in seconds versus hours has been provided for easy understanding. initially, charged battery feeds power to load. In the morning, loads were fed by both battery and solar PV. At midday time and afternoon, when the battery was off and load is small, the PV unit feeds power to a small load as well as an external grid through a transformer. Then after evening, when the load is high and there is no PV generation, the load is fed by the battery.

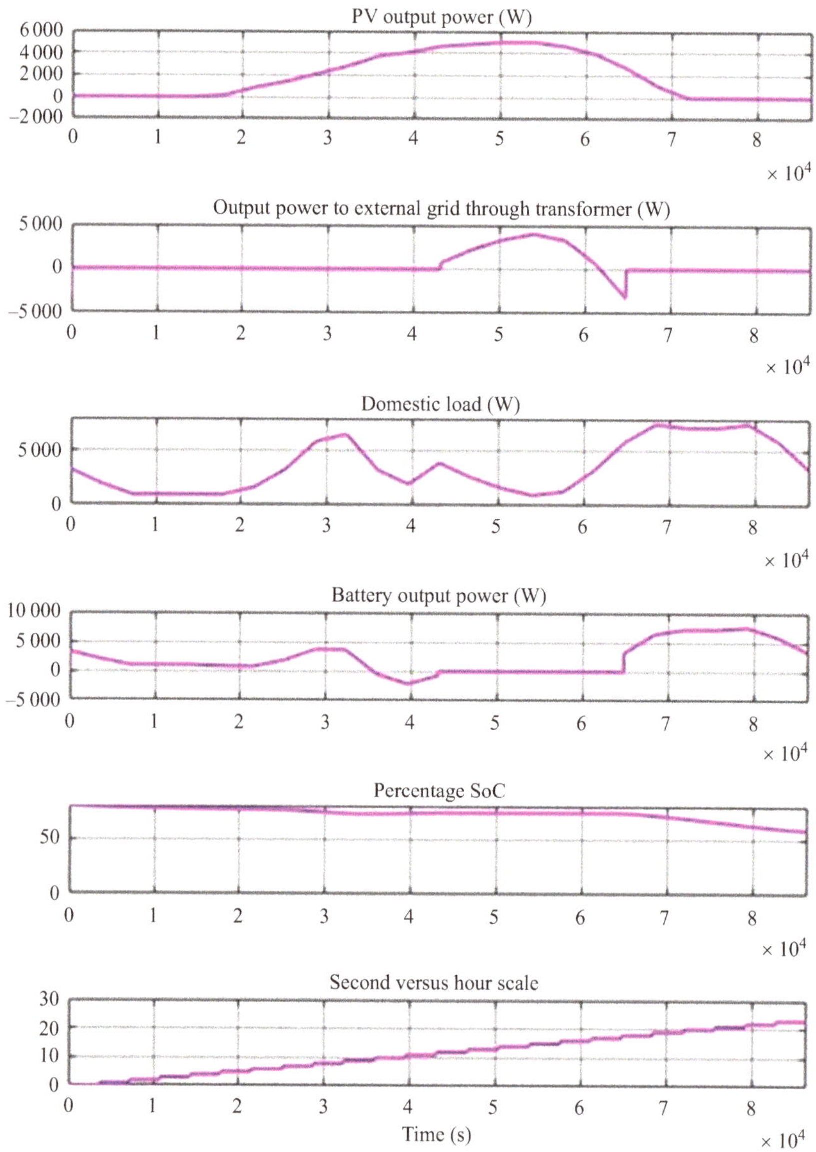

Figure 7.12 The result with a battery of 1 200 Ah capacity

7.10.6.1 SoC

SoC initially decreases slightly as shown in Figure 7.12. Before noon when the load is small, PV feeds power to the battery, and SoC increases. After the evening, when the load is high and there is no PV generation, SoC reduces drastically.

The same case study was carried out with a battery of lesser capacity. The result of the nano grid system with a battery of 400 Ah is shown in Figure 7.13. It has shown a more drastic reduction of SoC after evening as presented in this figure.

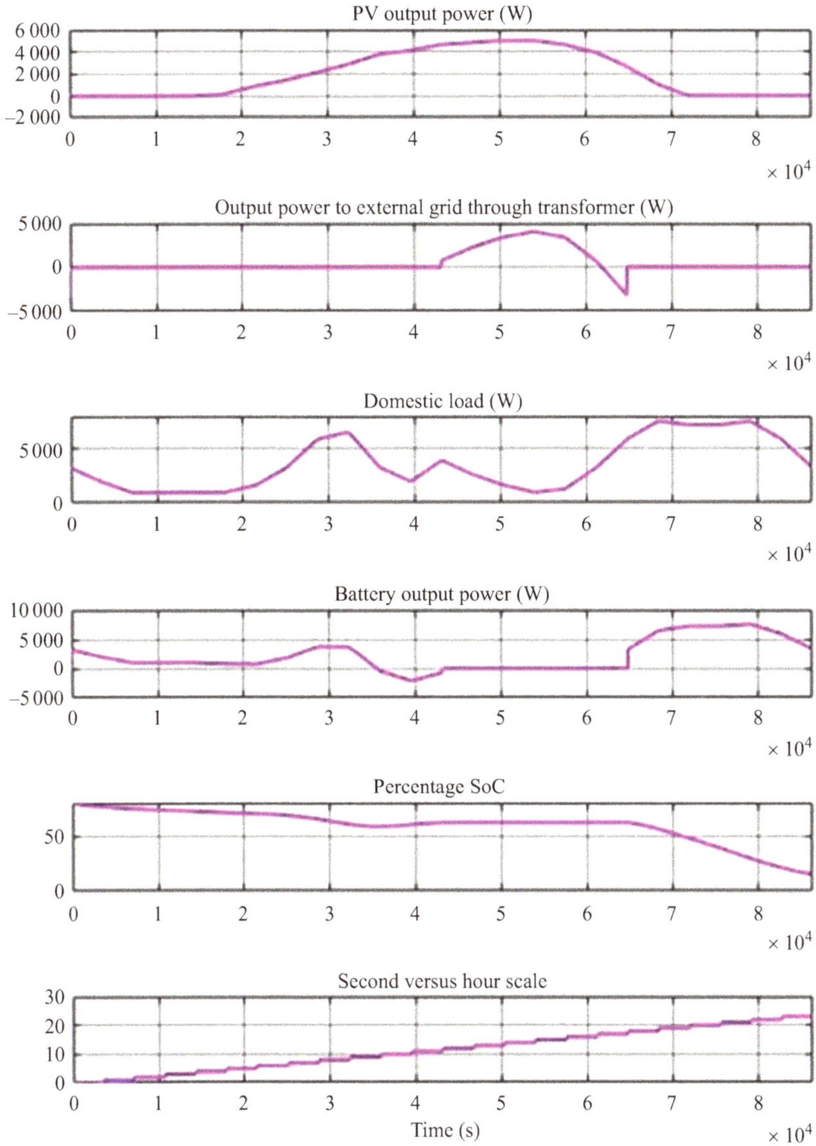

Figure 7.13 Result with 400 Ah battery

It indicates that proper selection of the ampere–hour capacity of the battery is very important.

7.10.6.2 Effect of short circuit at PV unit

A short circuit was considered in the PV network. Duration of the short circuit was considered for 30 s. The result is shown in Figure 7.14. It shows a sudden reduction of PV output. The fault results in a sudden rise of battery power as the load is on and a small oscillation in the external grid path.

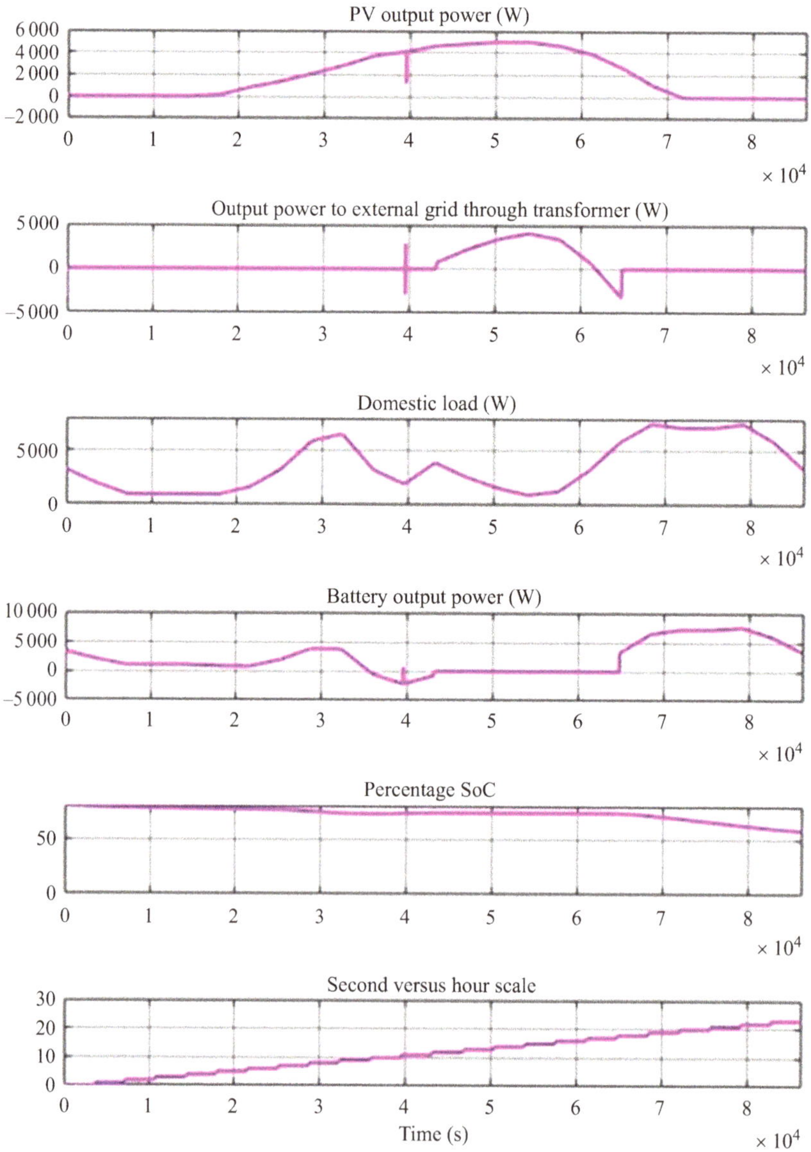

Figure 7.14 Result with a short circuit in PV network

7.10.6.3 Effect of a long open circuit in some part of PV unit

A long open circuit in some parts of the PV unit was considered. The open-circuit was considered in some parts of the PV network for the 4-h duration in the daytime. The result is shown in Figure 7.15. It shows a considerable reduction of PV output. The fault results in power flow from the external grid to meet the local load demand.

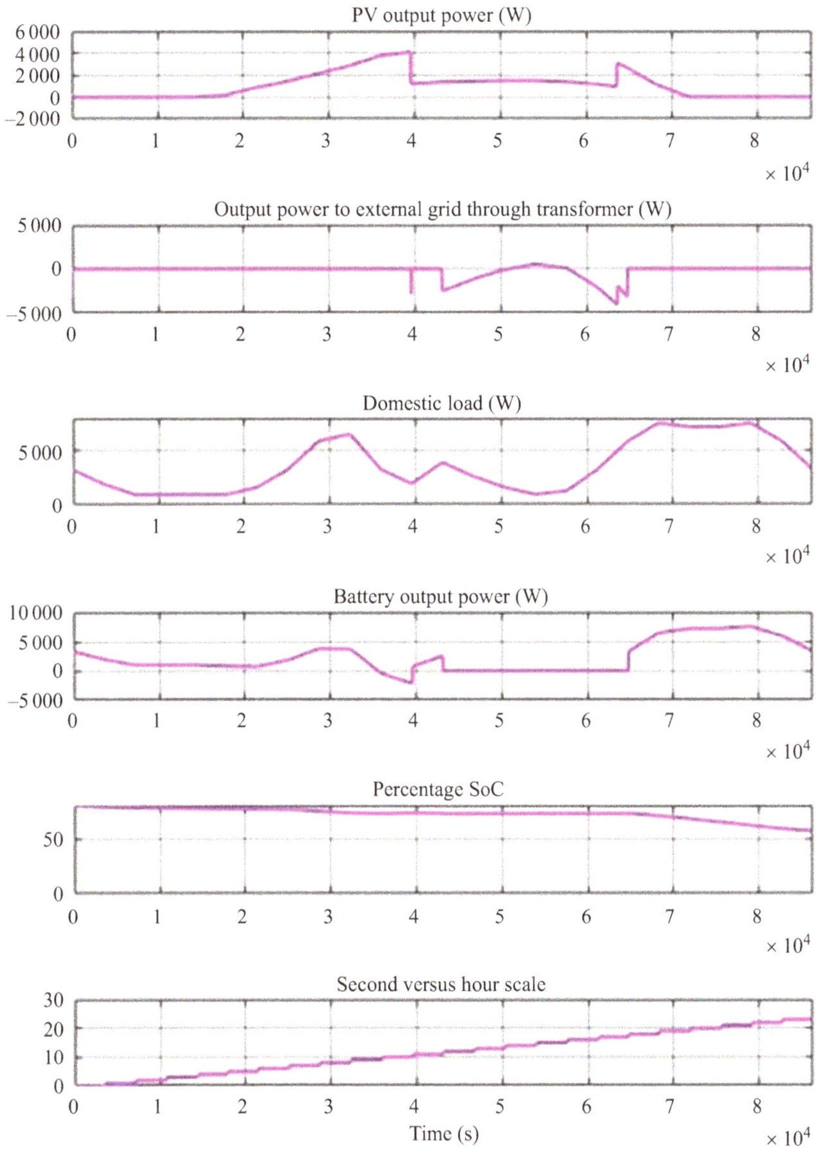

Figure 7.15 Result with a long open circuit in some part of PV network

7.10.6.4 Effect of sudden load cut off

Sudden load cut-off in one of three units was considered. Load cut-off was considered for 10 min at peak load at night. The result is shown in Figure 7.16. It shows a sudden drop in power output; however, no effect is considerably observed in SoC.

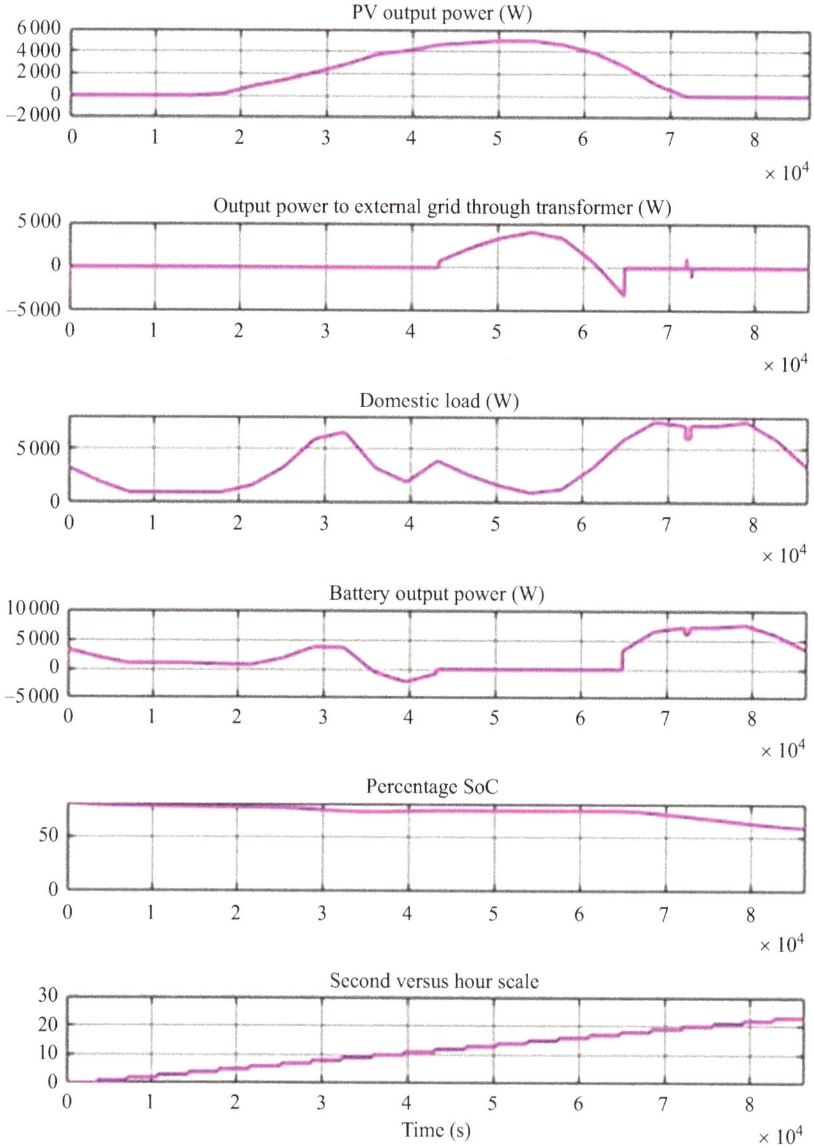

Figure 7.16 Result with sudden load cut off at load side

7.11 Design features of solar-wind hybrid picogrid-based pole-mounted lighting system

The use of solar and wind energy in lighting, security monitoring, etc. is very attractive. A solar-wind hybrid picogrid-based pole-mounted lighting system is shown in Figure 7.17. Equipment needed for this is listed below:

(i) Pole.
(ii) One wind turbine set.
(iii) Solar PV panel.
(iv) Light arrangement.
(v) Battery.
(vi) Solar-wind hybrid control.
(vii) Cables and connecting wires.

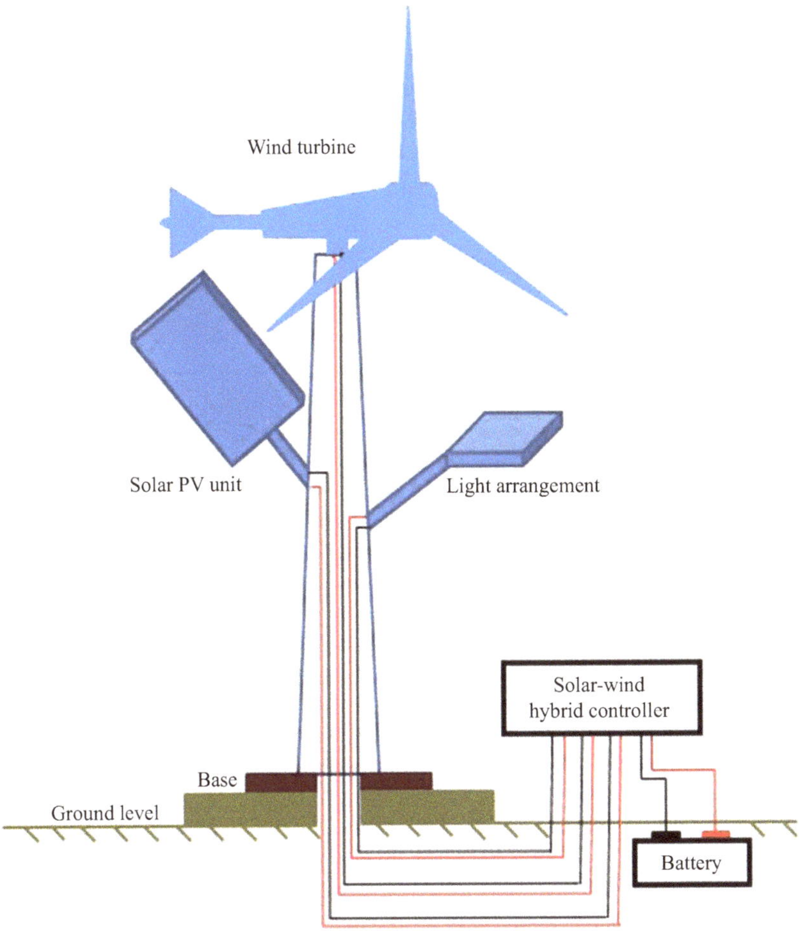

Figure 7.17 Solar-wind hybrid picogrid-based pole-mounted lighting system

Table 7.8 Some sample specifications of the main grid and solar PV parameters (frequency—50/60 Hz±2.5%, power factor—>0.92, configurable solar cell no. —72)

Grid				Solar PV			
Power (kVA)	Voltage (V)	Output current (A)		Power	V_{OC} (V)	Panel no.	String no.
		1 phase—2 wire	3 phase—4 wire				
5	220–240±10% (1 phase), 380–415±10% (3 phase)	17	6	5	400–740	12, 16	1
7.5	Do	26	9	7.5	400–740	16, 12	1, 2
10	Do	34	12	10	400–740	16, 16	2, 2
15	Do	52	17	15	400–740	12	4
20	Do	70	23	20	400–740	16	4
25	Do	87	23	25	540–810	16	5
30	Do	104	35	30	540–810	16	6
40	Do	139	46	40	540–810	16	8
50	Do	–	58	50	540–810	16	10
60	Do	–	70	60	540–810	16	12
80	Do	–	93	80	540–810	16	16
100	Do	–	116	100	540–810	16	20
120	Do	–	139	120	540–810	16	24

Design steps:

Step 1: The pole is mounted on a strong base foundation. Cables are passed through the pole for interconnections.

Step 2: Wind turbine is installed at the top. It may be flexible in any direction.

Step 3: Solar PV panel is installed below the wind turbine set.

Step 4: The lighting system is installed normally below the PV panel. Distance between ground level and lighting system is set to achieve maximum illumination. LED-based street light systems are very common. The use of LED reduces energy consumption.

Step 5: Battery is placed attached with the pole at mid hight or bottom or below ground level. Maintenance-free long-life batteries are used for this purpose. The battery may be placed at a remote distance increasing cable length. A 12 V or 24 V battery is common for this purpose. However, a battery of other voltage ratings can also be used.

Step 6: A solar-wind hybrid controller is used for operation. It is connected with the PV panel, wind turbine, lighting system, and battery unit. Solar power is available during the daytime and is used to charge the battery. Sufficient wind may be available both day and night. Rectified DC from the wind turbine is used to charge the battery. The battery feeds power to the lighting system. Operation of the control unit may be fully automatic or may be regulated remotely by software or mobile apps.

7.12 Some specifications useful for modeling, design, and selection of NPG-based applications

Some specifications useful for modeling, design, and selection of NPG based applications are provided in Tables 7.8, 7.9, 7.10, 7.11, and 7.12.

7.13 Summary

Different types of utilization of nanogrids and picogrids are discussed in this chapter. Based on different factors, various types of utilizations have been

Table 7.9 Some sample specifications of important inverter parameters

Output voltage (V)	220
Load current (A)	4.5, 9, 18, 36, 64, etc.
Output phase	1 phase, 2 or 3 wire
Output waveform	Sine or square wave
Control	PWM
Switching device	MOSFET, IGBT
Overload capacity	100–110%: 10 min 110–120%: 2 min 120–160%: 30 sec 150–200%: 2 sec 200–300%: 1 sec >300%: 2 ms
Cooling system	Controlled fan
Operating temperature	0–50°C
Max. relative humidity (25°C)	95%

Table 7.10 Some sample specifications of solar PV useful for application with MPPT and PWM base charge controller using IGBT switching device with efficiency greater than 95%

Sample no.	**P_{max} (kW)**	**V_{max} or V_{OC} (V)**	**Operating voltage range (V)**	**Low cut voltage (V)**	**Low cut recovery voltage (V)**	**High cut voltage (V)**	**High cut recovery voltage (V)**
1	0.15–0.5	22.5	18–22	12±1	16±1	NA or up to 10% high	NA or up to 10% high
2	0.25–1	45	30–44	22±1	26±1	Do	Do
3	0.5–1	90	72–89	68±1	72±1	Do	Do
4	1–3	180	68–176	48±1	68±1	Do	Do
5	3–5	360	136–352	96±1	136±1	Do	Do
6	5–15	450	170–440, 255–450, 340–450	120±1,180±1, 192±1	170±1,255±1, 272±1	Do	Do

Table 7.11 Some sample specifications of battery useful for application with MPPT and PWM base charge controller using IGBT switching device with efficiency greater than 95%

Sample no.	Nominal voltage (V)	Charging current (A)	Bulk/boost voltage (V)
1	12	10–20	14±0.5
2	24	10–20	29±0.5
3	48	10–60	58±1
4	96	30–60	116±1
5	120	30–60	145±1
6	180	30–60	217±1
7	192	30–60	232±1
8	240	30–60	290±1

Table 7.12 Some sample specifications of battery for e-rickshaw

Battery charger	12/24/48–60 V, 15 A
BLDC motor controller	12/24/48 V/20/30/50 A
DC–DC converter	12/24/48 V, 15 A
Battery capacity	140–250 Ah

classified. Picogrid utilizations are discussed with different examples of application areas. Domestic utilizations have been discussed. Grid efficiency has been defined and its dependency on the number of conversions in grid operation has been pointed out. Other aspects like mobile grid utilization, etc. have been discussed. Some interesting case studies have been presented for better understanding.

Further reading

[1] "IEEE Standard for Local and metropolitan area networks – Part 21: Media Independent Services Framework," in *IEEE Std 802.21-2017 (Revision of IEEE Std 802.21-2008 as amended by IEEE Std 802.21a-2012, IEEE Std 802.21b-2012, IEEE Std 802.21c-2014, and IEEE Std 802.21d-2015)*, pp. 1–314, 28 April 2017, DOI: 10.1109/IEEESTD.2017.7919341.

[2] "ISO/IEC/IEEE International Standard – Information Technology – Telecommunications and Information Exchange Between Systems – Local and Metropolitan Area Networks – Specific Requirements – Part 21: Media Independent Services Framework," in *ISO/IEC/IEEE 8802-21:2018(E)*, pp. 1–318, 5 June 2018, DOI: 10.1109/IEEESTD.2018.8376063.

[3] "IEEE DC Quick Charger for Use with Electric Vehicles Test Suite Specification – Version 1.0," in *IEEE DC Quick Charger for Use with Electric Vehicles Test Suite Specification – Version 1.0*, pp. 1–10, 17 September 2019.

[4] "IEEE Draft Standard Technical Specifications of a DC Quick and Bi-directional Charger for Use with Electric Vehicles," in *IEEE P2030.1.1/D1*, February 2021, pp. 1–159, 9 February 2021.
[5] "IEEE Recommended Practice for Performance Testing of Electrical Energy Storage (EES) System in Electric Charging Stations in Combination with Photovoltaic (PV)," in *IEEE Std 2836-2021*, pp. 1–27, 8 June 2021, DOI: 10.1109/IEEESTD.2021.9447313.
[6] "IEEE Grid Vision 2050 Reference Model," in *IEEE Grid Vision 2050 Reference Model*, pp. 1–15, 30 April 2013, DOI: 10.1109/IEEESTD.2013.6577591.
[7] G. Simard, "IEEE Grid Vision 2050," in *IEEE Grid Vision 2050*, pp. 1–93, 15 April 2013, DOI: 10.1109/IEEESTD.2013.6577603.
[8] Continental Automated Buildings Association, "Impact of Smart Grid on Connected Homes Landmark Research Study," in *Impact of Smart Grid on Connected Homes*, pp. 1–106, 17 May 2013, DOI: 10.1109/IEEESTD.2013.6786299.
[9] K. Bhat, V. Sundarraj, S. Sinha, and A. Kaul, "IEEE Cyber Security for the Smart Grid," in *IEEE Cyber Security for the Smart Grid*, pp. 1–122, 16 September 2013, DOI: 10.1109/IEEESTD.2013.6613505.
[10] "IEEE Standard for Utility Industry Metering Communication Protocol Application Layer (End Device Data Tables)," in *IEEE Std 1377-2012 (Revision of IEEE Std 1377-1997)*, pp. 1–576, 10 August 2012, DOI: 10.1109/IEEESTD.2012.6264065.
[11] "IEEE Draft Standard for Synchrophasor Measurements for Power Systems," in *IEEE PC37.118.1/D4.2*, May 2011, pp. 1–56, 23 June 2011.
[12] "IEEE Standard for an Architectural Framework for the Internet of Things (IoT)," in IEEE Std 2413-2019, pp. 1–269, 10 March 2020, DOI: 10.1109/IEEESTD.2020.9032420.
[13] "IEEE/ANSI Draft Recommended Practice for Electromagnetic Compatibility Limits and Test Levels," in *ANSI PC63.12/D12g*, May 2015, pp. 1–49, 1 January 2015.
[14] "IEEE Draft Standard for DC Microgrids for Rural and Remote Electricity Access Applications," in *IEEE P2030.10/D11*, February 2021, pp. 1–56, 10 March 2021.
[15] "IEEE Approved Draft Standard for Test Methods for Surge Protectors and Protective Circuits Used in Information and Communications Technology (ICT) Circuits, Including Smart Grid Data Circuits," in *IEEE PC62.36/D38*, February 2016, pp. 1–146, 1 January 2016.
[16] "IEEE Standard Conformance Test Procedures for Equipment Interconnecting Distributed Energy Resources with Electric Power Systems and Associated Interfaces," in *IEEE Std 1547.1-2020*, pp. 1–282, 21 May 2020, DOI: 10.1109/IEEESTD.2020.9097534.
[17] "IEEE Guide for Safety in AC Substation Grounding," in *IEEE Std 80-2013 (Revision of IEEE Std 80-2000/ Incorporates IEEE Std 80-2013/Cor 1-2015)*, pp. 1–226, 15 May 2015, DOI: 10.1109/IEEESTD.2015.7109078.

[18] "IEEE Draft Recommended Practice for Personnel Qualifications for Installation and Maintenance of Stationary Batteries," in *IEEE P1657/D9*, July 2017, pp. 1–114, 1 January 2017.

[19] "IEEE Standard Glossary of Stationary Battery Terminology," in *IEEE Std 1881-2016, pp. 1–42, 30 August 2016, DOI: 10.1109/IEEESTD.2016.7552407.*

[20] "IEEE Guide for Technology of Unified Power Flow Controller Using Modular Multilevel Converter—Part 2: Terminology," in *IEEE Std 2745.2-2021*, pp. 1–25, 1 September 2021, DOI: 10.1109/IEEESTD.2021.9525319.

[21] "IEEE Recommended Practice for Monitoring Electric Power Quality," in *IEEE Std 1159-1995*, pp. 1–80, 30 November 1995, DOI: 10.1109/IEEESTD.1995.79050.

[22] "IEEE Recommended Practice for Power Quality Data Interchange Format (PQDIF)," in *IEEE Std 1159.3-2019 (Revision of IEEE Std 1159.3-2003)*, pp. 1–185, 1 May 2019, DOI: 10.1109/IEEESTD.2019.8697192.

[23] "IEEE Recommended Practice for Monitoring Electric Power Quality," in *IEEE Std 1159-2019 (Revision of IEEE Std 1159-2009)*, pp. 1–98, 13 August 2019, DOI: 10.1109/IEEESTD.2019.8796486.

[24] "IEEE Recommended Practice for Monitoring Electric Power Quality," in *IEEE Std 1159-2009 (Revision of IEEE Std 1159-1995)*, pp. 1–94, 26 June 2009, DOI: 10.1109/IEEESTD.2009.5154067.

[25] "IEEE Guide for Cloud Portability and Interoperability Profiles (CPIP)," in *IEEE Std 2301-2020, pp. 1–62, 13 August 2020, DOI: 10.1109/IEEESTD.2020.9169938.*

[26] "IEEE 802 Nendica Report: The Lossless Network for Data Centers," in *IEEE 802 Nendica Report: The Lossless Network for Data Centers*, pp. 1–29, 17 August 2018.

Chapter 8
Electromobility

This chapter deals with electromobility. It has been defined and the different types of electric vehicles (EVs) have been classified. The need for electromobility has been pointed out and different advantageous features of electromobility have been presented. Different technical aspects of electromobility have been discussed. Scopes for electromobility in smart city applications and rural transportation have been discussed.

8.1 Introduction

Electric mobility or electromobility or E-mobility is an electric energy-driven sustainable climate-friendly solution to the transport system. It may be defined as a state of mobility enabling electric power to run a vehicle. It uses the powertrain concept and in-vehicle communication technologies. It also involves other various technologies and non-technical aspects. Electric mobility has been globally accepted as the best choice to replace conventional vehicles that are non-polluting, eco-friendly, and sustainable. Therefore, electric mobility together with renewable-based grid utility can help to resolve many environmental issues to a great extent.

8.2 Need for E-mobility

The energy consumption scenario depends on geography, availability of resources, and economic strength of the country. In the global scenario, the transport sector takes around 18–25% of total energy consumption. According to Intergovernmental Panel on Climate Change (IPCC), 23% of worldwide CO_2 emissions is due to transport sectors. A major part of the transport sector survives based on the consumption of crude oil, the price of which varies with the international market. This makes the transport sector vulnerable due to the volatile nature of oil prices. In addition to that, in the last few decades, statistics are showing that one of the major sources of carbon emission causing climate change is the transport sector. On the other hand, the transport system is an unavoidable component for rapid economic growth, rising urbanization, and better travel and communication needs of a country.

All these demands for an alternate solution of transportation that will give climate-friendly sustainable solutions fulfilling our needs. E-mobility has been globally accepted as this alternate solution. It also helps in balancing energy demand improving the security of energy of a country.

8.3 General benefits of E-mobility

Electric mobility offers the following main benefits to society:

- It helps in the reduction of carbon emissions.
- It gives zero or ultra-low tailpipe emissions.
- It creates low noise or sound pollution.
- It reduces running cost or fuel cost.
- It reduces maintenance cost.
- It improves the cleanliness of the vehicle.
- It is becoming more competitive in the market.

8.4 The main requirement for the growth of E-mobility

Electric vehicle (EV) manufacturing industries in most of the countries are far behind the level of requirement. In many countries, the EV industry shares 1–10% of total vehicles, and routes are dominated by conventional vehicles. This indicates that there is big demand and opportunity for this new industry throughout the world. To utilize this opportunity, the following are important:

- awareness of climate change,
- promotion of E-mobility,
- formation of regulations (nation-wise),
- formation of global standardization, and
- infrastructure development.

8.5 Infrastructure development

Adequate infrastructure development is the key requirement for accelerating the EVs industry. Infrastructure development includes follows:

- ensuring adequate EV charging station;
- modernization of battery industry;
- modernization of vehicle-motor industry;
- establishment of adequate service centers;
- inclusion of IoT and communication industries.

8.6 Classification of E-vehicles

Different types of EVs are shown in Figure 8.1. E-mobility provides three modes of operation as follows:

- Hybrid vehicles.
- Full EVs.
- Hydrogen fuel cells generated electricity-driven vehicles.

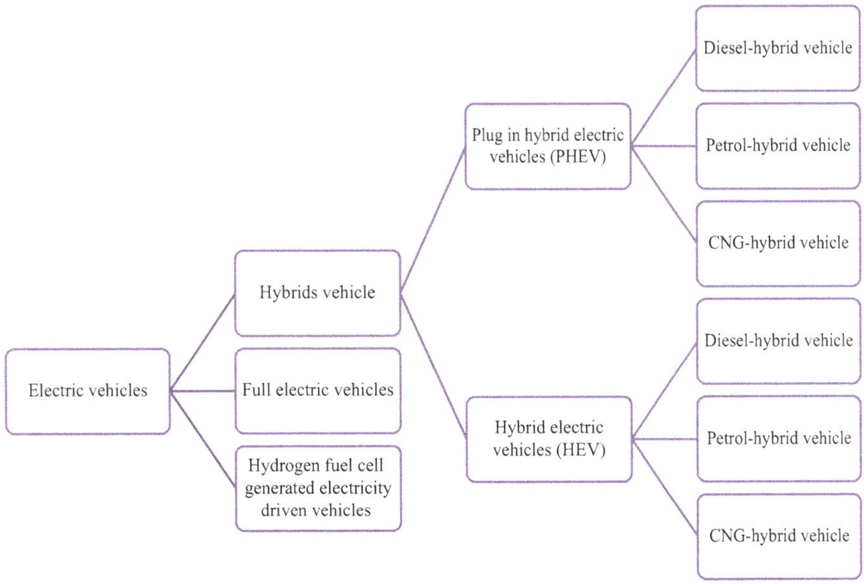

Figure 8.1 Classification of different types of EVs

Two major types of hybrid vehicles are available:

- *Plug-in hybrid EVs (PHEV)*: Batteries are charged by plug arrangement from externals sources.
- *Hybrid EVs (HEV)*: Regenerative braking-based energy conversion is utilized for charging the battery.

Based on the type of fuel used for internal combustion, hybrid vehicles may be of three categories as follows:

- Diesel-hybrid vehicle.
- Petrol-hybrid vehicle.
- Compressed natural gas (CNG)-hybrid vehicle.

Based on the application area, E-vehicles may be divided into three categories as follows (shown in Figure 8.2):

- E-vehicle for surface transport: e.g. 2, 3, 4 wheeler vehicles.
- E-vehicle for water transport: e.g. boats, water scooter, etc.
- E-vehicle for aerial application: e.g. manned or unmanned aerial vehicles.

8.7 Main components of EVs

Main components of different types of an EV can be understood by comparing with a conventional vehicle. The conventional vehicle is driven by oil or gas and an

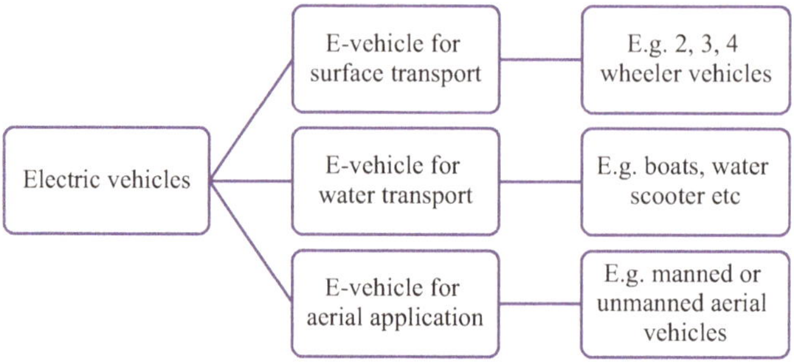

Figure 8.2 Classification of E-vehicles based on the area of application

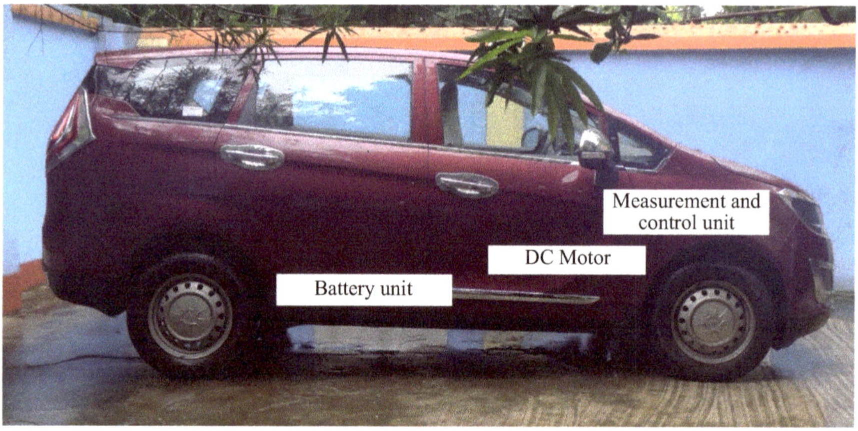

Figure 8.3 Basic components of EVs

internal combustion engine along with other compatible measurement and control units. These are replaced partially or fully by new means in e-vehicles as shown in Figure 8.3. EVs have three basic components: (a) battery unit, (b) DC motor, and (c) measurement and control unit.

Different types of semi or fully EVs are as follows:

(a) *Hybrid vehicle*: Oil or compressed natural gas (CNG)-based internal combustion engines are used in addition battery and DC motors are provided to run in electric mode (e-mode).
(b) *Full EV*: Here oil-based internal combustion engines are replaced fully by battery and DC motors to run in electric mode (e-mode).
(c) *Hydrogen cell run EV*: Oil-based internal combustion (IC) engines are replaced by hydrogen cells and DC motors are provided to run in electric mode (e-mode). Hydrogen cells are used to generate electricity that is used to run motors in the vehicle.

The basic components of different types of EVs are presented in Table 8.1.

Table 8.1 Basic components of different types of EVs

Main parts	**Hybrid vehicle**	**Full EV**	**Hydrogen cell-based vehicle**
Oil/CNG tank	Present	Absent	Absent
IC engine	Present	Absent	Absent
Battery	Present	Present	Absent
Hydrogen cell	Absent	Absent	Present
DC motor	Present	Present	Present
Measurement and control unit	Present	Present	Present
Fuel pump, fuel line, fuel tank, etc.	Present	Absent	Absent
Comfort and safety features	Present	Present	Present

Table 8.2 Benefits of electric motors of e-vehicles over IC engine

Issues	**Benefits**
Economical	• Less moving parts • Less maintenance • Less noise • Higher efficiency • Lower fuel cost • Lower operational cost
Improved air quality	• No tailpipe emission • Reduced local air pollution • Reduced greenhouse gas emissions
Convenience	• No gears • Much easier to drive • Much quieter driving • Charging provision at home as well as charging station outside • Incentives and other benefits from Government bodies in many countries • Less dependency on fossil fuel

8.8 Benefits of electric motors of E-vehicles over internal combustion engine

The benefits of electric motors of e-vehicles over internal combustion (IC) engines are presented in Table 8.2.

8.9 EV charging

EVs are driven by electric energy. Batteries are used to store and deliver this energy to the vehicles at run time. Therefore, these batteries need to be charged regularly. EV charging refers to charging batteries used to run an EV.

8.9.1 Charging time

Charging time (T_C) refers to the time required to get the battery fully charged. It depends on battery capacity and charging power. Mathematical, charging time may approximately be written as follows:

$$\text{Charging time } (T_C) = \frac{\text{battery capacity}}{\text{charging power}}$$

8.9.2 Different factors influencing charging time

The relation of charging time with power is not linear. In reality, charging time is higher than the calculated value. In the early days charging time was too long. But with advanced technologies, it has reduced significantly. Time of charging depends on the following factors:

- *Charging speed*: charging speed is not constant throughout the charging time.
- *Power during charging*: charging is not done at maximum power at all times.
- *Nonlinear nature of fast charging*: fast charging is done with a direct DC supply fed to the battery. The speed remains very fast up to 80–90% of maximum charging then the speed reduces for the next part of charging.
- *Battery temperature*: charging speed varies with the temperature of the battery. Optimum speed is normally observed at 20–30°C.
- *Dependency on charging control method*: charging time depends on method and algorithm that vary with different types of charge controllers.
- Type of charger.
- Source capacity.
- Control efficiency.
- Battery capacity.
- Battery state of charge (SOC) and chemical properties.

8.9.3 Classification based on EV charging

Based on charging time, charging modes may be classified as follows:

- Slow charging.
- Medium/moderate charging.
- Fast charging.

At present, EV charging for different segments (2/3/4-wheelers) takes 1–5 h to get charged from 0% to 80% by slow or medium charging mode and takes less than 1 h by fast charging mode.

8.9.4 Range of fully charged battery

The range (km) of EV batteries has increased significantly. For two-wheeler vehicles, it is around 80–90 km per charge. and for four-wheeler vehicles, it is around 150–200 km per charge. The range is improving with the advancement of technology.

8.9.5 EV battery swapping

The range of EVs is being increased by introducing the provision of extra battery units. If one battery capacity decreases, it may be replaced by another charged battery taking from the extra unit or the charging station. It also saves the time required for battery charging.

8.9.6 Lifetime of EV batteries and replacement

In the early days, battery lifetime was comparatively less. But now it has improved. Battery industries are now offering EV battery lifetime ranging from 3 to 8 years. Most of the batteries are now maintenance-free. They are also offering discounts for replacement old batteries and operational guaranty.

Lithium-ion batteries show long life. It offers high energy density and can be recharged many times. EV batteries have 20–60 kWh capacity in most cases.

8.9.7 EV charging stations

The number of charging stations is still not adequate. Proper planning and regulations are needed. Many governments are providing subsidies in establishing charging stations. Delicencing for setting up EV charging stations is an encouraging step that may be adopted to attract more private sectors and individuals to facilitate this.

8.9.8 Specification parameters of charging stations

The main specifications of the charging station are presented in Table 8.3.

8.9.9 EV supply equipment (EVSE)

EVSE refers to the electrical equipment that is external to the EV used for providing charging power for the battery units. These are equipped with an energy source, grids, converters, charge controllers, smart meters, communication networks, etc. EVSE forms the main parts of EV charging stations (EVCS).

Table 8.3 Main specification parameters of a charging station

Parameters	**Description**	**Example**
Charger type	Fast	Less than 1 h
	Moderate	3–5/6 h
	Slow	Greater the 5/6 h, etc.
Power capacity	kW ratings	20 kW, 50 kW, etc.
Output voltage levels	DC voltage levels	24 V, 48 V, 72 V, 200–500 V, etc.
Number of charging points	Number of vehicles that may be charged at a time	
EV type	Vehicle's segment type	2/3/4 wheeler, SUV, MUV, light motors, etc.

8.10 EV safety

Mechanical safety features are being taken as like conventional vehicles. Safety concerning battery units are also being improved. Lots of tests and research are going on to reduce the probability of the explosion of a battery unit. Advanced vehicles are providing a special battery management section (BMS). BMS considers the following issues:

- optimum and efficient use of the battery,
- reduction of explosion possibility,
- limiting heat generation,
- proper cooling
- proper thermal and electrical insulation, and
- shielding from vibrational effect and short circuit.

8.11 EV speed, efficiency, and price

8.11.1 Speed

EVs' speeds are improved now and fulfills the limits of urban and rural road regulation. It also offers faster acceleration. Breaking facilities in EV are also improving.

8.11.2 Efficiency

Car engine efficiency may be defined as the ratio of output power to the power fed to the engine. Gasoline engines show around 35% and diesel engines show around 45% of efficiency. On the other hand, battery-fed electric power trains show around 90% efficiency. Loss component feeds mainly heat loss. In hybrid models, power is feedback to the battery using recuperation. Recuperation refers to the reuse of energy by feedback path during braking to charge the battery in electric cars, acceleration, braking, i.e. change of torque can be made faster than conventional vehicles.

8.11.3 Price

Prices of different types of EVs are reducing rapidly. It must be considered that the running cost i.e. the cost of charging electricity is much less than the fuel cost. Also, cost maintenance is less than that of conventional vehicles. The cost of the battery is a matter of concern. However, the effective cost of the battery is reducing with the increase of life span of the battery and reuse of some parts of battery materials.

8.12 EV charging tariff

Mainly cost (c_{EV}) is defined as energy consumption (kWh) multiplied by EV charging tariff (T_R). Mathematically,

$$c_{EV} = \text{kWh} \times T_R$$

where c_{EV} is the cost of charging; T_R is the charging tariff; kWh is the energy consumption.

Let d is the distance/full charge; D is the total distance covered by the car; $\text{kWh}_{\text{full charging of battery}}$ is the amount of energy consumption for full charging of the battery.

Therefore, energy consumption will be as follows:

$$\text{kWh} = \frac{\text{distance covered}}{\text{distance per full battery charge}} \times \text{energy consumption for full charging of battery}$$

$$\text{kWh} = \frac{D}{d} \times \text{kWh}_{\text{full charging of battery}}$$

Therefore, the cost of a run may be determined as follows:

$$c_{EV} = \text{kWh} \times T_R = \frac{D}{d} \times \text{kWh}_{\text{full charging of battery}} \times T_R$$

Tariff may be a flat rate or two part-variable type (one fixed part for the cost of infrastructure and another part for energy consumption) or multi-parts variables considering other business conditions, like a rebate, penalty, etc. Different types of tariff and billing systems are seen in different countries. Mainly tariff depends on the following factors:

- charging time: it refers to the amount charged for energy consumption-based on an amount per kilowatt–hour (kWh);
- basic fee (fixed cost);
- starting fee (optional) based on the type of service.

Different types of payment options are also seen as follows:

- payment by charging card;
- payment by debit or credit card;
- paying by smartphone app;
- paying by net-banking;
- payment by cash.

Example of energy consumption:

- Very small e-cars can consume less than 7 kWh per 100 km.
- Small ranged e-car can consume 10–20 kWh per 100 km.

Example 1 A car claims approximately 20 km per liter. Let us assume a run for 100 km. Therefore, it consumes around 5 l of petrol. Assuming the cost of petrol of $1 per liter.

Total price,

$$p_{\mathrm{PETROL}} = \frac{\text{distance traveled}}{\text{kml}} \times \text{petrol price per liter}$$

$$p_{\mathrm{PETROL}} = \frac{100}{5} \times 1 = \$20$$

Let the car runs 100 km per full charge. Let the fully charged energy consumption is 21 kWh and the electricity price is $0.28:

$$c_{EV} = \text{kWh} \times T_R = \frac{D}{d} \times \text{kWh}_{\text{full charging of battery}} \times T_R = \frac{100}{100} \times 21 \times 0.28$$

$$= \$5.88$$

8.13 Present scenario

The vehicle industry has gone through development for a long time. Two incidences are remembered in the time frame:

- 1867: Before the advent of the combustion engine, Werner von Siemens introduced his electric generator using the dynamo-electric principle at World's Fair in Paris.
- 1899: Belgian Camille Jenatzy presented the first cars with an electric motor. It was the first on-road vehicle showing a speed of 100 km/h.

Over years, many countries have shown lots of progress. In the E-vehicle industry, Norway is the forerunner. The average travel (approximate) in km in three countries is shown in Table 8.4.

The number of charging stations at present condition (2018) is shown in Table 8.5.

Table 8.4 Average travel (approximate) in km

Country	Average travel (approximate) in km
Germans	Less than 40
Americans	26 km (during winter)–32 (during fall)
Norwegians	47 km

Table 8.5 Number of charging stations

Country	Number of charging stations
Norway's capital Oslo	One per 488 inhabitants
Amsterdam (Dutch capital)	One per 650 inhabitants
Germany	Total 5 000 (March 2018)
China	190 000 (2017)

The upcoming scenario is changing very rapidly increasing the challenge. According to report of "The UNO's World Urbanization Prospects 2014", urban will be occupied by 70% of the world's population by 2050. On the other hand, according to "Statistical Review of World Energy 2017," oil reserves will exist for 50 years considering current levels of oil consumption.

8.14 Smart city application

Smart cities are being designed considering all aspects of the causes of climate change. The main target of smart cities is to provide the best needs of human beings in an ecofriendly way. Along with the applications of the Internet of Things and communication technology, all civil works and planning are framed to reduce the level of pollutions without hampering modern facilities. This has catalyzed to choice options of E-mobility as the first choice in upcoming days. Hundred percent E-vehicles in a smart city is the ultimate goal. This requires the use of EVs special in the segments of two, three, and four-wheeler vehicles. In addition to that, adequate charging stations are required. This can be fulfilled by keeping space at the time of planning. Availability of service centers in the smart city must be incorporated. Vehicle garages and community parking areas can be equipped with charging points. Charging stations will be fed by solar energy. Aerial unmanned e-vehicles can be utilized for surveillance and other purposes. Boats and vehicles in children's parks will be e-vehicles only. Thus, the main features of e-mobility in the smart city may be pointed out as follows:

- setting aim to replace all vehicles inside the city with e-vehicle;
- adequate charging stations inside the smart city area;
- vehicle garages and community parking areas equipped with charging points;
- use of solar energy utility power as well as for charging EV batteries in charging station;
- use of E-boats in a water park;
- use of E-vehicle in children parks;
- use of unmanned E-vehicles for surveillance.

8.15 Rural utility

Planned utilization of EVs in rural areas will help villages to become smart villages. In many ways, applications of e-mobility in rural areas are easier. Some natural facilities can be obtained in rural areas as follows:

- The pollution level in rural areas is comparatively low.
- Adequate space is available for setting up charging stations.
- Adequate road area that can be developed to the level of standards for running different types of e-vehicles.
- The cost of land is low.
- The cost of electricity is low.
- Enough space is available for solar PV installation.

However, many challenges are to be overcome. Some are as follows:

- lack of awareness of climate change and increasing pollution level;
- poor economy;
- poor road conditions;
- lack of communication with cities;
- lack of advanced technologies.

To overcome these limitations, initiatives are to be taken. Some are as follows:

- rural road and other infrastructure development;
- spread of awareness;
- strengthening rural economy;
- providing subsidies in setting solar-based EV charging station;
- providing subsidies in purchasing E-vehicles;
- developing a community model for planning and development.

Some applications of rural e-mobility are as follows:

- human transportation;
- growing agricultural products;
- E-boat for fishing;
- E-ambulance;
- E-pull cars;
- E-rickshaws.

Some benefits of utilization of rural e-mobility are as follows:

- faster and cheaper transportation;
- better communication with schools, colleges, hospitals, etc.
- help in creating jobs;
- help in fishing and marketing;
- help in growing agricultural product and their marketing.

8.16 Unmanned aerial vehicle (UAV)

UAVs are rapidly growing industries. It can be used to monitor as well as to provide some service in remote areas. UAVs are normally battery-driven and made of DC–DC or DC–AC converter. As actuators, permanent magnet DC (PMDC) or brushless DC (BLDC) motors are widely used. Based on the number of actuators, UAVs may be classified as a helicopter, tri-copter, quadcopter, hexacopter, etc. The main advantage of UAV is that they can be controlled remotely. The main limitations of UAVs are ensuring stable flying and low battery run time, etc.

8.17 Surveillance

Battery-driven unmanned aerial e-vehicles can be utilized for surveillance. It can be also utilized in monitoring remote and difficult areas. It can be used on the river, on the sea, and even on the hilly areas. Crowded areas can be monitored by a UAV system. However, it requires battery swapping of charging. Lots of research are going on to improve the run time of UAV by one full-charged battery unit.

8.18 Challenges

There are many challenges to be overcome to leverage electric mobility in the coming days. Some are as follows:

- lack of world awareness about climate change and pollution;
- lack of infrastructure;
- inadequate charging stations;
- inadequate service centers;
- high cost of EVs;
- battery longevity factors;
- distance range per full charge of battery;
- emergency failure-based service;
- lack of Government subsidies in some countries;
- lack of adequate models of all vehicles' segments.

8.19 Summary

This chapter has presented the concept of electro-mobility. The basic needs of e-mobility have been described. Then different features of e-mobility have been described. Classifications of e-vehicles have been made. Battery-related different issues have been discussed. The tariff has been presented along with an example. Different aspects of rural e-mobility like application area, limitations, benefits, etc. have been presented. Then UAV and its surveillance application have been highlighted. Lastly, challenges based on the present scenario have been mentioned that are to be overcome.

Further reading

[1] "IEEE Standard for Local and Metropolitan Area Networks – Part 21: Media Independent Services Framework," in *IEEE Std 802.21-2017 (Revision of IEEE Std 802.21-2008 as amended by IEEE Std 802.21a-2012, IEEE Std 802.21b-2012, IEEE Std 802.21c-2014, and IEEE Std 802.21d-2015)*, pp. 1–314, 28 April 2017, DOI: 10.1109/IEEESTD.2017.7919341.

[2] "ISO/IEC/IEEE International Standard – Information Technology – Telecommunications and Information Exchange Between Systems – Local and Metropolitan Area Networks – Specific Requirements – Part 21: Media Independent Services Framework," in *ISO/IEC/IEEE 8802-21:2018(E)*, pp. 1–318, 5 June 2018, DOI: 10.1109/IEEESTD.2018.8376063.

[3] "IEEE DC Quick Charger for Use with Electric Vehicles Test Suite Specification – Version 1.0," in *IEEE DC Quick Charger for Use with Electric Vehicles Test Suite Specification – Version 1.0*, pp. 1-0, 17 September 2019.
[4] "IEEE Draft Standard Technical Specifications of a DC Quick and Bi-directional Charger for Use with Electric Vehicles," in *IEEE P2030.1. 1/D1*, February 2021, pp. 1–159, 9 February 2021.
[5] "IEEE Smart Grid Vision for Vehicular Technology: 2030 and Beyond Roadmap," in *IEEE Smart Grid Vision for Vehicular Technology: 2030 and Beyond Roadmap*, pp. 1–17, 30 June 2015.
[6] "IEEE Vision for Smart Grid Controls: 2030 and Beyond Reference Model," in *IEEE Vision for Smart Grid Control: 2030 and Beyond Reference Model*, pp. 1–10, 12 September 2013, DOI: 10.1109/IEEESTD.2013.6598993.
[7] "IEEE Smart Grid Vision for Computing: 2030 and Beyond Roadmap," in *IEEE Smart Grid Vision for Computing: 2030 and Beyond Roadmap*, pp. 1–14, 11 January 2016.
[8] S. F. Bush, S. Goel, and G. Simard, "IEEE Vision for Smart Grid Communications: 2030 and Beyond Roadmap," in *IEEE Vision for Smart Grid Communications: 2030 and Beyond Roadmap*, pp. 1–19, 3 September 2013, DOI: 10.1109/IEEESTD.2013.6690098.
[9] "IEEE Recommended Practice for Performance Testing of Electrical Energy Storage (EES) System in Electric Charging Stations in Combination with Photovoltaic (PV)," in *IEEE Std 2836-2021*, pp. 1–27, 8 June 2021, DOI: 10.1109/IEEESTD.2021.9447313.
[10] "IEEE Grid Vision 2050 Reference Model," in *IEEE Grid Vision 2050 Reference Model*, pp. 1–15, 30 April 2013, DOI: 10.1109/IEEESTD.2013.6577591.
[11] G. Simard, "IEEE Grid Vision 2050," in *IEEE Grid Vision 2050*, pp. 1–93, 15 April 2013, DOI: 10.1109/IEEESTD.2013.6577603.
[12] Continental Automated Buildings Association, "Impact of Smart Grid on Connected Homes Landmark Research Study," in *Impact of Smart Grid on Connected Homes*, pp. 1–106, 17 May 2013, DOI: 10.1109/IEEESTD.2013.6786299.
[13] K. Bhat, V. Sundarraj, S. Sinha, and A. Kaul, "IEEE Cyber Security for the Smart Grid," in *IEEE Cyber Security for the Smart Grid*, pp. 1–122, 16 September 2013, DOI: 10.1109/IEEESTD.2013.6613505.
[14] "IEEE Standard for Utility Industry Metering Communication Protocol Application Layer (End Device Data Tables)," in *IEEE Std 1377-2012 (Revision of IEEE Std 1377-1997)*, pp. 1–576, 10 August 2012, DOI: 10.1109/IEEESTD.2012.6264065.
[15] "IEEE Draft Standard for Synchrophasor Measurements for Power Systems," in *IEEE PC37.118.1/D4.2*, May 2011, pp. 1–56, 23 June 2011.
[16] "IEEE Standard for an Architectural Framework for the Internet of Things (IoT)," in *IEEE Std 2413-2019*, pp. 1–269, 10 March 2020, DOI: 10.1109/IEEESTD.2020.9032420.

[17] "IEEE/ANSI Draft Recommended Practice for Electromagnetic Compatibility Limits and Test Levels," in *ANSI PC63.12/D12g*, May 2015, pp. 1–49, 1 January 2015.

[18] "IEEE Draft Standard for DC Microgrids for Rural and Remote Electricity Access Applications," in *IEEE P2030.10/D11*, February 2021, pp. 1–56, 10 March 2021.

[19] "IEEE Approved Draft Standard for Test Methods for Surge Protectors and Protective Circuits Used in Information and Communications Technology (ICT) Circuits, Including Smart Grid Data Circuits," in *IEEE PC62.36/D38*, February 2016, pp. 1–146, 1 January 2016.

[20] "IEEE Standard Conformance Test Procedures for Equipment Interconnecting Distributed Energy Resources with Electric Power Systems and Associated Interfaces," in *IEEE Std 1547.1-2020*, pp. 1–282, 21 May 2020, DOI: 10.1109/IEEESTD.2020.9097534.

[21] "IEEE Guide for Safety in AC Substation Grounding," in *IEEE Std 80-2013 (Revision of IEEE Std 80-2000/Incorporates IEEE Std 80-2013/Cor 1-2015)*, pp. 1–226, 15 May 2015, DOI: 10.1109/IEEESTD.2015.7109078.

[22] "IEEE Draft Recommended Practice for Personnel Qualifications for Installation and Maintenance of Stationary Batteries," in *IEEE P1657/D9*, July 2017, pp. 1–114, 1 January 2017.

[23] "IEEE Standard Glossary of Stationary Battery Terminology" in *IEEE Std 1881-2016*, pp. 1–42, 30 August 2016, DOI: 10.1109/IEEESTD.2016.7552407.

[24] "IEEE Guide for Technology of Unified Power Flow Controller Using Modular Multilevel Converter—Part 2: Terminology," in *IEEE Std 2745.2-2021*, pp. 1–25, 1 September 2021, DOI: 10.1109/IEEESTD.2021.9525319.

[25] "IEEE Recommended Practice for Monitoring Electric Power Quality," in *IEEE Std 1159-1995*, pp. 1–80, 30 November 1995, DOI: 10.1109/IEEESTD.1995.79050.

[26] "IEEE Recommended Practice for Power Quality Data Interchange Format (PQDIF)," in *IEEE Std 1159.3-2019 (Revision of IEEE Std 1159.3-2003)*, pp. 1–185, 1 May 2019, DOI: 10.1109/IEEESTD.2019.8697192.

[27] "IEEE Recommended Practice for Monitoring Electric Power Quality, " in *IEEE Std 1159-2019 (Revision of IEEE Std 1159-2009)*, pp. 1–98, 13 August 2019, DOI: 10.1109/IEEESTD.2019.8796486.

[28] "IEEE Recommended Practice for Monitoring Electric Power Quality," in *IEEE Std 1159-2009 (Revision of IEEE Std 1159-1995)*, pp. 1–94, 26 June 2009, DOI: 10.1109/IEEESTD.2009.5154067.

[29] "IEEE Guide for Cloud Portability and Interoperability Profiles (CPIP)," in *IEEE Std 2301-2020*, pp. 1–62, 13 August 2020, DOI: 10.1109/IEEESTD.2020.9169938.

[30] "IEEE 802 Nendica Report: The Lossless Network for Data Centers," in *IEEE 802 Nendica Report: The Lossless Network for Data Centers*, pp. 1–29, 17 August 2018.

[31] "IEEE Standard for Adoption of Open Fog Reference Architecture for Fog Computing," in *IEEE Std 1934-2018*, pp. 1–176, 2 August 2018, DOI: 10.1109/IEEESTD.2018.8423800.

[32] "Cloud Computing Innovation in India: A Framework and Roadmap – 2.0," in *Cloud Computing Innovation in India: A Framework and Roadmap – White Paper 2.0*, pp. 1–206, 31 December 2014, DOI: 10.1109/IEEESTD.2014.7039186.
[33] "IEEE Approved Draft Guide for the Use of Artificial Intelligence Exchange and Service Tie to All Test Environments (AI-ESTATE)," in *IEEE P1232.3/D3.2*, June 2014, pp. 1–96, 10 October 2014.
[34] "IEC 62243 Ed. 1 (IEEE Std 1232(TM)-2002): Standard for Artificial Intelligence Exchange and Service Tie to All Test Environments (AI-ESTATE)," in *IEC 62243 First edition 2005-07 IEEE 1232*, pp. 1–128, 20 November 2002, DOI: 10.1109/IEEESTD.2002.8684709.
[35] "IEEE Standard for Artificial Intelligence and Expert System Tie to Automatic Test Equipment (AI-ESTATE): Overview and Architecture," in *IEEE Std 1232-1995*, pp. 1–32, 15 March 1996, DOI: 10.1109/IEEESTD.1996.80816.
[36] "IEEE Standard for Software Interface for Maintenance Information Collection and Analysis (SIMICA)," in *IEEE STD 1636-2009 (New Trial-Use Standard)*, pp. 1–38, 14 August 2009, DOI: 10.1109/IEEESTD.2009.5210055.
[37] "IEEE Standard for Software Interface for Maintenance Information Collection and Analysis (SIMICA)," in *IEEE Std 1636-2009 (Upgraded to Full Use)*, pp. 1–37, 14 August 2009, DOI: 10.1109/IEEESTD.2009.8654231.
[38] "IEEE Draft Standard for the Classification of Adaptive Instructional Systems," in *IEEE P2247.1/D20*, December 2020, pp. 1–28, 25 January 2021.
[39] Q. Yang, L. Fan, R. Tong, and A. Lv, "IEEE Federated Machine Learning," in *IEEE Federated Machine Learning – White Paper*, pp. 1–18, 15 June 2021.

Chapter 9
Nanogrid and picogrid integration with electric mobility

This chapter deals with nanogrid and picogrid integration with electromobility. Charging stations are important components of electromobility. This chapter provides the application of nanogrid and picogrid for feeding power to charging stations used for electric vehicles (EVs). Different aspects of charging stations have been provided. Design features have been described and then factors of grid connections have been described. Different advancements occurred in this field have been presented. One case study has been described. Some useful standards have been provided followed by a summary.

9.1 Introduction

The market for electric vehicles (EVs) is growing rapidly along with the spread of nano and pico grids (NPGs). With the growth of the EV sector, the need for sufficient numbers of charging stations has increased. In China, growth of EV market is growing at a very high rate (recently it has grown by 111%). On the other hand, Europe is going to be the second-largest marketplace for EVs. It is expected that 15% of global sales of a passenger vehicle will be in Europe in 2025. At present, around 185 000 charging points are existing publicly in this region that provide one charging point for an average of 7 EVs. World wide requirement of EV charging stations is high and increasing with the increase of EVs. It has increased the demand for electric power. Conventional energy resources are at present meeting the demand for electric power in EV charging stations.

9.2 Need for integration of EV with NPG

EVs need electric power for charging their battery units. This need can be fulfilled by conventional grid supply. However, concerning pollutions, environmental aspects, and other factors, renewable and green energy sources are the best solution to meet this demand aiming for sustainable development. Second, where grid supply is not reliable or continuously not available or unavailable, an alternate and a reliable solution is necessary for EV charging. These aspects have directed the use of dedicated NPG systems for charging. NPG integration for EV charging gives the following advantages:

- reliable supply of power;
- economic energy solution;
- provision of better flexible tariff for charging;
- use of alternate pollution-free solar energy;
- enabling charging stations to be built up at remote places where conventional grid supply is not available;
- enabling charging stations to be built up at remote places where conventional grid supply is poor;
- enabling charging stations to be built up at a nearby place like parking areas, garages, etc.;
- reduce the cost of transportation;
- reduce the necessity of looking for petrol pumps;
- good supervisory monitoring;
- reduce fire hazards;
- easy to set up, safe in operation, and easy to operate.

9.3 Generalized model of EV-NPG integration

Modeling of NPG integration with EVs considers both electrical aspects and nonelectrical aspects as follows:

(a) Planning of grid-connected charging stations.
(b) Planning of logistics and other business at premises of charging stations.

The above aspects should be considered from the initial stages of planning and installation of integration of EV with NPG.

9.3.1 Planning of grid-connected charging stations

Planning of grid-connected charging stations considers resources, grids, and charging loads. Specifications/capacity of the following units are important:

(a) the capacity of the solar photovoltaic (SPV) unit,
(b) connectivity with the utility grid,
(c) the capacity of the battery storage unit,
(d) number of charging points,
(e) EV charging power load,
(f) non-EV or other electrical loads, and
(g) voltage levels.

Consider a model of charging stations equipped with all facilities (SPV, utility grid, battery storage unit) as shown in Figure 9.1. The general model power network with both AC and DC charging facilities for EVs is shown in Figure 9.2.

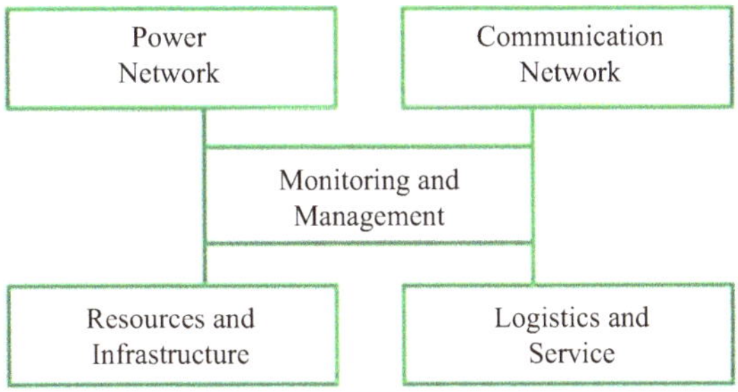

Figure 9.1 Block model of charging stations equipped with various facilities

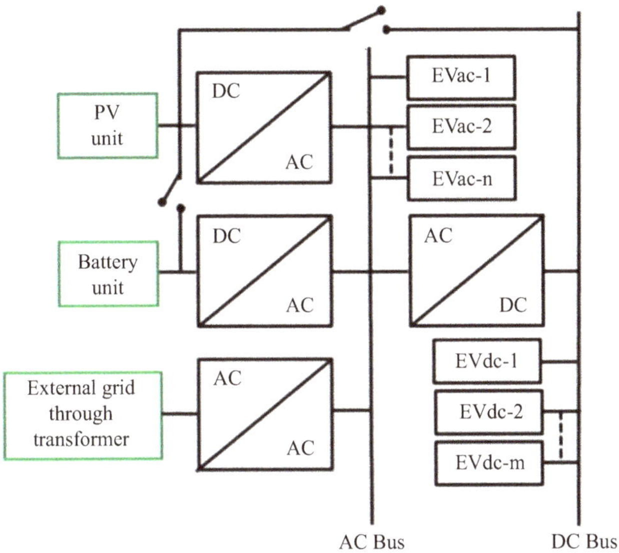

Figure 9.2 Model power network with both AC and DC charging facilities for EV

Let us define the following parameters:

P_{SPV} = power input from SPV
P_{BT} = power input from battery storage unit of charging station
P_{UG} = power input from external utility grid
P_{IN-NPG} = total power input to NPG
$P_{OUT-NPG}$ = power output from NPG
$P_{Loss-NPG}$ = power loss in NPG
$P_{EV\ Load}$ = electric vehicle charging load
$P_{Non-EV\ Load}$ = electric load in charging stations other than EV charging

Therefore, the total input power can be written as follows:

$$P_{IN-NPG} = P_{SPV} + P_{BT} + P_{UG}$$

Considering grid loss,

$$P_{IN-NPG} = P_{OUT-NPG} + P_{Loss-NPG}$$

Therefore,

$$P_{OUT-NPG} = P_{IN-NPG} - P_{Loss-NPG}$$

Therefore, NPG efficiency can be written as follows:

$$\eta_{NPG} = \frac{P_{OUT-NPG}}{P_{IN-NPG}} = \frac{P_{IN-NPG} - P_{Loss-NPG}}{P_{IN-NPG}} = 1 - \frac{P_{Loss-NPG}}{P_{IN-NPG}}$$

Total NPG output power can be written as follows:

$$P_{OUT-NPG} = P_{EV\ Load} + P_{Non-EV\ Load}$$

EV charging load can be written as follows:

$$P_{EV\ Load} = P_{OUT-NPG} - P_{Non-EV\ Load}$$

or,

$$P_{EV\ Load} = P_{IN-NPG} - P_{Loss-NPG} - P_{Non-EV\ Load}$$

Thus, the object of modeling will be to

(a) *Maximize* $P_{EV\ Load}$ = *maximize* $(P_{OUT-NPG} - P_{Non-EV\ Load})$
(b) *Minimize* $P_{Loss-NPG}$

with subject to – $P_{IN-NPGmin} \leq P_{IN-NPG} \leq P_{IN-NPGmin}$

9.4 Optimization

Modeling of EV integration with charging stations is done considering the optimization of different issues and parameters. Common aspects of optimization are as follows:

(a) operation, utilization, and business optimization;
(b) charging optimization;
(c) fleet management optimization;
(d) energy consumption optimization.

9.4.1 Operation, utilization, and business optimization

Operation, utilization, and business optimization aim in a single supervisory and control system that can manage multiple units in the whole vehicle network area. It covers

- communities,
- smart cities,

- municipalities,
- parking garages,
- charging infrastructures, and
- customer choices.

Supervisory control enables integration of

- highway charging,
- public charging,
- private charging to the best extent,
- vehicle depots, and
- E-mobility ecosystem.

9.4.2 Charging optimization

It enables the following benefits:

- optimization of rolling out of EV fleet;
- optimization of fleet vehicles charging;
- optimization of energy consumption;
- minimization of CO_2 emissions;
- optimization of return on investment (ROI).

Some common challenges and solutions are given in Table 9.1.

9.4.3 Fleet management optimization

Major challenges and common solutions for fleet management optimization are given in Table 9.2.

9.4.4 Energy consumption optimization

Major challenges and common solutions are given in Table 9.3.

9.5 Issues in EV integration with grids

Some important issues must be considered in the planning of the integration of EV networks and grid systems. The major issues are as follows:

(i) resource availability;
(ii) grid type;
(iii) public charging network;
(iv) private charging station;
(v) smart cities/communities;
(vi) vehicle density;
(vii) charging at parking, depot, and garages;
(viii) corporate parking;
(ix) very important person (VIP) parking;
(x) car sharing/fleet;

Table 9.1 Some challenges in charging optimization

Challenges	Solutions
Rolling out charging	Simple registration process
Customer and employee satisfaction	Intuitive app based on map and user-friendly features
Maximization of vehicles for charging	Maximization of charging points, Fast charging Adequate space Communication with vehicles Refreshment and other facilities in charging stations for customer
Data analysis	Data collection Cloud-based communication and computations Use of apps Feedback collection
Maximization ROI	Lowering charging prices Offering time-based discount Flexible tariff Use of different modes of payment and use of offers with them
Interaction with other charging networks, if needed	Load sharing during peak hour Maintain ecosystem Share best practices

Table 9.2 Some challenges in fleet management optimization

Challenges	Solutions
Charging-customer interface	Easy user-friendly interface Transparent metering
Cope with grid limitations	Both-way monitoring and control Proper load management
Maximization of vehicles' number	Enhancing charging capacity Increase of charging points Space management
Fast charging	Technical infrastructure development
Employee satisfaction	Use of apps and software tools Feedback and facility arrangement
Collecting and analyzing data	Track usage patterns and tariffs
Business or corporate objectives	Maximize ROI

Table 9.3 Some challenges in consumption optimization

Challenges	Solutions
Management charging points	Capacity-based charging management
Local energy resource and grid utilization	Utilization of local resource Minimization of grid intake Minimization of battery discharge Effective operation of hybrid vehicles
Maximize usage and ROI	Different payment possibilities, clearing & roaming as well as the web portal/app with real-time availability information

(xi) battery swapping;
(xii) existing fuel (gas/oil) pumps.

9.5.1 Resource availability

The use of local resources makes charging most economic. SPV-based charging stations are rapidly growing around the world. Both off-grid and on-grid models of charging stations based on nanogrid and picogrid are used. Different types of business models for power trading and utilization are available depending upon the regulatory bodies in the country or state. Based on the choice of the business model of grids, reduced or flexible charging tariffs can be set.

9.5.2 Grid type

The type of grid providing energy to charging points is important and must be decided at the beginning. Grid type is mainly chosen based on locally available resources, type of EVs and other electric loads, and business model. Efficiency and economic operation depend on grid loss which varies with the type of grid.

9.5.3 Public charging station

Public charging stations are normally planned on roadsides. Based on average vehicle density on the road and length of the road, the number and locations of roadside public stations are planned. Public charging stations can also be built at public parking areas and depots.

9.5.4 Private charging stations

Private charging stations are built mainly in personal or private garages and parking areas. For daily users of EV, if the charging point is provided at their place, then, it reduces the need of going out for charging saving much time. However, private installations and tariffs (whether they should be considered as commercial of domestic or other types) should be guided by the regulatory bodies. Adequate safety measures are to be added for installing private charging stations.

9.5.5 Smart cities/communities

The utilization of EV in smart cities and communities deserves smart management for grid-based charging stations. It aims at three main aspects:

- maximizing use of renewables,
- minimizing energy consumption, and
- minimizing carbon emission and pollution.

Smart city planning leverages parking, building or home automation, e-charging, web portal or apps-based services, etc.

9.5.6 Vehicle density

Statistics on vehicle density in the area helps in EV load management. Vehicle density is a dynamic parameter and varies with the time, days and month. Prior information on an average and maximum vehicle density can be considered during the planning and design of a charging station to set the number of charging points as well as to decide the power capacity of the charging station. During operation, based on vehicle density and the number of charging points available at charging stations, EV load sharing can be done to avoid congestion and to save time.

9.5.7 Charging at parking, depot, and garages

The best utilization of space for parking, depots, and garage is to plan, design, and build charging stations in these areas. It saves time. In urban mobility, this is a good solution in public areas as well as in public places. However, safety measures should be checked for installing charging stations in such areas. Monitoring, metering, and commercial dealings are to be set for charging facilities in open or public areas.

9.5.8 Corporate parking

Corporate parking model provides parking facilities incorporate areas where a large number of the vehicle comes regularly. Therefore, if corporate parking areas are upgraded with grid-based charging points, the demand for charging a large number of vehicles can be met in one place. Solar photovoltaic (PV)-based nanogrid and picogrid-integrated EV charging stations provide an environment-friendly economic solution to corporate needs.

9.5.9 VIP parking

Parking areas for use of vehicles of VIPs are normally zones of high alert and security. Therefore, if VIP parking areas are upgraded with grid-based charging points, the demand of going out for charging can be eliminated. Again, SPV-based nanogrid and picogrid-integrated EV charging stations provide environment-friendly economic solutions to corporate needs. Along with other common factors, security is an important issue to be considered while setting charging stations at VIP parking areas.

9.5.10 Car sharing/fleet

A fleet of cars often involves the business of car-sharing. This business model should be planned with charging provisions, grid-connected charging stations at different places, parking, and depots. Service stations for vehicles near parking or charging areas give an added advantage. Flexible tariff model, quick and secure payment, roaming service, and information sharing system help to attract this business model.

9.5.11 Battery swapping

To provide fast service to vehicles, battery swapping is a good solution. It needs the coordination of battery type and available charged battery. On-road app-based information sharing system among vehicles in need of charged battery and battery swapping station make this model more attractive to the customer. It also reduces the space requirement and saves time.

9.5.12 Existing fuel (gas/oil) pumps

For safety issues, planning of charging station at or near fuel gas or oil (petrol/diesel) pumps is not preferred. As petrol, diesel, etc. are highly inflammable, in normal practice, electrical circuits should not be installed at the pump premises. If in any unavoidable circumstances, oil pumps and charging stations are located in common premises, high safety measures must be taken that will increase the cost of installation, operation, and maintenance. For this research, general guidelines advise not to install any charging stations at the oil pump premises.

9.6 Design of EV charging stations

EV integration with nanogrid and picogrid is done with charging point(s) in EV charging stations. The design of EV integration with NPG starts with user need and also ends with user satisfaction. User demands following needs to be fulfilled concerning design aspects:

(a) adaptive charging points,
(b) easy charging, and
(c) reliable charging.

Different technical aspects for EV integration with grids are discussed in the following subsections.

9.6.1 Charging mode

Based on the requirement, the user can demand for particular charging mode. Design of the charge controller can be made for charging in the following modes:

(a) fast charging,
(b) medium charging, and
(c) long charging.

Anyone facility or all facilities can be offered to the customer. Charging mode can be selected considering the following aspects:

(a) customer's choice,
(b) matching with available charge controller, and
(c) considering available time for charging.

Fast charging is chosen during peak demand hours. Long charging occupies space for a long period. This charging mode may be selected when demand is less and flexible tariffs can be provided.

9.6.2 Voltage levels

Design of rectifier and charge controllers are done based on the output voltage requirement. DC voltage levels can broadly be categorized as follows:

(a) low-voltage levels: $\leq$ 12 V or 24 V;
(b) medium-voltage level: > 24 V and $\leq$ 48 V;
(c) large voltage level: > 48 V.

9.6.3 Connector shape

Shapes of contact points are important in integration. It ensures connection with the correct polarity of battery units. It should match with the battery points of the vehicles of different segments and models. Based on the shapes of the contact point, connectors may broadly be divided into the following categories:

(a) Open contact type.
(b) Plug-in type.

9.6.3.1 Bare-type connector

Sometimes battery terminals are available in the bare condition. In these cases, two separate conducting wires or cables are used to connect battery terminals for charging (shown in Figure 9.3). There is no dedicated plug-in type socket for charging purposes. Some old versions of EV use this; however, at present, bare charging is not recommended.

9.6.3.2 Plug-in type

The plug-in type is safe and easy to use. Varieties of plug shapes are used. Broadly, they may be classified into two main groups as follows:

(i) *Conventional socket type plug-in connector*: This type of connection looks like conventional electrical plug-in types (shown in Figure 9.4).
(ii) *Fuel injector type plug-in connector*: This type of connection looks like a fuel injector used in a conventional petrol pump (shown in Figure 9.5(a and b)).

9.6.4 Continuous availability of power at charging station

As the charging station is expected to serve 24 h, power should be available at all times. In NPG-based system, solar power is available in the daytime and be directly

Figure 9.3 Bare-type charging connector without any dedicated inlet

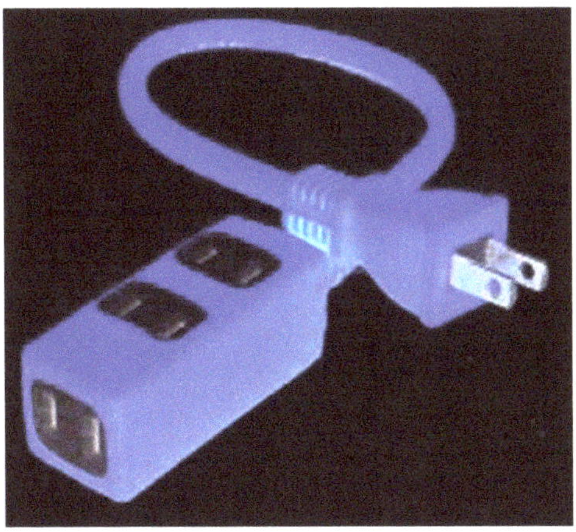

Figure 9.4 Conventional bare socket-type connector for EV charging

utilized for charging. But, for charging at night time, it can be designed and controlled to take power from other sources. In normal design and operation practice, at night, it can take power either from the utility grid or from the battery storage unit of the station.

9.6.5 Smart metering

Smart meters are used as best practices in charging stations. It monitors energy consumed by individual vehicles during charging and also monitors power inputs from solar PV, utility grid, and battery storage unit if any. Energy data can be

(a) (b)

Figure 9.5 (a) Fuel injector gateway used in petrol or diesel car and (b) fuel injector type plug-in connector for EV charging

monitored and stored for further analysis. In some places, cloud-based supervision is made by the owner of the charging station and/or regulatory bodies.

9.6.6 Payment

Two types of payment methods are used as the methods commonly followed in petrol or other fuel pumps:

(a) *Cash payment*: customers pay by cash.
(b) *E-payment*: customers pay by debit card, credit card, net banking, mobile banking, etc.

E-payment options are incorporated in two ways as follows:

(a) The payment method that is not connected with smart meters: meter reading is done and e-payment is done.
(b) The payment option that is connected with smart meters: meter reading is automatic and then e-payment is done.

Following features are found attractive and may be adopted:

(a) advanced booking,
(b) monthly billing system,
(c) prepaid payment, and
(d) offering a discount in connection with a flexible tariff system.

9.6.7 Interactive information sharing and booking facility in the vehicle network

Availability and services may be shared by the internet and apps to the drivers of the vehicle networks. On-demand, drivers can select and book charging points at nearest or optimum locations. Four layered Internet of Things (IoT) topology and cloud-based communication and computation may be adopted for the sharing of information among different charging stations, vehicle owners, and other stakeholders.

9.6.8 Vehicle load sharing and traffic management

If any charging station is overloaded with a large number of vehicles, it can transfer some vehicles to nearby other charging stations instead of keeping them waiting for long. On the other hand, lightly loaded charging stations charging less number of vehicles than capacity, can take extra vehicle loads of other nearby charging stations. This can be done by information sharing and the application of cloud-based apps.

Charging stations are required by main by following two categories of vehicles:

- Full EV (FEV): It is also known as battery electric vehicle (BEV) or pure EV (PEV).
- Plug-in hybrid EVs (PHEVs): It utilizes a battery as well as an internal combustion engine for an energy source.

Both of these above two types require external sources charging. A comparison of batteries used in BEV and PHEV-based primary aspects are given in Table 9.4.

9.6.9 Main EV charging components

The main charging components of EV are as follows:

(i) the electric motor in the EV,
(ii) battery,
(iii) onboard charger, and
(iv) inlet for charging.

9.6.10 Charging inlet

Two types of charging inlet are used as follows:

(a) AC charging inlet.
(b) DC charging inlet.

Table 9.4 Comparison of batteries used in BEV and PHEV

The battery used in BEV	The battery used in PHEV
It feeds power to the vehicle all run time	It feeds power to the vehicle initially, then an oil/gas-driven internal combustion engine provides power to the vehicle
The average run by full charge is around 140–180 km	The average run by full charge is around 30–50 km
Battery capacity and size are high	Battery capacity and size are small
When the battery state of charge (SoC) is insufficient, the vehicle stops; it requires charging or replacement	When battery SoC is insufficient, IC engine drives the vehicle
BEV does not have greenhouse gas emission	PHEV emits greenhouse has when it is driven by IC engine

Based on the type of external source to be used for charging, the following two types of charging systems are available:

- *AC charging system*: It uses a board charger with an AC charging inlet. External supply feeds AC power to the onboard charger that converts AC to DC for charging the battery.
- *DC charging system*: It does not need a board charger. DC type external source is directly connected to feed power for charging the battery. For fast charging, this type is used.

Onboard charger

An onboard charger is installed inside the EV. It receives electric power from an external supply system through an AC inlet. It then converts AC to DC that is fed to the battery for charging. Onboard charger provides necessary control of charging current, battery voltage, and power. For fast charging, DC is directly fed to the battery, therefore, in fast charging mode, the onboard charger is bypassed and not used for charging.

9.6.11 EV connector for charging inlet

EV connectors make the contact between the external supply or charging point of the charging station and the charging inlet of the EV. Inlets are made as per various international standards.

IEC 62196-2 refers to AC connectors and IEC 62196-3 refers to DC connectors. Different types of commonly known connectors are presented in Table 9.5. EV connectors for charging inlets commonly used in different regions of the world are presented in Table 9.6. The pin positions of different connector connectors are shown in Figure 9.6. The meaning of different connecting pins of CCS – Type 2 is shown in Figure 9.7.

9.6.12 Different modes of charging connections for EV charging

EV charging is classified into four modes based on charging connection by IEC 61851-1. Four modes of charging are presented in Table 9.7.

9.6.13 EV charging classification based on charging time

Based on charging time, EV charging is classified into three main categories as follows:

- Slow charging.
- Moderate charging.
- Fast charging.

In some cases, charging time is reduced to a very minimum time that is called ultra-charging.

An example of charging time for different types of charging is given in Table 9.8.

Table 9.5 Different types of EV connectors for charging inlet and charging points of charging station

Type 1	• The J1772 connector is used in AC single phase charging station • It has three large pins like a common power outlet used at home • Three large pins serve for connections with phase, neutral, and ground • It has additional and two smaller pins. • Smaller pins help in communication between charger and EV • This type is used up to It can deliver 10 kW (approx.) • It has a current capacity normally up to 32 A • Often it has a lock so that the line connection does not break or disconnected during charging
Type 2 (IEC 62196-2)	• Pins of type 2 are of a similar type to type 1 and are used both for single-phase and three-phase AC charging stations • It is round with a flat edge at the top • For three-phase charging system, it has two extra pins for two additional phases • Current is up to 32 A for single-phase applications and up to around 60 A for three-phase applications • It is used normally up to 50 kW • Often it has a lock so that line connections do not break or disconnected during charging
Type 3	• Type 3 connector is coming obsolete because of attractive features of another type
Type 4 (SAE J1772)	• This type is used for DC fast charging
CHAdeMO	• This type is used for the DC charging station • CHAdeMO comes from the Japanese words "O cha demo ikaga desuka" which means "You will have tea while the car is charging." • It is used for fast charging • It carries direct current • Normally it is used up to around 150 A • It is used charging station of capacity up to 400 kW
Combo 1: Combined Charging System (CCS-1)	• CCS Combo 1 consists of features Type 1 (J1772) and DC fast charging • This type can be used both for AC charging and DC fast charging. • This type is normally used up to 350 kW
Combo 2: Combined Charging System (CCS-2) (IEC 62196-3)	• CCS Combo 2 consists of features of type 2 and DC fast charging • This type can be used both for AC charging and DC fast charging • This type is normally used up to 350 kW

Table 9.6 EV connectors for charging inlet commonly used in different regions of the world

DC/AC	Plug	Region
DC	Type 1 (J1772)	Japan, America
	GB/T	China
	Type 2	Europe and other parts
DC	CCS Type 1	America
	CCS Type 2	Europe and other parts
	GB/T	China
	CHAdeMO	Japan

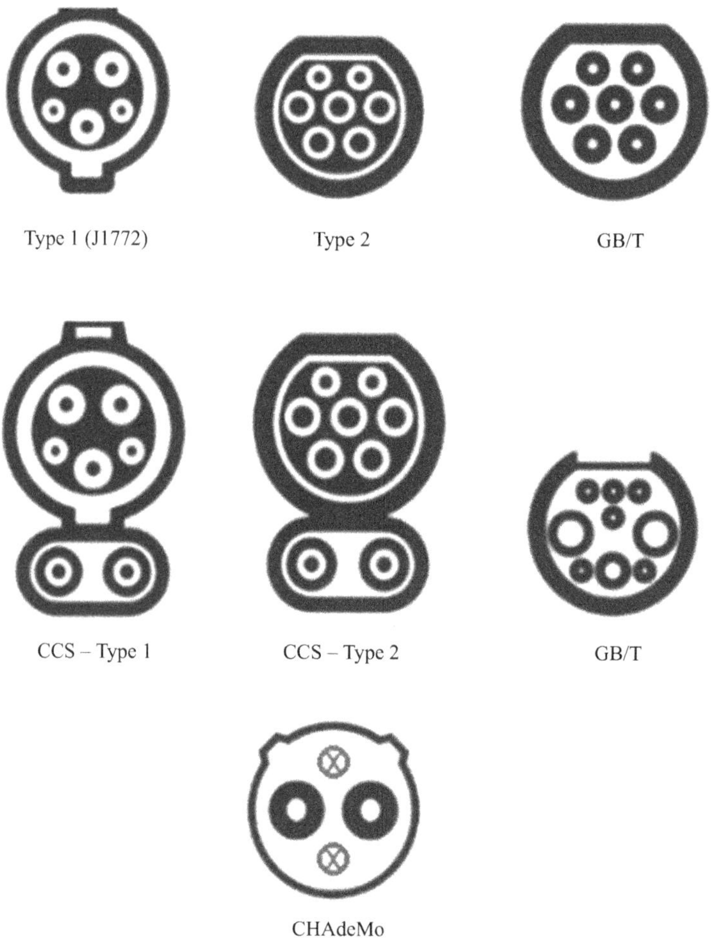

Figure 9.6 Pin positions of different connector connectors (as mentioned in Table 9.6) for EV charging inlets

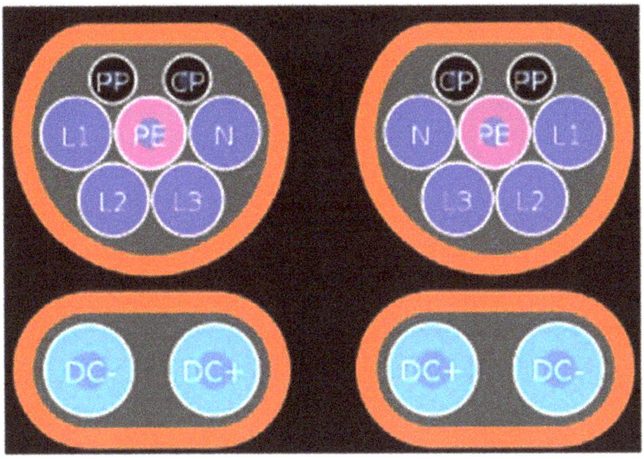

Male type Female type

Figure 9.7 Meaning of different connecting pins of CCS—type 2

9.6.14 Main advantages and limitations of fast charging

Different types of charging methods like slow, moderate, and fast are used. Fast charging saves time and offers a quick solution for charging. However, frequent fast charging (say, 4–7 times per month or more) reduces the span of battery life.

9.6.15 Mode selection

As per international standard IEC 61851, mode 3 and/or mode 4 are normally selected. This selection can be followed for setting up charging stations both in indoor and outdoor applications.

9.6.16 Environmental aspects: altitude and temperature

As per IEC 61851-1, it should be compatible

- to operate up to 2 000 m of altitude;
- to operate minimum temperature of −25 °C outdoor environments;
- to operate minimum temperature of −5 °C for indoor environments.

In most cases, the operating temperature lies between −30 °C and +50 °C and relative humidity lies between 5% and 95%.

9.6.17 Number of charging output

The output of the charging station is of two categories based on connecting facility:

- *Socket-outlet*: It has no cable attached to the output point.
- *EV connector*: It has a cable attached to the output point.

Table 9.7 Different modes of charging connections for EV charging

Mode 1	• It uses a standard socket used in domestic applications with an AC network. The charging point outlet is connected with an EV charging inlet by standard plug and cable without any additional equipment. Common specifications used for this connection is as follows: ○ For single-phase system: 16 A and 250 V AC ○ For three-phase system: 16 A and 480 V AC • Related international standard is IEC 61851-1 • The main advantages are that it offers a solution at home premises and is simple for setting up • It offers slow charging which is its main disadvantage
Mode 2	• It uses standard socket-outlet • In addition, protection against electric shock and safety measures are taken • Common specifications used for this connection is as follows: ○ For single-phase system: up to 32 A and 250 V AC ○ For three-phase system: up to 32 A and 480 V AC • Related international standard is IEC 61851-1 • It is used slow and moderates charging
Mode 3	• Instead of using only a standard socket-based charging connection, this mode uses a dedicated charger connected with the AC supply network • Adequate measures for protection are taken • Advances control techniques are used for charging • It offers a higher power rating • It offers moderate charging time i.e., less than mode 1 or 2 • It offers additional communication facilities between the vehicle and the charging system • It also allows the implementation of control functions • Remote monitoring and control can be integrated verification that the EV is correctly connected to the EV supply equipment • It offers optimized energy consumption • It offers safe charging for the battery enhancing its life span • It reduces risk and provides safe operation • Smart meters and flexible tariffs and e-transactions can be integrated
Mode 4	• It provides direct charging by DC • If the supply is DC, it is connected directly with the battery. If the supply is AC, then it is converted to DC that is directly fed to the battery without using a charge onboard controller • It provides fast charging than modes 1, 2, and 3 • DC EV supply equipment • The power rating for DC charger is normally kept high, e.g., 20 kW or above • Relevant international standard is IEC 61851-24

Table 9.8 Example of charging times for different types of charging

Charging mode	Charging time	Battery capacity	Charging system
Slow charging	10–11 h	40 kWh	4 kW home charging station
Moderate charging	6–7 h	40 kWh	7 kW onboard charging system connected with external 15 kW AC charging system (charging with being limited by the upper limit of onboard charger)
Fast charging	40–50 min	40 kWh	60 kW DC type fast system
Ultra-fast charging	10–15 min	40 kWh	200 kW DC type ultra-fast charging system

In domestic applications, there may be a single of few outputs. Normally, it serves the need of a single-family or person.

But, public charging stations have multiple outputs. Both types of outputs i.e., socket outlet and EV connector with attached cables can be used in public charging stations. The main advantage of multiple outputs is that several EVs can be charged at a time. These are suitable for public places like corporate houses, institutes, market complexes, parks, roadside, etc. However, the charging station must have adequate power delivering capacity to charge multiple EVs.

9.6.18 Authentication

In public charging stations, authentication is followed before charging. This helps in access management, EV management, and secure payments. Often it helps in priority management for charging to EV of VIPs, ambulance, etc. RFID technology and smart cards are commonly used for this purpose. Authentication management can be done in the following modes:

- locally;
- remotely by online or apps;
- advanced by online or apps.

9.6.19 Communication system in charging station

Communication among charging stations and EV networks as well as with other charging stations is important for EV charging management. It helps in many ways like:

- selection of charging stations by EV drivers remotely;
- authentication process;
- booking for charging point and time slots;
- payment process;
- monitoring;
- EV load sharing with other charging stations;
- peak demand management.

There are several ways of communication methods to be adopted like Ethernet, Bluetooth, Wi-Fi, NFC, 3G/4G/next generation, etc. Some typical examples of AC charging station's power ratings: 3.7, 7.4, 11, or 22 kW. Current ratings: 16 and 32 A.

9.6.20 Type of charging point installation

Different types of installations of charging points may be broadly be classified into two categories:

- wall-mounted installation and
- floor installation.

Wall-mounted installation: this method installs charging points on walls, mainly by socket-type outlets. There is no fixed cable attached to it. This type is used for charging points or stations in the home or residential applications.

Floor installation: this type uses a pedestal stand set up on the floor on which output charging points are installed. Fixed cables are attached with output points. It looks like an oil pump station. This type of installation technique is used mainly at public charging stations at outdoor locations.

9.6.21 Fixed cable for charging

Cables are used for connecting output points with EV. Floor-type installations normally use fixed cables on a pedestal. Dedicated and attached cable applications are found mainly at outdoor public charging stations. But, at indoor home charging stations, fixed cables are not generally found.

A cable connector for charging should match with EV vehicle inlets. By one particular cable type, the same type of vehicles can be connected. For different types of vehicles, different types of cable connectors may be compatible. For these reasons, outdoor public stations having multiple charging points, keep different types of cable arrangements different charging points to charge EV vehicles of different inlet points. Charging stations without attached cable (with type 1 or 2 socket outlets) do not consider this issue and can charge any type of EV and is more flexible in this aspect.

9.6.22 Maximum charging current, number of charging points, and mode

The maximum charging current is set based on the capacity of the charging station. Modes are selected based on the number of charging points. Normally AC charging station of sing charging point uses either mode 1 or 2. Both single-phase and three-phase AC charging stations with multiple charging points use mode 3 and DC-type charging stations use mode 4. Different current ratings used for maximum charging current at different modes of charging stations are presented in Table 9.9.

Table 9.9 Different settings for maximum current for different application

Number of charging points	Preferred charging mode	Number of phases (P)	Maximum current
1	Mode 1, Mode 2	1P	16 A
Multiple	Mode 3	1P, 3P	16 A, 32 A

9.6.23 Earthing in EV charging

EV charging system must have proper earthing for safety. Earthing is referred by IEC 60364-7-722 (Clauses 314.01 and 312.2.1). It recommends that

- At the charging station, a dedicated path must be provided for energy transfer of energy from/to EV.
- EV connector and socket-outlet must have an earthing connection and it must phase-earth-neutral (PEN) structure.

9.6.24 Electric shocks and their protection in the EV charging system

9.6.24.1 Sources and causes

The main sources and causes of electric shock are given in Table 9.10.

Table 9.10 Major sources and causes of electric shocks in the EV charging system

Source	Cause
Cable	Mechanical damage, over-stress on insulator, etc.
Plug	Disconnection in earthing connection
EV	Failure or damage in basic safety and protective systems
External environment	Wet or high level of moisture content

9.6.24.2 Protection

Other than routine check-ups and safety measures, residual current device (RCD) and Residual Direct Current Detecting Device (RDC-DD) are used for protection. Recommended protection for different modes is presented in Table 9.11. The tripping range normally used for Options 1, 2, and 3 of mode 3 considering the aging effect and on insulation and humidity is presented in Table 9.12.

9.6.25 Surge or transients overvoltage protection

Different types of surges or transients due to lightning, switching effects, and faults may appear in grid-connected charging stations. These surges can damage the controllers, converters, chargers, measuring instruments as well as EV connected with the charging points. Overvoltage protection, lightning arrestors, etc. are used to protect against surges. Causes and protection schemes used for surge/transient overvoltage that can propagate and dame EC charging systems are presented in Table 9.13.

Table 9.11 Recommended protection for different modes of charging

Mode	Protection type	Standard
Mode 1	RCD 30 mA type A	For RCDs:
Mode 2	RCD 30 mA type A	IEC 61008-1,
Mode 3	Option 1: RCD 30 mA type B, Option 2: RCD 30 mA type A + 6 mA RDC-DD,Option 3: RCD 30 mA type F + 6 mA RDC-DD	IEC 61009-1, IEC 60947-2, IEC 62423
Mode 4	Not applicable as no AC points are there. Adequate separation is recommended	For RDC-DD: IEC 62955

Table 9.12 Tripping range normally used for options 1, 2, and 3 of mode 3 considering the aging effect and on insulation and humidity

Option	Tripping range	Consideration
1	(15–60 mA)	Leakage current can be 5–7 mA in humid environments or due to the aging effect of insulation. This may lead to mal tripping
2	(3–6 mA)	
3	(3–6 mA)	

Table 9.13 Causes, protection schemes used for surge/transient overvoltage, and useful standards

Causes	Lightning, switching transients, faults
Protection schemes	Lightning rod: lightning protection (LPS) to be instated in the building of indoor and outdoor application) surge protective device: Type 1 or 2 for main low-voltage switchboard (MLVS) Type 2 for EV charging system and bus bar distribution panel Type 3 for electronics equipment Common practice: provide over-voltage protection if it exceeds 2.5 kV (IEC 17409). Sometimes, the limit is set at 1.25 kV for the charging system connected with the building
Useful standards	IEC 60664-1, IEC 60364-7-722, IEC 17409

9.6.26 EV charging and load management

EV load characteristics are different from other types of electrical loads. It varies with the type and location of charging stations and with time. Different types of EV charging and load management system are in use as follows:

- *Static LMS*: manage EV charging load by setting the fixed level of charging power.

- *Dynamic LMS*: manage EV charging load by optimization considering building and other load and local resources.
- *Artificial intelligence-based LMS*: use statistical data based deep and machine learning to optimize and manage EV load, other local loads, resources, and the structure of tariffs.
- *Predictive control*: model-based predictive control may be used for LMS.

Common solutions for two different cases are presented below:
Case 1: EV demand for charging is lower than installed charging power capacity.
Solutions:

- attempt for adding more EV;
- feed energy to local needs;
- feed energy storage;
- feed energy to the conventional grid, if the local power generation capacity is greater than local needs and EV demand.

Case 2: EV demand for charging is greater than installed charging power capacity.

Solutions:

- use local source;
- use storage;
- take power from the conventional grid;
- change charging time slot and perform load management;
- in long run, add charging capacity and add more charging points.

9.6.27 Charging station at dealer premises

For EV marketing, dealers should have their charging stations having multiple points. This is required mainly for the following reasons:

- for presenting demo cars,
- for test drive,
- for initial EV start,
- for EV service and repair, and
- for EV parking for customers and employees.

9.7 Smart supervision and challenges

9.7.1 Smart supervision and smart charging

Smart supervision is an advanced type of supervision incorporated with cloud-based communication and computation. It normally used a platform like Open Charge Point Protocol (OCPP) for EV charging, hardware connections, and software applications.

Smart charging refers to EV charging with optimum use of local resources, grids, EV demands, and best market solutions.

Table 9.14 Different useful topologies for EV charging

V1G	Step by step manage to charge power in a unidirectional way
V2X	Manages power step by step both for charging and discharging. Different vehicle to grid (V2X) topologies are • Vehicle-to-Grid (V2G): refers to EV battery interaction with electric grid (frequency control . . .) • Vehicle-to-Building (V2B): refers to EV battery operation with optimized consumption of building energy • Vehicle-to-Home (V2H): refers to optimizing the use of home energy for EVs keeping backup power for emergency • Vehicle-to-Load (V2L): refers to the operation of EV battery feeding energy to load

For smart supervision and charging, different protocols are used as presented in Table 9.14.

9.7.2 Challenges

Some barriers that may appear as major challenges for smart EV charging are as follows:

- lack of EV driver's willingness;
- lack of proper regulation in many regions;
- mismatch of dynamic tariff with regulations;
- lack of smart metering;
- lack of common topology and standards;
- possibility of cyber-attack;
- data privacy issues;
- grid disagreement on many technical and commercial issues.

9.8 Case study on the V2X configuration

A case study was carried out on the V2X configurations. The utility of 60 EV vehicles was considered. EV users were grouped into five following categories:

- Type 1: EV users go to the office on regular basis by EV and charge EV at work premises.
- Type 2: EV users go for a long drive by EV.
- Type 3: EV users go to the office on regular basis by EV, travel a longer distance than type 1, and charge EV at work premises.
- Type 4: EV users stay most of the time at home and have very little utility of EV.
- Type 5: EV users go to the office at might shift on regular basis by EV and charge EV at work premises.

In each category, 12 EVs were considered and the SoC of their battery units were monitored in terms of high (charging time) or (low run time). The result is presented in Figure 9.8. This type of study may help charge scheduling and charging station management.

9.9 Some useful standards

Some useful standards with relevant issues are provided in Table 9.15.

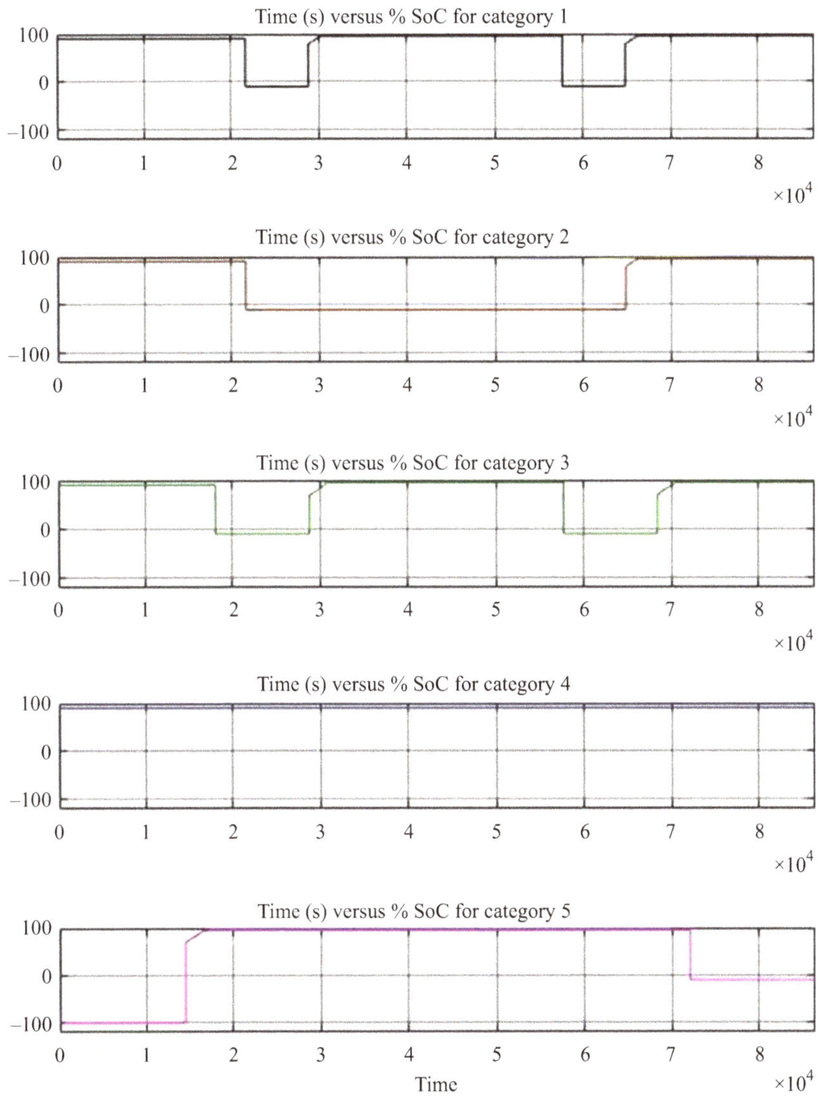

Figure 9.8 SoC of different types of vehicle usage

Table 9.15 Some useful standards

IEC standards	Issues
IEC 60364	Low-voltage electrical installations (Part 7-722: EV installation or location requirements)
IEC 61851	The EV charging station of modes 3 and 4 category • It is referred to for modes 3 and 4 • Part 1 refers to general requirements • Part 21-2: refers to EMC requirements for off-board EV charging systems; it defines two classes as follows: ○ Class A: suitable for all locations apart from residential (it can also directly be connected to a low-voltage residential power network) ○ Class B: suitable for residential building areas (it can also directly be connected to a low voltage residential power network) • Part 23 refers to DC charging stations' requirement Part 24 refers to digital communication among EV and DC charging station
IEC 62955	RDC-DD
IEC 60309-1, IEC 60309-2, IEC 62196-1, IEC 62196-2, IEC 62196-3, IEC TS 62196-4	Socket outlet and EV connector
IEC 60947-2, IEC 60947-6-2,IEC 61009-1, IEC 60898 IEC 60269	Overcurrent protection
IEC 60364-7-722.311	Maximum charging current configuration

9.10 Summary

Nanogrid and picogrid integration with electromobility has been discussed. Different features of charging stations have been presented and then design considerations have been described. Different types of charging schemes along with their merits and demerits have been presented. Different connectors used in charging stations have been shown. Operational aspects of NPG integrated charging stations have been discussed. Advancements introduced in the design and operations of charging stations connected with the grid have been discussed. The study recommends that installations of EV charging stations should follow international standards to be compatible with vehicles' standards and should follow guidelines of regional regulatory bodies.

Further reading

[1] "IEEE DC Quick Charger for Use with Electric Vehicles Test Suite Specification – Version 1.0," in *IEEE DC Quick Charger for Use with Electric Vehicles Test Suite Specification – Version 1.0*, pp. 1–0, 17 September 2019.

[2] "IEEE Draft Standard Technical Specifications of a DC Quick and Bi-Directional Charger for Use with Electric Vehicles," in *IEEE P2030.1. 1/D1*, February 2021, pp. 1–159, 9 February 2021.

[3] "IEEE Smart Grid Vision for Vehicular Technology: 2030 and Beyond Roadmap," in *IEEE Smart Grid Vision for Vehicular Technology: 2030 and Beyond Roadmap*, pp. 1–17, 30 June 2015.

[4] "IEEE Vision for Smart Grid Controls: 2030 and Beyond Reference Model," in *IEEE Vision for Smart Grid Control: 2030 and Beyond Reference Model*, pp. 1–10, 12 September 2013, DOI: 10.1109/IEEESTD.2013.6598993.

[5] "IEEE Smart Grid Vision for Computing: 2030 and Beyond Roadmap," in *IEEE Smart Grid Vision for Computing: 2030 and Beyond Roadmap*, pp. 1–14, 11 January 2016.

[6] S. F. Bush, S. Goel, and G. Simard, "IEEE Vision for Smart Grid Communications: 2030 and Beyond Roadmap," in *IEEE Vision for Smart Grid Communications: 2030 and Beyond Roadmap*, pp. 1–19, 3 September 2013, DOI: 10.1109/IEEESTD.2013.6690098.

[7] "IEEE Recommended Practice for Performance Testing of Electrical Energy Storage (EES) System in Electric Charging Stations in Combination with Photovoltaic (PV)," in *IEEE Std 2836-2021, pp. 1–27, 8 June 2021, DOI: 10.1109/IEEESTD.2021.9447313.*

[8] Continental Automated Buildings Association, "Impact of Smart Grid on Connected Homes Landmark Research Study," in *Impact of Smart Grid on Connected Homes*, pp. 1–106, 17 May 2013, DOI: 10.1109/IEEESTD.2013.6786299.

[9] K. Bhat, V. Sundarraj, S. Sinha, and A. Kaul, "IEEE Cyber Security for the Smart Grid," in *IEEE Cyber Security for the Smart Grid*, pp. 1–122, 16 September 2013, DOI: 10.1109/IEEESTD.2013.6613505.

[10] "IEEE Standard for Utility Industry Metering Communication Protocol Application Layer (End Device Data Tables)," in *IEEE Std 1377-2012 (Revision of IEEE Std 1377-1997)*, pp. 1–576, 10 August 2012, DOI: 10.1109/IEEESTD.2012.6264065.

[11] "IEEE Draft Standard for Synchrophasor Measurements for Power Systems," in *IEEE PC37.118. 1/D4.2*, May 2011, pp. 1–56, 23 June 2011.

[12] "IEEE Standard for an Architectural Framework for the Internet of Things (IoT)," in *IEEE Std 2413-2019, pp. 1–269, 10 March 2020, DOI: 10.1109/IEEESTD.2020.9032420.*

[13] "IEEE/ANSI Draft Recommended Practice for Electromagnetic Compatibility Limits and Test Levels," in *ANSI PC63.12/D12g*, May 2015, pp. 1–49, 1 January 2015.

[14] "IEEE Draft Standard for DC Microgrids for Rural and Remote Electricity Access Applications," in *IEEE P2030.10/D11*, February 2021, pp. 1–56, 10 March 2021.
[15] "IEEE Approved Draft Standard for Test Methods for Surge Protectors and Protective Circuits Used in Information and Communications Technology (ICT) Circuits, Including Smart Grid Data Circuits," in *IEEE PC62.36/D38*, February 2016, pp. 1–146, 1 January 2016.
[16] "IEEE Standard Conformance Test Procedures for Equipment Interconnecting Distributed Energy Resources with Electric Power Systems and Associated Interfaces," in *IEEE Std 1547.1-2020*, pp. 1–282, 21 May 2020, DOI: 10.1109/IEEESTD.2020.9097534.
[17] "IEEE Guide for Safety in AC Substation Grounding," in *IEEE Std 80-2013 (Revision of IEEE Std 80-2000/ Incorporates IEEE Std 80-2013/Cor 1-2015)*, pp. 1–226, 15 May 2015, DOI: 10.1109/IEEESTD.2015.7109078.
[18] S. Chattopadhyay, M. Mitra, and S. Sengupta, *Electric Power Quality*, 1st ed., Springer, The Netherlands, 2011.
[19] S. Karmakar, S. Chattopadhyay, M. Mitra, and S. Sengupta, *Induction Motor Fault Diagnosis*, 1st ed., Springer, Singapore, 2016.
[20] S. Chattopadhyay and A. Das, *Overhead Electric Power Lines – Theory and Practice,* 1st ed., IET, Stevenage (UK), 2021.
[21] "IEEE Standard Glossary of Stationary Battery Terminology," in *IEEE Std 1881-2016, pp. 1–42, 30 August 2016, DOI: 10.1109/IEEESTD. 2016.7552407.*
[22] "IEEE Guide for Technology of Unified Power Flow Controller Using Modular Multilevel Converter—Part 2: Terminology," in *IEEE Std 2745.2-2021*, pp. 1–25, 1 September 2021, DOI: 10.1109/IEEESTD.2021.9525319.

Chapter 10

Nanogrid-integrated EV charging with IoT-enabled cloud computing for smart time and space management

The mode of mobility in both rural and urban areas is changing from conventional oil-based vehicles to hybrid and electrical electric vehicles (EVs). With the increase in the number of EVs, it has become a challenging job to provide proper charging facilities throughout the vehicle network. On the other hand, considering the rooftop space available in a city environment, nanogrid (NG) instead of a microgrid is becoming an alternate or parallel choice to meet the power demand. With this perspective, in this chapter, an attempt has been taken to propose nanogrid-integrated EV charging with the Internet of Things (IoT)-enabled smart time and space management. First NG integration has been presented and then NG has been incorporated with the EV charging system. IoT-enabled information sharing system (IEISS) has been introduced to the link between consumers and NG stations. Selection of charging station has been done. Then, standby battery charging station (SBCS) has been added to the system, and the selection process has been modified accordingly for space and time optimization. Comparative case studies have been presented.

10.1 Introduction

Smart city planning includes proper mobility management. In the last few years, a paradigm shift has been observed in the vehicle category. The rate of increase of conventional oil-based vehicles has decreased. The hybrid vehicle has been initially considered as an intermediate solution. Then, with the advancement of battery technology, the EV has become the most preferred solution for future mobility. This demands pollution-free, cost-effective, and time-optimized management of EV charging.

To engineer the world in a better way from threats of pollution, the whole world is looking for alternate solutions of energy resources and green transportation to achieve carbon minimization and sustainable development. The mode of mobility in both urban and rural areas is changing from conventional oil-based vehicles to hybrid and electrical EVs. With the increase in the number of EVs, it has become a challenging job to provide proper charging facilities throughout the vehicle network. On the other hand, considering the rooftop space available in a city environment and open space available in rural areas, NGs are spreading rapidly for different utilities. From this perspective, a concept of the topological model will be developed by NG-integrated EV charging with the IoT-enabled cloud computing-based smart time and space management. First NG integration has been presented and then NG has been incorporated with the EV charging system (EVCS). IEISS is used to link consumers and NG stations. Selection of charging station has been done. Then, SBCS is added to the system, and the selection process has been modified accordingly for space and time optimization. Comparative case studies have been performed to validate the model with some real data. The NG (or NPG) integrated model offers the following benefits:

(a) Effective use of green solar energy in NG networks enable wide distributed generation in the capacity of a small scale to medium-scale.
(b) Effective integration of in NG and EV charging station (EVCS) both in DC and AC format.
(c) Effective integration of IoT in NG-fed EVCS.
(d) Effective use of cloud computing for time and space management considering upcoming large numbers of EVs.
(e) Effective use of digital fog computing for faster execution.
(f) It can reduce the cost of energy generation by use of solar-based power for EV charging that can also be utilized in applications increasing revenue to a great extent.
(g) *Aim of carbon zero*: The model has a direct impact on the reduction of carbon by promoting NG with solar resources and EVs in the transport sector.

The NPG and EVCS integration offers the following opportunities:

(a) a business model of small- to medium-scale power generation;
(b) a business model of EVCS;
(c) a business model of mobile apps for booking, charging, load management, and supervision benefiting owners and customers;
(d) a business model of a mobile-apps-based third-party servicing.

Additionally, it offers the Internet of NG-integrated EVCSs, EVS, vehicle owners, charging station owners, and NG owners having the potential to benefit a large number of people globally. The model can be implemented both in rural and urban areas and helps in carbon reduction to a great extent (of around 60–80%). The model offers four types of business models benefiting the small to the large business sector. It can reduce the cost of energy generation by use of solar-based power for EV charging that can also be utilized in applications increasing revenue to a great extent.

It can reduce the cost of energy consumption by use of solar-based power for EV charging that can also be utilized in applications increasing revenue to a great extent. It also reduces fuel or energy costs for the vehicle. Considering the operation of the model for 5–10 years, it can improve GDP by reducing energy costs for utility and EV by 80%. Also, the opportunity of apps-based management can be offered remotely for other countries increasing foreign earning. Additionally, it can boost up the rural economy.

The IoT-enabled cloud computation-based NPG-EVCS model can generate job opportunities in the following sectors:

- Energy sector.
- Transport sector.
- Service sector.
- Networking sector.
- Software and mobile apps sector.
- Data analytics.

The model can develop the following business, sales, and user opportunities:

- energy trading;
- battery trading;
- business for maintenance service;
- business for service to EV owners, drivers;
- business opportunities in different locations like roadsides, parking areas, shopping mall areas, rural areas, smart city areas, and many more;
- networking business;
- data business.

A typical 12 kW NG-integrated model can trade $12 \times 24 \times 365 = 105\ 120\ \mathrm{kW-h}$ energy, corresponding EVs, services, etc.

From a global perspective, it helps especially in the following sectors:

- can benefit small businessmen to rich businessmen;
- boosting rural economy;
- smart city design;
- green rural and urban environment;
- creates large numbers of jobs in various sectors, energy, automobile, service, IT, data analytics, etc.;
- serves a large number of people.

10.2 Internet of Things (IoT)

10.2.1 Definition

IoT is advanced networking among many “Things.” It applies intelligent technologies and analytics systems for sensing, interfacing networking, and controlling to improve performance and service. It has wide application in industries and home utilities with enhanced data analytics, automation, power savings, optimization, and smart solutions.

10.2.2 History of IoT

The history of the initial stages of the development of IoT is shown in Figure 10.1. The thought of connecting devices by wireless mode started in 1970. After

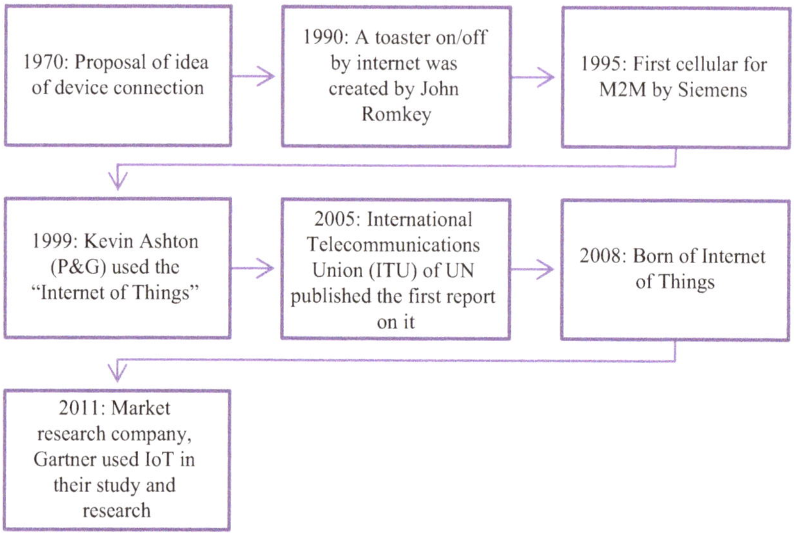

Figure 10.1 History of initial stages of development of IoT

20 years, in 1990, John Romkey successfully switched on and off a toaster by Internet. In 1995, the first machine-to-machine cellular communication was introduced by Siemens. The new concept of networking of things with the title "Internet of Things" was used first by Kevin Ashton (P&G) in 1999. International Telecommunications Union (ITU) of the United Nations (UN) published its first report on it in 2005. IoT got recognition in 2008 and was first marketized by Gartner, a market research company that used IoT in its study and research. in 2011.

10.2.3 IoT protocol layers and topology

IoT is incorporated by smart integration of advanced technologies in the field of different areas like measurement, communication, data science, control, power management, etc. The operating protocol of IoT is based on four main layers (shown in Figure 10.2) as follows:

- Layer 1: Sensing layer.
- Layer 2: Networking layer.
- Layer 3: Data processing layer.
- Layer 4: Application and interfacing layer.

10.2.4 Sensing layer

The sensing layer is the bottom-most layer of IoT. It connects various devices and systems through sensors and creates the interface between various systems (Things) and IoT frameworks. Sensors perform the jobs of smart measurement and identifications of relevant parameters. The operation of smart sensors is based on compatibility with advanced communication technologies like GPS, RFID, WiFi, etc.

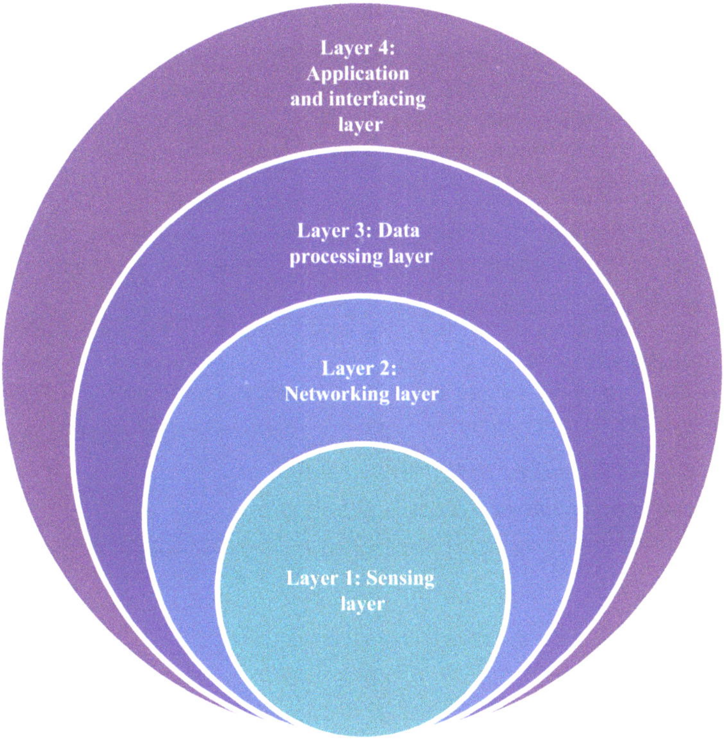

Figure 10.2 Layers of IoT

10.2.5 Networking layer

The networking layer acts as an interconnection among all other layers including various systems, data processing units, the application layer. It may be developed by wired networking or by wireless networking or both depending on the systems' location, behavior, and objective of IoT. Some commonly used network technologies in IoT are WLAN, WPAN, WMAN, ethernet, optical fiber, etc.

10.2.6 Data processing layer

The data processing layer collects data from the network layer and processes them through different data processing tools. Data processing can take place at a localized center or remote center. Various types of servers, clouds normally act as a place for data processing. Different tools of artificial intelligence (AI), cloud computing, machine learning, etc. are used for data processing. Lots of advancements are observed in the field of data processing and data science. Quantum communication and quantum computing have great future aspects in the field of IoT.

10.2.7 Application and interfacing layer

The application layer is the topmost of IoT and serves the main objectives of IoT. Based on data processing results, it provides the necessary information to the consumers and signals to various object systems. It can control the operation of various units to achieve desired outputs. Thus, the application layer consists of the following sublayers:

(a) *Data collection unit*: It collects relevant data from the data processing layer.
(b) *Control unit*: It performs control operations based on feedback, reference signals, and control logic.
(c) *Actuator operation unit*: It operates based on control signals to achieve the desired output.
(d) *User interaction*: This layer also serves interfacing and interaction with users.

10.2.8 Key hardware components of IoT

Key hardware components of IoT are as follows:

- sensors,
- server,
- router,
- dashboard,
- power supplies,
- actuators,
- transmitter,
- receiver, and
- network modules.

10.2.9 IoT—sensors

Sensors are the basic and mandatory component of IoT. They sense, identify, and measure various parameters from different devices, systems, environments, etc. With the advancement of sensing technology, many smart sensors are available and are coming in near future. Considering grid applications and EV integration, sensors are classified into two groups: electric parameter sensors and nonelectrical parameter sensors, as presented in Table 10.1.

A sensing layer used in IoT is shown in Figure 10.3 that consists of four modules as follows:

- sensing modules,
- energy modules,
- power management modules, and
- RF modules.

Widely used communication modules involve RF communications and use technologies like a radio transceiver, ZigBee, Bluetooth, WiFi, BAW, duplexer, etc.

Table 10.1 Different types of electric parameter sensors and nonelectrical parameter sensors

Sensor type	**Sensors**
Electrical parameter sensors	Smart meters Voltage sensor Current sensor
Nonelectrical parameter sensors	Temperature sensor Proximity sensor Image sensor Light sensor RFID gas sensor Microflow sensors Accelerometer Magnetometer Gyroscope Acoustic sensor Pressure sensor Humidity sensor

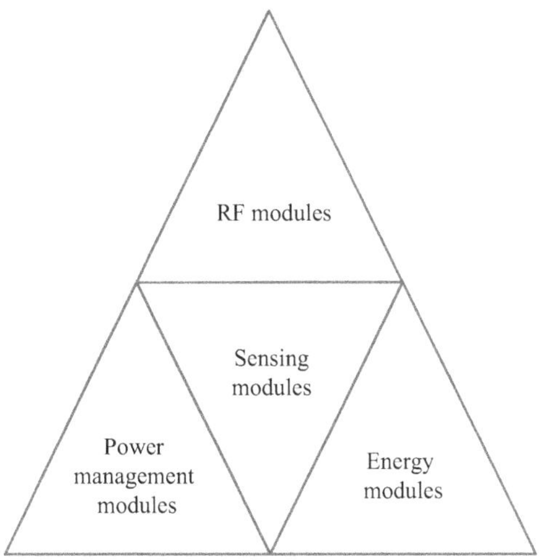

Figure 10.3 Sensing layer

10.2.10 Networking protocols for IoT

Communication technology offers different networking topologies. Commonly used protocols are

- radio-frequency identification (RFID),
- near-field communication (NFC),

- low-energy Bluetooth,
- low-energy wireless,
- low-energy radio protocols,
- LTE-A, and
- WiFi-direct.

The main features of different networking protocols are presented in Table 10.2.

10.2.11 IoT application software

Application software presents useful information and other operational features to the users of IoT-based systems. Through application software, real-time IoT

Table 10.2 Main features of different networking protocols

Networking protocol	Features
Near-field communication (NFC)	It uses standard protocols for sensing modules with standard electronic devices, like a mobile device, etc. It offers • Simple configuration • Low-energy usage • Flexible options like identity access • Bootstrapping, payment-based gateway, etc.
RFID	It uses radio transmitter–receivers for two-way communication and performs identification, tracking, and transmission. Like NFC, it offers • Simple configuration • Low-energy usage • Flexible options like identity access • Bootstrapping, payment-based gateway, etc.
Low-energy Bluetooth	It gives low-power communication It can be chosen to fulfill the need for long-use of IoT across various systems
Low-energy wireless	It provides a wireless solution It offers reduces energy consumption It can be operated for long use of IoT It offers an enhanced life span of IoT devices
Radio protocols	Some radio protocols are ZigBee, Z-Wave, and thread It is useful for a low-cost solution It can fulfill the need for IoT in a private area It offers low-power
LTE advanced (LTE-A)	It offers advanced features of LTE It offers reliable solutions for the application of IoT in mobility involving vehicles, UAVs, etc.
WiFi-direct	It provides peer to peer (P2P) communication at high speed It does not require a fixed access point.

functions are executed by including sensing, data process, analytics, and communication. It performs the following main roles:

- *Device integration*: It is done with the help of communication protocol, data process output, and user's objective.
- *Collection of data*: It is done with the help of sensors and communication protocol. Data may be stored locally or at a remote place.
- *Real-time data analytics*: It is done with the help of cloud computing, software algorithms. Many advanced tools like AI, machine learning, etc. are commonly used for this purpose.
- *Process, control, and application*: It is done by sending command or control signals to the systems based on the result of the data process. Different parameters' set points, reference signals, desired output, etc. are used through application software.

10.2.12 IoT—key features and advantages

IoT offers many benefits to users. Some important key features of IoT are as given below:

- *Smart sensing*—IoT enables all smart features of sensing and metering technology. Both useful electrical and nonelectrical parameters are sensed to gather data and then they are processed, shared, and applied for different purposes. Smart sensing has made IoT suitable for real-time applications.
- *Connectivity*—IoT enables interconnections among different systems located at a far distance. This networking of things provides access, monitoring, control, data sharing from remote locations. By this, the benefit of one system may be transferred to others. Connectivity can be established by a small area or by a large area.
- *AI*—IoT enables application software and clouds to apply the latest analytical tools in an effective way to analyze data collected through different sensing devices. AI has become part of IoT. It makes IoT more user-friendly, reliable, and "smart." Intelligence level is strengthened by the use of modern tools. Machine learning cloud computing, quantum computation, etc. find more and more use in IoT.
- *Active interconnection and engagement*—IoT has enabled the interconnection of various systems. The system of systems is operating with the help of IoT. Both similar type and dissimilar type systems are engaged actively to share information and operate keeping coordination with each other. For example, IoT has made it possible and easy to build a system of systems like vehicle to vehicle, vehicle to grid, etc.
- *Smaller devices*—IoT has reduced the size and cost of individual utility devices, controllers, etc., and made individual items more compact and distributed in a wide area.

10.2.13 IoT—advantages

Offering various attractive features of IoT gives many advantages. Some major advantages are as follows:

- Remote access.
- Improved customer engagement.

- Intelligent solution.
- Technology optimization.
- Reduced waste of time.
- Reduced waste of space, property, and infrastructure.
- Reduced energy.
- Wide data coverage.
- Advanced data analytics.

10.2.14 Major challenges with IoT

IoT has many challenges and limitations to overcome. Some are given below:

- *Security*—Device access, data handling, and processing are to be made more secure.
- *Privacy*—There is a chance to lose the privacy of an individual system or device or operator. Maintaining privacy is a challenging task in IoT applications.
- *Lack of privacy may raise the following issues*:
  - Users' or owners' consent.
  - "To be left alone" right.
  - Indistinguishable data.
  - User's comfort.
- *Complexity*—To give better facilities, IoT applications sometimes make the simple system more complex. Complex applications may reduce consumer satisfaction.
- *Flexibility*—IoT applications are very much dependent on various types of protocol. Too much dependency on particular protocols may reduce flexibility.
- *Compliance*—IoT should comply with systems of various nature to provide a sustainable solution.
- *Data misuse*—IoT devices collect different data that include system data, environmental data, and human data. There is the probability of misuse of these data and bringing them in danger.
- *Device malfunction*—For reliable operation of IoT applications, secure and stable operation of each layer is important. Malfunction of any IoT layer or device may lead to unreliable operation and may result in total failure.

10.2.15 Cyber-attacks

As physical and human beings (users and owners) are involved through communication networking in IoT, there is always the risk of cyberattack. Attention should be paid to the following:

- in-built-security,
- encryption,
- regular risk assessment, and
- authentication and authorization for access of devices and data.

10.2.16 Data theft

As IoT is involved with lots of data, there is a risk of data theft. Again, attention should be taken to maintain security at collection, storing, and accessing data.

10.2.17 Other challenges

Other challenges are the following:

- no update for long time,
- minimum testing,
- complexity of software,
- data size,
- limitations of AI,
- continuity of energy supply, and
- communication failure.

10.2.18 Industrial IoT (IIoT)

IIoT is a dedicated framework of the IoT for industry solutions. Major differences between IoT and IIoT are presented in Table 10.3.

10.2.19 Cellular topology for IoT

Cellular topology is used in many places. This topology uses a cellular network for Internet connection. The topology diagram is shown in Figure 10.4.

Like other IoT frameworks, the base of cellular topology is a sensor that collects data. Then through different IoT devices data flow to the device gateway by PAN/LAN, WiFi cellular, etc. Then through router data get connected with a wide area of network-based Internet. Then through the Internet, different IoT devices and interfacing devices get access based on the type of application.

10.2.20 Wearable electronics

Sometimes wearable electronic devices become useful for operators or users in IoT applications. Electronics devices that are small in size and worn on their clothes, caps, head (helmets, glass), neck (collars, jewelry), torso (clothing), arms (rings, wristband, watches), and feet (shoes, shocks) are commonly known as wearable electronics device.

Table 10.3 Major differences between IoT and IIoT

Issues	**IoT**	**IIoT**
Significance	The framework of the general IoT	The framework of IIoT for dedicated industry solution
Benefit object	Human-centric	Machine-centric
Customer	Individuals	Industry applications
Network communication	Wireless devices	Either wire or wireless or both wired and wireless
Data quality	Medium to high	High to very high

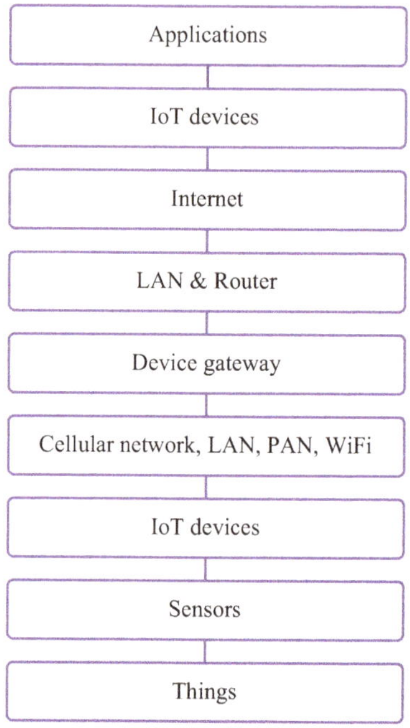

Figure 10.4 Cellular model for IoT

10.2.21 Difference between machine to machine (M2M) network and IoT

M2M network establishes interconnection between machines locally through the embedded system. The major difference between the M2M network and the IoT is presented in Table 10.4.

10.2.22 Factors to be considered in applying IoT-based solution

Before applying an IoTs based solution following points should be considered:

- dashboard features,
- security features,
- compatibility of IoT protocols,
- data management features, and
- energy efficiency.

10.2.23 IoT in energy management

IoT is becoming popular in the energy management system. Through load data supervision and analytics, need-based energy usage is implemented by IoT. This reduces energy waste and saves energy. Smart metering is one important

Table 10.4 Major difference between M2M and IoT

Issues	M2M	IoT
Relevance	M2M networking	Internetworking between things (maybe machines and other devices)
Communication	Localized at client site by embedded system	Small area or wide area networking and not localized
Device type	Interconnected devices are mainly of the same type and standard	Interconnected devices are mainly of multiple types and standards
Upgradation	Upgradation is difficult	Upgradation is easy
Commonly preferred connectivity	Cellular or wired network	Internet connection
Flexibility	Less (multiple bidirectional) communication is not possible at a time	High (multiple bidirectional) communication is possible at a time

component of IoT-based load management system. Through resource data collection and critical demand analysis, IoT can result in better and efficient energy usage. Besides, residential and commercial energy management, IoT can provide a better solution for grid management. IoT-based energy management is broadly divided into three categories:

- IoT-based residential load management.
- IoT-based load energy management.
- IoT-based grid and energy distribution management.

IoT helps energy monitoring much easier than conventional approaches. Additionally, it provides user-friendly operation and makes energy consumption more economic.

10.2.24 IoT in mobility

IoT finds wide application in mobility. Some application areas of IoT in mobility are mentioned below:

- *IoT in rails*: IoT can be used for better customer facility, passengers' comfort, supervisory monitoring, maintenance, etc.
- *IoT in mass transit*: IoT can be applied in tracking mass transit, traffic congestion management, load & delivery schedule, etc.
- *IoT in the road*: IoT can be used for road condition monitoring, traffic control, parking, etc.
- *IoT in an automobile application*: IoT can be utilized in various ways in the automobile industry. Vehicle condition monitoring, tracking, fuels monitoring, battery monitoring, etc. are now being incorporated with IoT applications. Driver and owner access and control can be incorporated with IoT-based

solutions. Complete battery management for an EV can be done through IoT. Identification of nearby charging stations, charging monitoring, payment, etc. can be incorporated with IoT.

- *IoT for connecting vehicles*: IoT can be utilized to handle billing, fuelling, charging, parking, insurance, and other needs.
- *IoT for vehicle booking*: IoT can be utilized for booking cars and other vehicles.
- *IoT for booking charging points* (CPs): IoT can be utilized for booking CPs and monitoring them.
- *Activity trackers*: IoT can be utilized to monitor the activity of vehicle drivers.
- *IoT in smart outlets of charging stations*: Effective use of IoT in charging stations can make them smarter and customer-friendly. Booking, charging, monitoring, etc. can be done through IoT.
- *IoT in parking*: Complete vehicle parking management can be done through IoT. Searching of the parking area, parking spot, parking time, payment, etc. can be done.
- *IoT in EV and charging stations health monitoring*: Through IoT, different health conditions involving electrical and nonelectrical parameters can be monitored.
- *IoT in water transport*: IoT can be used for water transport for better monitoring, management, and customer service.
- *IoT in unmanned aerial vehicles (UAVs)*: IoT can be used for water transport for better monitoring and control of UAVs.

10.2.25 Best IoT practices

Some best practices that may be adopted in IoT applications are as follows:

- consider reliability and security issues during the design of products;
- use strong authentication and security protocols;
- disable non-essential services;
- ensure secured Internet, IoT management hubs, and services;
- use energy-efficient algorithms for the system to be active longer.

10.2.26 Testing of IoT devices

Before application, IoT devices should be undergone different tests. Some commonly used tests for IoT devices are mentioned below:

- *Usability testing*: For a utility point of view, a usability test is carried out on the same setup.
- *Compatibility testing*: This test is done to see the compatibility of communication and networking protocols and configurations of software and hardware.
- *Reliability and scalability testing*: This is done for performance assessment mainly concerning sensors.
- *Data integrity testing*: Data accuracy and performance of data analytics are tested.

- *Security testing*: Data security and privacy in IoT applications are tested.
- *Performance testing*: Overall performance is tested with data collected from some random objects.
- *Other tests*: Pilot testing, regulatory testing, and upgrade testing.

Different tools are used for IoT tests and they are broadly classified into two categories:

- software tools and
- hardware tools.

Some commonly used tools are the following:

- *Software tools for testing*: Wireshark, Tcpdump, etc.
- *Hardware tools for IoT testing*: Digital Storage Oscilloscope, JTAG Dongle, etc.

10.3 Artificial Intelligence (AI)

10.3.1 Definition

AI refers to adding intelligence to machines to improve performance. In recent years, AI has undergone tremendous advancement. Considering human intelligence as the highest level of intelligence, different new mathematical tools are being introduced to improve machine intelligence that a machine can present to result in improved performance.

10.3.2 Limitation

AI can fail. Besides many constraints, it should be remembered that AI does not have different virtues of the human mind like, feeling, emotion, social value, etc.

10.3.3 Classification

Based on the functionality, AI is classified into four major categories as follows:

(a) Reactive machines (RMs).
(b) Limited memory (LM).
(c) Theory of mind (TM).
(d) Self-aware (SA).

Based on the technology, AI is classified into three major categories as follows:

(a) Artificial narrow intelligence (ANI).
(b) Artificial general intelligence (AGI).
(c) Artificial super intelligence (ASI).

The main features of the above seven types of AI are presented in Table 10.5.

Table 10.5 Main features of seven types of AI

AI type	Features
RMs	• The oldest type of AI • Limited technical capabilities • Memory base is not used with function • Cannot use past experiences for the present task • No or very little power to 'learn' things for future reaction
LM	• LM access for implementing AI in machines • Can react as well as "learn" through past experiences • Can use Deep Learning, for training to react and learn • Need a large amount of data set for training • Can adopt new learning • Example, fingerprint scanner
TM	• Advanced AI • Try to understand different entities • Can show some emotions, beliefs, thinking, etc.
SA	• Try to follow human brain functioning • Try to create self-awareness • Try to develop self-consciousness • Still at the developing stage
ANI	• A most common type of AI • Can serve one or two tasks • Can be trained and can learn from past experiences can decide on a predefined set under a limited set of constraints (also called the "Weak AI")
AGI	• Use functions with "Theory of Mind" • Still at the developing stage • The aim is to make the ability to establish relationships with various domains independently like a human brain
ASI	• Still at the developing stage • Can access and process large memory data set rapidly • Can make decision • Execution time will be very less • The concept is much theoretical at the present stage

10.4 Artificial neural network (ANN)

10.4.1 Definition

ANN has become a powerful method for information processing to achieve and implement AI. Attempts are taken to follow the neuron structure of the human body and their way of reactions. The human brain consists of around 86 billion nerve cells, neurons, thousands of connections, axons, sensory organs dendrite to take stimuli input from the external environment. The first neurocomputer was invented by Dr Robert Hecht-Nielsen.

ANN refers to an advanced computation network designed by several interconnected elements, which takes inputs from external information sources, processes and computes them to provide outputs by dynamic state response of network elements.

10.4.2 Classification of ANN topologies

Topologies used for ANNs are broadly classified into two groups as follows:

(a) Feed-Forward ANN.
(b) Feed-Back ANN.

10.4.2.1 Feed-Forward ANN

Feed-Forward ANN maintains the unidirectional flow of information through its network (shown in Figure 10.5). One unit sends information to another from which it does not receive information. Thus, this type of ANN operates without any feedback path and normally, has a fixed set of input and output. Feed-Forward ANN is commonly used for recognition, classification, pattern generation, etc.

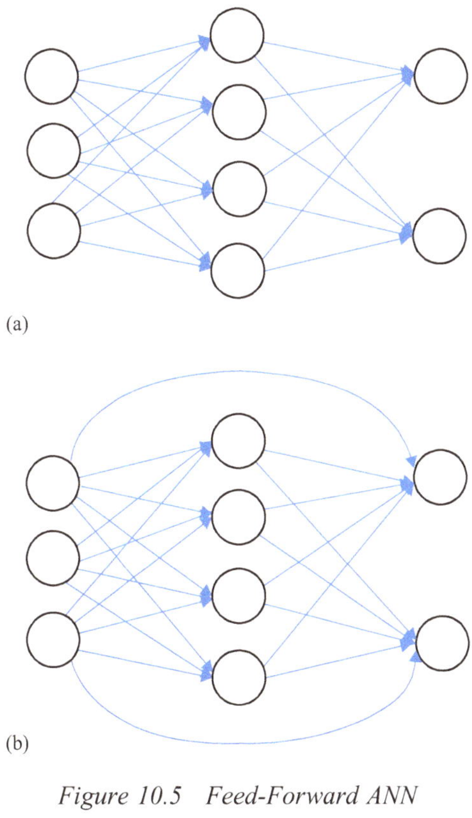

Figure 10.5 Feed-Forward ANN

10.4.2.2 Feed-Back ANN

Feed-Back ANN allows the bidirectional flow of information through its network (shown in Figure 10.6). One unit sends information to another from which it can receive information. Thus, this type of ANN operates with a feedback path and has memories.

10.4.3 Working topology of ANNs

ANN has input, output, and intermediate layers called hidden layers as shown in Figure 10.7. The network is made of neurons and connections. Neurons are called

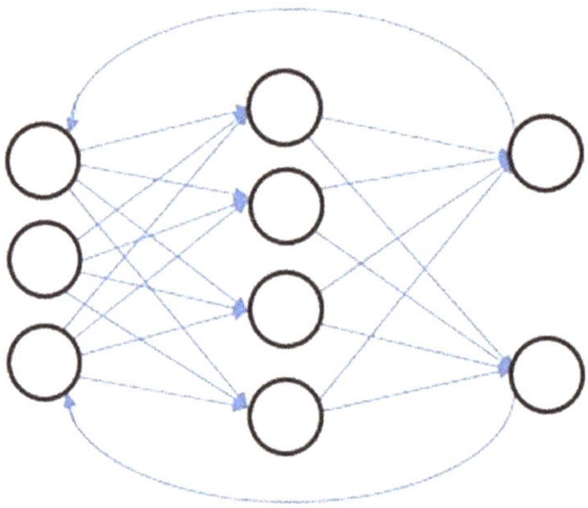

Figure 10.6 Feed-Back ANN

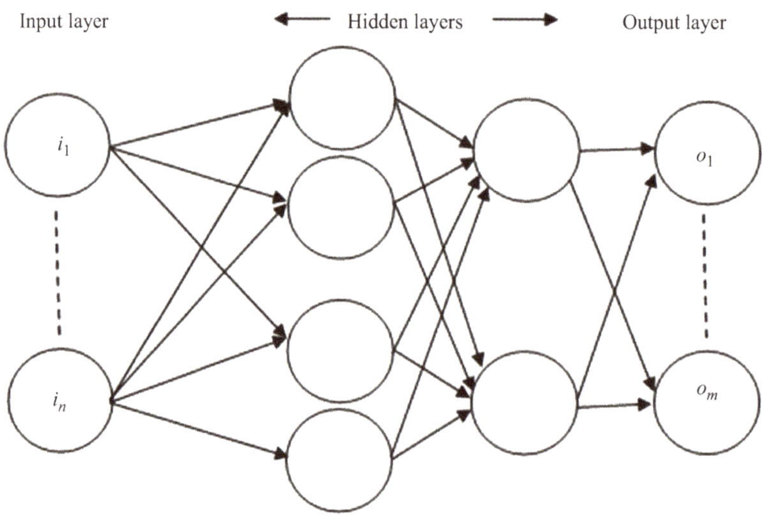

Figure 10.7 Basis structure of ANN

nodes. A neuron may have multiple inputs but produces a single output. Each arrow serves as a connection between one neuron and another neuron. The direction of each arrow indicates the direction of the flow of information through that connection. Thus, each connection transfers the output of one neuron to another. One neuron may be connected with multiple connectors or arrows. Each arrow of connection is specified by weight. The weight of the ANN connector refers to an integer number used for that controlling flow signal between two neurons. Each network link is involved with weight. Each node output is called the node value or activation. The operation of ANN needs learning from a known set of input and output data. The weight of each connection indicates the influence of one node or neuron on another.

10.4.4 Propagation function

Computation of input to a neuron is done by propagation function using outputs of its predecessor neurons with the help of weight sum of connections.

10.4.5 Machine-learning ANNs

The performance of ANN depends on network structure and the learning process. Different types of ANN and machine-learning methods have been introduced. Machine-learning methods may be broadly classified into three groups as follows:

(a) Supervised learning.
(b) Unsupervised learning.
(c) Reinforcement learning.

- *Supervised learning*—It needs an external source other than ANN itself to feed data from examples with known output/answers to supervise the learning process.
- *Unsupervised learning*—It does not need any example of input data with known output.
- *Reinforcement learning*—Based on observation, strategy is set and connections' weights are adjusted to achieve better output next time.

10.5 Cloud computing

10.5.1 Definition

Cloud computing is all means of providing hosting services to clients through the Internet, involves configuration, manipulation, and access of online applications with an online data store and infrastructure. It can offer service at a large scale and does not involve up-from investment. Cloud computing is defined by NIST as follows:

"cloud computing is a model for enabling convenient, on-demand network access to a shared pool of configurable computing resources (e.g., networks, servers, storage, applications, and service) that can be rapidly provisioned and released with minimal management effort or service provider interaction"

Computing resources mainly include remote data storage, applications, networks, and service. The term cloud represents the internet providing remote data storage, client services.

10.5.2 Main advantages and limitations of cloud computing

The main advantages of cloud computing are as follows:

- access to utility applications over the Internet;
- access of infrastructure, platforms, software over the Internet;
- access of remote storage;
- access to remote resources;
- access on demand;
- cost-effective and energy-efficient computing service.

10.5.3 Major limitations of cloud computing are as follows:

The major limitations of cloud computing are as follows:

- security challenges;
- possibility of violating privacy;
- data or resource misuse;
- dependency on the internet and non-availability of the Internet in many areas.

10.5.4 Classification of cloud computing

Different models have been introduced for cloud computing. All models can be broadly classified into two groups as follows:

(a) Deployment model.
(b) Service model.

The deployment model indicated the type and access facility of the cloud. Deployment models are again divided into four categories as follows:

(i) public cloud,
(ii) private cloud,
(iii) hybrid cloud, and
(iv) community cloud.

Based on different types of services, cloud computing models are classified into three main categories as follows:

(i) Infrastructure as a Service (IaaS).
(ii) Platform as a Service (PaaS).
(iii) Software as a Service (SaaS).

10.5.5 Public, private, hybrid, and community clouds

The basic feature of public, private, hybrid, and community clouds is presented in Table 10.6.

Table 10.6 Basic feature of public, private, hybrid, and community clouds

Cloud	Features
Public cloud	Access to the public in general Less secure
Private cloud	Access to private in general More secure
Hybrid cloud	Allow access to both public and private based on need There is the threat of security
Community cloud	Access to a particular community More secure

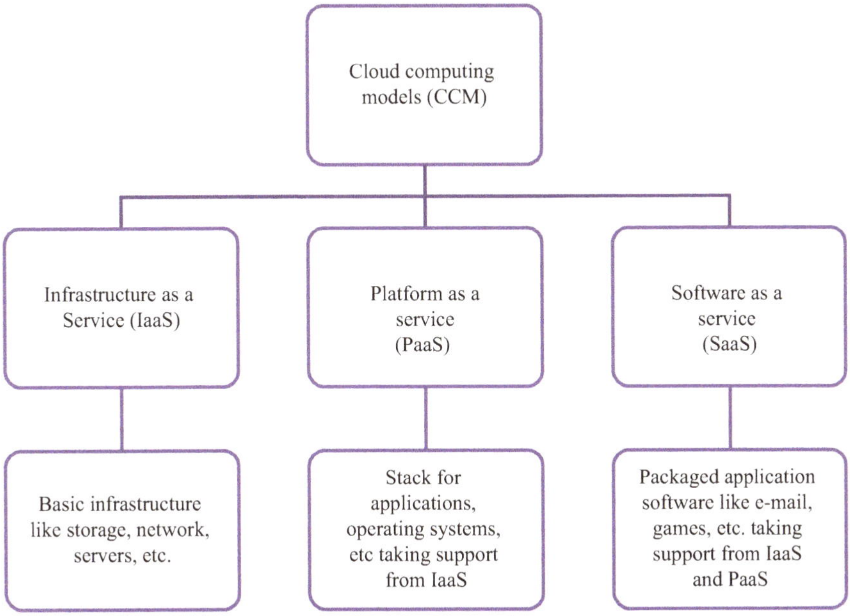

Figure 10.8 Basic features of IaaS, PaaS, and SaaS

10.5.6 IaaS, PaaS, and SaaS

Basic features of IaaS, PaaS, and SaaS are shown in Figure. 10.8. IaaS includes all features of basic infrastructure like storage, network, servers, etc. PaaS includes a stack for applications, operating systems, etc. It needs the support of IaaS. SaaS represents packaged application software like email, games, etc. It needs support from IaaS and PaaS.

Table 10.7 Basic feature of different types of technologies for cloud computing

Technologies	Features
SOA	It offers services to other technology, product, or client by providing applications to be used by them. They can use the program and get benefit from it without adding any program
GC	It offers distributed computing. The complex task is divided among multiple computers that are placed at multiple geographical locations, the main object remaining the same. It speeds up and makes the execution process of multiple tasks more simple
UC	It offers computing services on a utility basis and charges accordingly. "Pay per utility" basis service is provided to clients or users
Virtualization	It manages and acts as a pointer for presenting physics resources, applications, etc. to multiple clients or organizations virtually. It has three main parts: (i) storage and networks (ii) virtual infrastructure (iii) a virtual machine consisting of applications and operating system

10.5.7 Technologies for cloud computing

Different types of technologies for cloud computing have been introduced. Broadly, cloud computing technologies can be divided into four groups as follows:

(a) Service-oriented architecture (SOA).
(b) Grid computing (GC).
(c) Utility computing (UC).
(d) Virtualization.

Basic features of different types of technologies for cloud computing are presented in Table 10.7.

10.6 Fog computing

Fog computing refers to edge computing performed in the distributed network at the periphery of the cloud. It reduces the burden of cloud computing and makes overall computing faster. Fog computing devices are distributed at the layer of internet and IoT devices. Sensors generate lots of voluminous data which are normally sent to the cloud in raw form. In fog computing, raw data coming from sensors are processed and useful data are sent to the cloud. The data processed through fog computing can also be utilized locally to give an additional benefit. Fog computing reduces bandwidth requirements for data transfer to the cloud. The fog layer is also known as fog networking. It consists of two main parts: a control plane and a data plane. The control part handles the selection of sensor interfacing and data transfer whereas the data plane handles computation. Fog computing thus provides faster computation, reduced bandwidth, and improved quality of service.

10.7 Model for NG and internet of EV-things integration

10.7.1 Cluster of NGs

NG has been modeled where solar PV arrays feed power to three units as shown in Figure 10.9; one connected with the battery unit for charging, second is connected with the DC utility system, and third is connected with the inverter system. The inverter converts it into AC. The output of the inverter is either after stepping up or directly goes to the local grid system and feeds power to the AC utility system. Battery charging unit can take power from either solar PV or conventional grid after rectification.

NGs may be located in various locations and can feed to a common conventional grid system. A small contribution from each NG may be summed up to contribute a significant amount of power meeting the power demand of a large area as shown in Figure 10.10.

10.7.2 NG-integrated EV charging

With the increase of EVs, there has been an increasing demand for stations for charging an EV. At present, most of the charging stations are incorporated with conventional power supplies. Here EV charging station (EVCS) has been connected with NG as shown in Figure 10.11.

The EV charging unit (EV-CU) has the following options:

1. it can take power directly from solar PV units,
2. it can take power from the inverter unit after rectification at desired (controlled) level,
3. it can take power from a conventional power grid after rectification at the desired level, and
4. it can take power from the storage battery system.

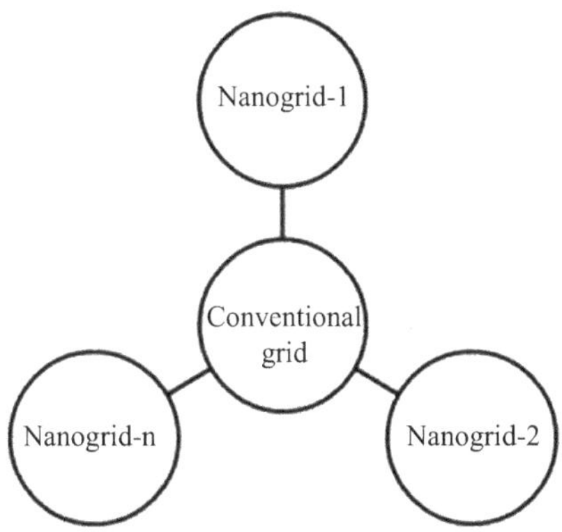

Figure 10.9. NG system connected with conventional grid system

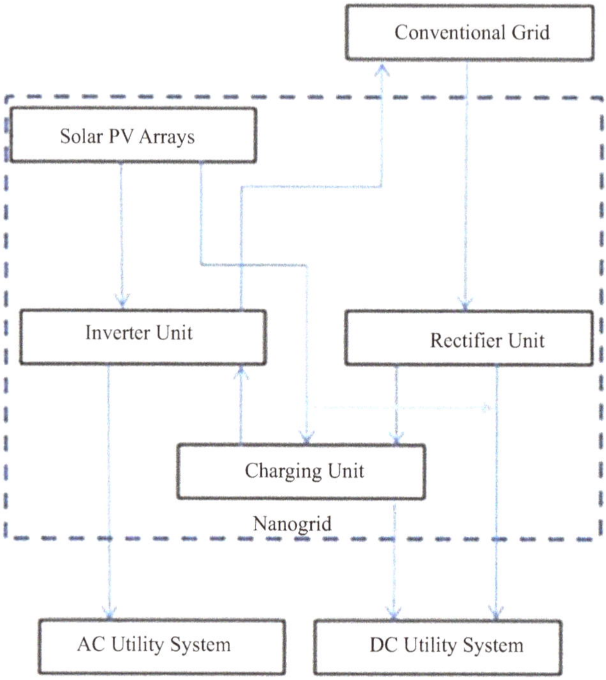

Figure 10.10 NG system without storage unit

Among four options, option 1 is relatively slow and depends on the availability of solar power. Options 2 and 3 are being commonly used. During an emergency or power cut-off, option 4 may be chosen. For all these four options, the time of charging depends on the charging characteristics of the vehicle battery.

Now when a vehicle needs charging, it will have to perform the following:

(i) Search and choose a charging station.
(ii) Reach nearby charging station.
(iii) Wait in a queue.
(iv) Check availability of CP.
(v) If a CP is available choose one (out of four options, if all available) option of charging, else go to 1.

The whole process is a lengthy one and there may be an unpredictable delay at any stage of execution.

10.7.3 IoT-enabled information sharing and selection of charging station

Here an attempt has been taken to reduce the overall execution time by incorporating the IEISS as shown in Figure 10.12. IEISS is virtually connected between the

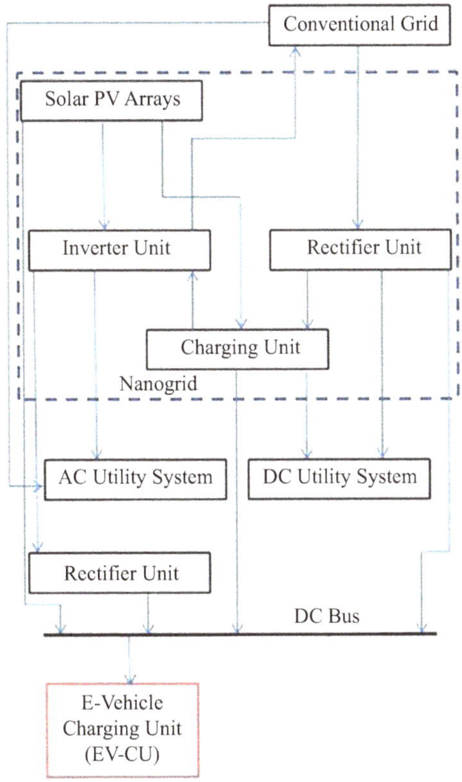

Figure 10.11 NG system integrated with EV charging system

multiple numbers of charging stations and multiple numbers of vehicles. All charging stations are sharing their information including location, charging capacity, number of available CPs. If there is no CP available, then the estimated waiting time is required to get a CP available.

Whenever charging is required, a vehicle can search charging stations located nearby and check the availability of CPs. In addition, IEISS can share traffic information. Accordingly, e-vehicle can choose the best solution. This scheme is a good one; however, one will have to spend minimum searching time, time to reach available CPs, and time of charging.

10.7.4 NG-fed EVCS and SBCS

Here the scheme described in the previous section has been modified by connecting SBCS with NG as shown in Figure 10.13. It consists of a standby EV battery charging and storing unit (SEVB-CSU). The NG may or may not be connected with EVCS. In SBCS, different types of vehicle batteries will be charged and stored with ampere–hour specifications and dimensions.

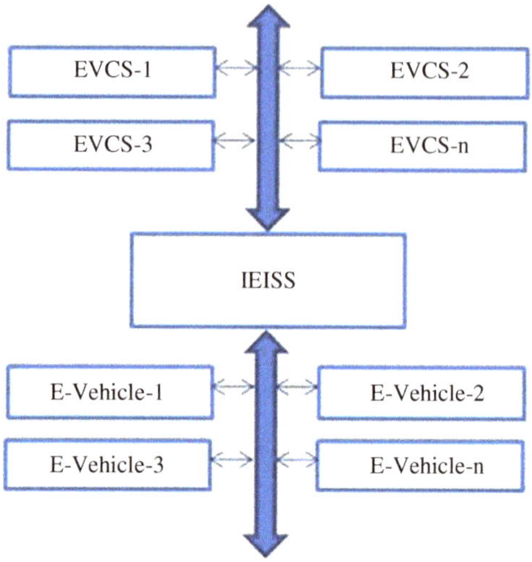

Figure 10.12 Cloud based information sharing system among E-vehicles and EVCS

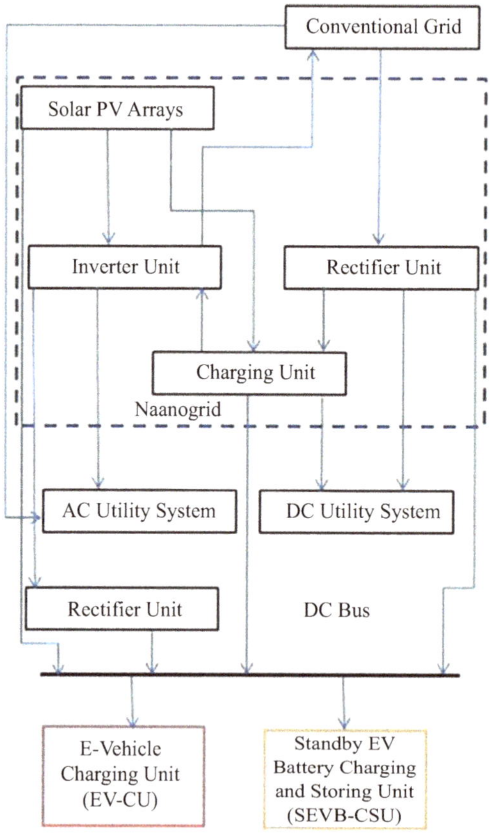

Figure 10.13 NG system with EV charging unit and standby battery charging and storing unit

10.7.5 Modified IoT-enabled information sharing and selection of EVCS and SBCS

IEISS has been virtually connected (as shown in Figure 10.14) with between multiple numbers of charging stations with and without SBCS facility and multiple numbers of vehicles. All charging stations will share data of location, charging capacity, the number of available CPs, if there is no CP available, then, the estimated waiting time is required to get a CP available. In addition, the NG-fed charging station will send the status data of SBCS like the number of available charged batteries, their ampere–hour ratings, and other specifications.

Whenever charging is required, a vehicle can search charging stations located nearby and check the availability of CPs. In addition, IEISS can share traffic information. Accordingly, e-vehicle can choose the best solution from EVCU and SEVB-CSU. This scheme is a good one. If one opts for the second one, only, he will spend time searching through IEISS, time to reach available CP, and time of replacement of the battery.

Let,

A = status of EVCS.
B = status SBCS.

Two binary status logics have been defined as follows:

$1 \cong$ available .
$0 \cong$ not available.

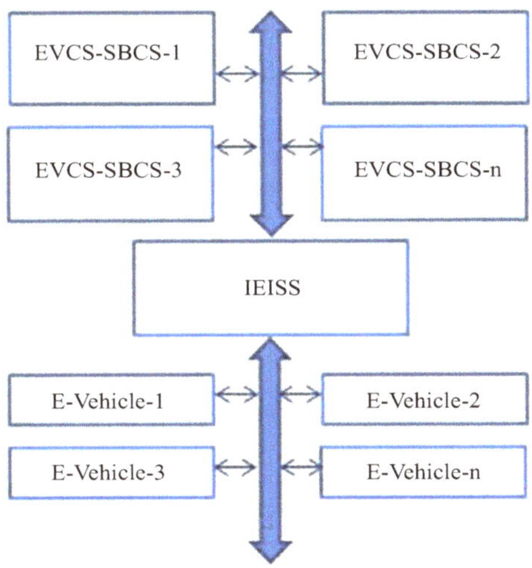

Figure 10.14 Cloud based information sharing system among E-vehicles and EVCS-SBCS

Based on the above, four categories of property values of NG-integrated charging station have been defined as follows:

Property value $= p = f(A, B)$

$$p = \begin{bmatrix} p_0 \\ p_1 \\ p_2 \\ p_3 \end{bmatrix} = \begin{bmatrix} \overline{A}\,\overline{B} \\ \overline{A}B \\ A\overline{B} \\ AB \end{bmatrix} = \begin{bmatrix} 00 \\ 01 \\ 10 \\ 11 \end{bmatrix}$$

$$p = \begin{bmatrix} \text{NG with not available EVCS and not available SBCS} \\ \text{NG with not available EVCS and available SBCS} \\ \text{NG with available EVCS and not available SBCS} \\ \text{NG with available EVCS and available SBCS} \end{bmatrix}$$

Now, NG concerning EV charging may be specified by two parameters: EVCS capacity and SBCS capacity where EVCS capacity refers to the maximum number of EV CP and SBCS capacity refers to the maximum number of fully charged standby batteries (SB) are available for replacement. Mathematically let,

$$\begin{aligned} i_{\max} &= \text{EVCS capacity} \\ &= \text{the maximum numbers of EV charging point} \\ j_{\max} &= \text{SBCS capacity} \\ &= \text{the maximum numbers of standby battery} \end{aligned}$$

Thus, NG-integrated charging station has two units: EVCS and SBCS; EVCS is specified by $i_{\max}$ i.e. EVCS capacity and SBCS is specified by $j_{\max}$ i.e., SBCS capacity.

If one or a few CPs are used, the number of available EV CPs (CP) will decrease and becomes less than EVCS capacity.

If one or a few SBs are used, the number of available fully charged SBs available for replacement will decrease and becomes less than SBCS capacity.

Let,

$i =$ numbers of EV charging point (CP) available.
$j =$ numbers of fully charged SB available.

NG utility for charging and use of SB has been shown by flow chart in Figure 10.15. A and B will remain at logic "1" if i and j remain nonzero. When i becomes zero, A will be converted to "0," and when i becomes zero B will be converted to "0." Initially,

$i = i_{\max}$
$j = j_{\max}$
A = 1
B = 1

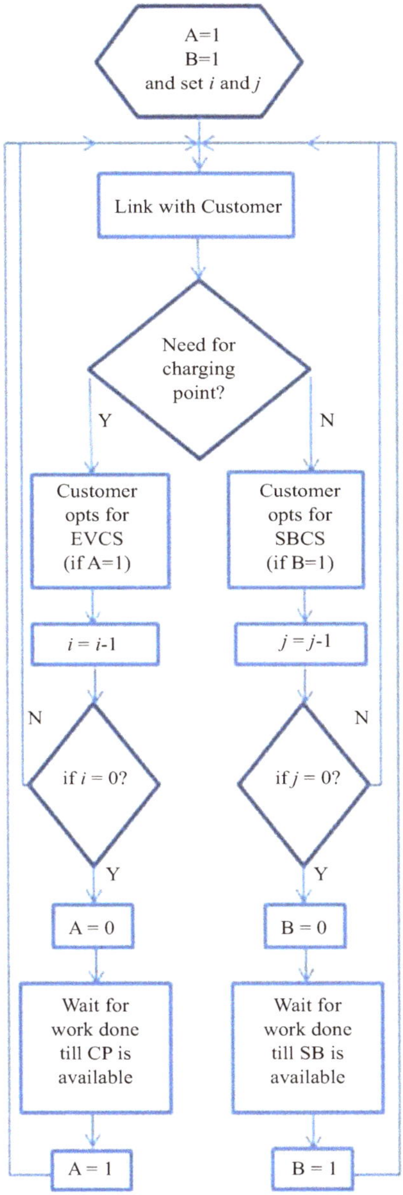

Figure 10.15 Flow chart for EVCS and SBCS utilization

Through IoT-based information sharing system, customers will be linked to one NG where one can opt either for EV CP or replacement of battery by SB.

Thus, depending on the increase of use of CPs and SB, AB value may be converted from "00" to "01" or "10" or "00." Thus, the property value of a NG for EV charging may not be constant and may vary among p_3, p_2, p_1, and p_0.

The probability of NG having property values p_3, p_2, and p_1 will increase with the increase of EVCS capacity and SBCS capacity. More space and time are involved to increase EVCS capacity, whereas less space and time are required for the improvement of SBCS capacity. Thus, instead of increasing SBCS capacity, more effort and attention may be given to improve SBCS capacity. The working flow is shown in Figure 10.15.

10.7.6 Case study

Space–time requirements for different property values are presented in Table 10.8. For "00," property value, time requirement is infinite which indicates it is not suitable at all for EV and will not be chosen. For "01" property value, both time and space requirements are small as it involves with replacement of EV battery only. For "10," both space–time requirements are maximum as it involves direct charging which requires the presence of EV for as long as charging is being done. For the "11" property value, the space–time requirement may vary from minimum to maximum as it may involve either battery replacement or direct charging.

Different possible properties of NGs and selection probability are presented in Table 10.9. The property of NG may be variable. NG with property "00" is not suitable for EV and will not be chosen.

Table 10.8 Space–time requirements for different property values

	$\overline{B}$	B
$\overline{A}$	00 $T=\infty$ $S=0$	01 $T=T_{min}$ $S=S_{min}$
A	10 $T=T_{max}$ $S=S_{max}$	11 $T=(T_{min}-T_{max})$ $S=(S_{min}-S_{max})$

Table 10.9 Different possible properties of NGs and selection probability

NG_{ij}	p	Selection observation
$NG\ i=0$ $j=0$	00	–
$NG\ i=0$ $j\neq 0$	01	This probability increases when j_{max} increases
$NG\ i\neq 0$ $j=0$	10	This probability increases when i_{max} increases
$NG\ i\neq 0$ $j\neq 0$	11	This probability increases when i_{max} and j_{max} increase

Probability of selection of NG with property value "01" increases with the increase of j_{max} (SBCS capacity).

Probability of selection of NG with property value "10" increases with the increase of i_{max}(EVCS capacity).

Probability of selection of NG with property value "11" increases with the increase of i_{max}(EVCS capacity) and j_{max}(SBCS capacity).

Initial property value and next probable property value are presented in Table 10.10. The initial property value $p = p_0$ cannot change.

But, the property value $p = p_1$ can change into either $p = p_0$ or $p = p_1$.

The property value $p = p_2$ can change into either $p = p_0$ or $p = p_2$.

The property value $p = p_3$ can change into either $p = p_0$ or $p = p_1$ or $p = 2$ or $p = 3$.

Let,

Number of NGs with property value $p_0 = N_0$
Number of NGs with property value $p_1 = N_1$
Number of NGs with property value $p_2 = N_2$
Number of NGs with property value $p_3 = N_3$
Total number of NGs $= N$

Based on different numbers of NGs of different initial property values, five different cases have been considered.

Five different cases are as follows:

Case 1: $N = N_0$
Case 2: $N = N_1$
Case 3: $N = N_2$
Case 4: $N = N_3$
Case 5: $N_0 = N_1 = N_2 = N_3 = \frac{N}{4}$

The probabilistic study has been made among five cases as presented in Table 10.11. The selection probability of EVCS (x) and selection probability SBCS (y) are shown in Figure 10.16.

Table 10.10 Initial property value and next probable property value

Initial property value	Next probable property value
$p = p_0$	$p = p_0$
$p = p_1$	$p = p_0$
	$p = p_1$
$p = p_2$	$p = p_0$
	$p = p_2$
$p = p_3$	$p = p_0$
	$p = p_1$
	$p = p_2$
	$p = p_3$

Table 10.11 Selection probability for EVCS and SBCS

Condition	User choice	Probability	Remarks
Case 1	User having no option	100	$x = 0\%$
$N = N_0$	User having option EVCS	0	$y = 0\%$
	User having option SBCS	0	
	User having option EVCS and SBCS	0	
Case 2	User having no option	0	$x = 0\%$
$N = N_1$	User having option EVCS	0	$y = 100\%$
	User having option SBCS	100	
	User having option EVCS and SBCS	0	
Case 3	User having no option	0	$x = 100\%$
$N = N_2$	User having option EVCS	100	$y = 0\%$
	User having option SBCS	0	
	User having option EVCS and SBCS	0	
Case 4	User having no option	0	$x = 50\%$
$N = N_3$	User having option EVCS	0	$y = 50\%$
	User having option SBCS	0	
	User having option EVCS and SBCS	100	
Case 5	User having no option	25	$x = 50\%$
$N_0 = N_1 = N_2 = N_3 = \frac{N}{4}$	User having option EVCS	25	$y = 50\%$
	User having option SBCS	25	
	User having option EVCS and SBCS	25	

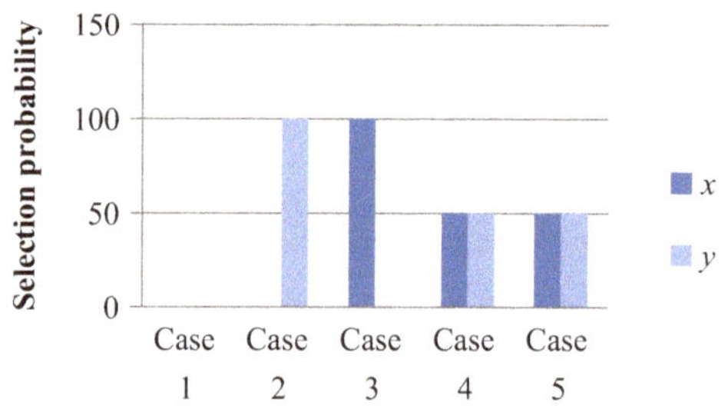

Figure 10.16 Selection probabilities of "x" and "y" EVCS and SBCS

- Case 1 results in zero probability for both EVCS and SBCS. Therefore, it becomes a very undesirable situation both from the service provider and consumer point of view.
- Case 2 results in 100% for SBCS. It indicates by replacing the old battery with a fully charged SB meets the demand with the lowest time and at the smallest area requirement for parking, charging, etc.
- Case 3 results in 100% for EVCS. It meets consumer demand by 100%. But it requires lots of space for parking, charging, etc. and requires vehicles to be at the station for a long time to get charged.

Table 10.12 Comparison concerning facility, space–time requirement, and option-based user satisfaction

If	Facility	Space–time requirement	Option-based user satisfaction
$N = N_0$	Poor	Not applicable as in this category no one is fit for EV	Not suitable
$N = N_1$	Good	Minimum time and space	Best
$N = N_2$	Moderate	Maximum time and space	Poor
$N = N_3$	Best	Moderate time and Space	Medium

- Case 4 results in a 50% chance for both EBCS and SBCS. Therefore, based on the consumer's choice and availability, the charging demand of multiple vehicles can be met.
- In case 5, 50% of consumers can be met by either EVCS or SBCS.

A comparison between cases 1 and 4 is presented in Table 10.12 for the facility, space–time requirement, and option-based user satisfaction.

- Case 1, no station is fit for EV resulting in "not suitable" concerning user satisfaction.
- Case 2 becomes best offering minimum space and time requirement for battery replacement.
- Case 3 gives poor user satisfaction as they can offer only EVCS which requires large space and time of charging.
- Case 4 becomes "medium" as here one can opt for either SBCS or EVCS.

In many cities, population density is the highest at the central position, and then it gradually decreases along the radius and, on the other hand, space availability increases at the outer side of the city. Along radial direction, city may be divided into three zones: 1 (central position), zone 2 (mid-position), and zone 3 (periphery). NG of a different category may be placed as shown in Figure 10.17:

- Zone 1: NG with SBCS facility (category 1) may be installed at the central position i.e., zone 1.
- Zone 2: NG with EVCS & SBCS facility (category 3) may be installed at the central position i.e., at zone 2.
- Zone 3: NG with EVCS facility (category 2) may be installed at the city peripheral position i.e., zone 3.

10.8 Research trend

With the shift of vehicle category, technology has observed its progress in the field of the grid system, starting from conventional to micrograd and then in recent years

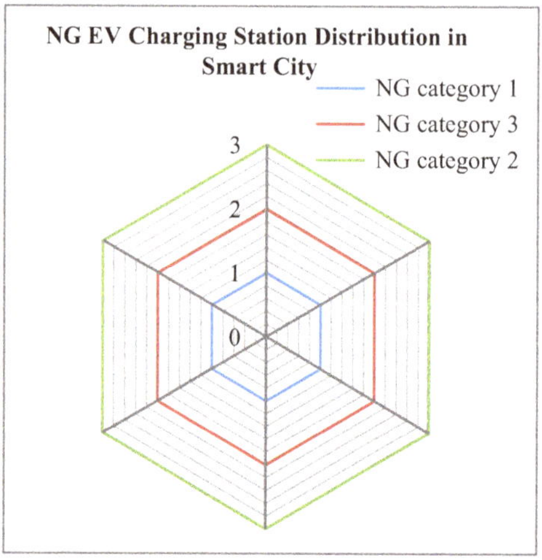

Figure 10.17 Distribution of NG-fed EVCS in a smart city

NG which effectively utilizes carbon-free solar energy. Lots of research chapters have been observed in the field of NGs and their effective use in EV charging [1–4]. The concept of NGs was described by Nordman (2013) et al. that suggests at least one load unit and one gateway to the outside [1]. Sometimes the term is also used for localized generation and distribution of loads less than 20 kW; however, no such rigid demarcation has been found in this regard. Haidar Samet et al. obtained full-selective contribution of sources in DC *NGs* [2] and AC *NG* [3]. Mehran Heidari et al. introduced an integrated cost-effective battery model in grid-connected smart *NG* [4].

Price-based power management has been done on an off-grid *photovoltaic* (PV) unit with a centralized DC bus [5]. Roman Uhlig et al. have presented a profitability analysis of *EV charging* supported by grid connection [6]. Manuela Sechilariu et al. presented an electromobility framework for urban planning for *EV charging stations* empowered by PV-based microgrids [7]. Kazem Khalkhali et al. utilized data envelopment analysis in the design of a plug-in hybrid EV *charging station* [8]. Marjan Gjelaj et al. introduced an optimal infrastructure planning for *EV* fast-*charging stations* using the prediction of user behavior [9].

Marjan Gjelaj et al. described some multifunctional applications of batteries with fast-*charging technologies* [10].

Maria Fazio et al. applied IoT, Edge, and Cloud Computing in Smart Cities [11]. Binod Vaidya et al. presented a vehicle charging system for smart cities [12]. Rahmani-Andebili et al. described planning and operation aspects of parking considering various traffic conditions [13]. Rigas et al. optimized congestion for urban EV charging systems [14]

Wang et al. introduced a novel charging method [15], and Moghaddass et al. introduced smart control of fleets of EVs [16] for smart communities.

Timpner et al. presented charging station scheduling strategies for EVs [17]. Multi-stage optimization has been carried out for coordinated electrical vehicles [18]. Cost efficiency maximization [1] and minimal waiting [20] have been achieved based on user convenience.

Cao [21] proposed a reservation system for EV battery switch service [21]. Online charging scheduling has been done with on-arrival commitment [22] and EV public parking infrastructures [23]. Cao et al. proposed for EV charging reservation under pre-emptive service [24].

Different smart charging strategies [25] and parking systems [26] have been introduced for EVCSs. IoT-enabled dependable control has been introduced for solar energy harvesting in smart buildings [27]. Agent-based control of EV charging [28] and off-board EV home charging [29] have been tested for residents in smart cities [28]. Hybrid optimization has been done for economic deployment of ESS in charging stations [30]. Smart load management has been done with plug-in EVs in distribution for loss minimization [31].

Wang et al proposed a two-stage mechanism for massive EV charging using renewable energy [32]. Renewable energy sources have been utilized for cost and emission reductions [33]. IoTs-based management of solar energy for smart homes has been done in Refs. [34,35]. Thus, the research trends are focusing on a more integrating and inclusive approach of new fields like IoT, NPG, AI, Fog, Cloud Computing, etc.

10.9 Useful information

Different IoT-related information is provided in Table 10.13. Different types of ANNs, learning algorithms, nomenclature, and useful information are provided in Table 10.14.

10.10 Summary

In this chapter, a NG-integrated EV (EV) charging system has been proposed which is enabled with the IoT-based information sharing system and meets consumer demands with time and space minimization. First integrated NG integration has been added with the EV charging system. IoT-enabled information sharing system has been utilized to make the bridge between consumer and NG-based charging stations. The selection of charging stations has been done through an information sharing system. A SBCS has then been added which utilizes the very small time required for replacement and the small area that is required during the replacement. The selection process has been modified accordingly for space and time optimization. Comparative case studies show that an increase of SBCS and its proper utilization with EVCS can improve consumers' satisfaction with the additional advantage of lowering time and area requirement which may be appropriate to be built up in city areas.

Table 10.13 Different IoT-related information

Item	Description
Sensor businesses and IoT	The operation of the sensor business does not require an Internet connection. IoT needs an Internet connection
Bluegiga APX4 protocol	It refers to support protocol for both WiFI and BLE platforms using normally 450 MHz ARM9 processors
Bluetooth low energy	It refers to technology for wireless Personal Area Network (PAN) using low power for data transmission over short and long-distance
MicroPython	It refers to a Python version having a library smaller than the standard library and compatible with MCU Mode using a microcontroller
Some commonly used IoT protocols	WiFi Zigbee Message Queue Telemetry Transport Protocol (MQTT) AMQP XMPP Data distribution service (DDS) Very simple control protocol (VSCP) Simple text-oriented messaging protocol (STOMP)
IoT publishers	It refers to sensors used for sending real-time data to intermediate IoT layers and devices
Salesforce IoT Cloud	It refers to an online platform capable of storing and processing IoT data. It can store various application software (known as lightning) and can collect useful information from various sources like devices, websites, customers, parallel company data centers, etc. partners
Replication	It refers to syncing data among multiple servers (2 or more)
IoT Thingworx	It refers to a platform useful for faster networking and application development for IoT-connected devices integrated with IoT tools for analysis and providing other features of IoT
IoT general electric (GE)-PREDIX	It refers to software for the collection of information collected from industrial measurements helpful for operation management and optimization
Types of CAN frame	Frame 1: data, Frame 2: request, Frame 3: error, Frame 4: overload
Floating CPU and fixed-point CPU	A floating CPU can directly handle floating values. Fixed CPU is useful for integer format; dealing with floating values by fixed point CPU may lead to loss of data/information and resolution
GPIO	It refers to a control pin to control input and output and the status of the pin is programmable
Some useful databases for IoT	Apache Cassandra, influx DB, RethinkDB, Sqlite, Mon-goDB, etc.
Android things	It refers to Android-based OS for embedded devices
Airflow sensors	It refers to sensors capable to measure the flow level of air in a different location, normally, in the soil
Scheduler	It refers to switch one task from another
IoT Contiki	It refers to software for program management and data processing useful for small Internet-connected devices

(Continues)

Table 10.13 (*Continued*)

Item	Description
Data in IoT	It refers to collect information from sensors and other virtual and physical sources
Some IoT antennas	PCB antenna Chip antenna Wire antenna Whip antenna Proprietary antenna
InfluxDB	It refers to the distributed database in time-series without any external dependencies
WSN	It refers to wireless sensor network (WSN) to deal with physical parameters in an application
Zigbee	It refers to a complex (like Bluetooth), low power, robust system with high security
Z-Wave	It refers to an IoT protocol with low power, RF communication useful for home automation
MQTT	It refers to Message Queue Telemetry Transport Protocol. This is a protocol for messaging useful for tracking IoT devices
Some IoT software	Blockchain, windows IoT, Predix, Microsoft Azure, Bluemix, Node-RED, etc.
Shodan	It refers to a tool for IoT testing useful for device connection checking with the Internet
Thing in IoT	It refers to any device-embedded and/or connected in IoT framework
Some MEMS sensor	MPU6050-Gyroscope, piezoelectric sensor, ADXL345, accelerometer, etc.
Sharding	It refers to a method for data splitting
Some IoT hardware prototypes	Raspberry Pi, Arduino, ARM Cortex Family, etc.
Raspberry Pi	It refers to a mini form of a computer with onboard WiFi, Bluetooth, GPIO pins, etc. capable of communicating with external things
IoT testing	IoT testing is a type of testing to check IoT devices. Today there is an increasing need to deliver better and faster services. There is a huge demand to access, create, use, and share data from any device. The thrust is to provide greater insight and control over various interconnected IoT devices. Hence, the IoT testing framework is important
Types of IoT	IoT and IIoT
Thingful	It refers to a search engine useful for IoT objects and data
WSN and IoT	WSN connects things, gather and monitoring some data IoT can connect things, collect and monitor data, share data, control application and interfacing, take a decision, etc. integrated with WSN, Internet, Cloud Storage, mobile, or web-based application

Table 10.14 Different types of ANNs, learning algorithms, nomenclature, and related information

Item	Remarks
Back-propagation algorithm (BPA)	• One type of training algorithm • It learns from the example and updates weight values useful for recognition of simple type patterns and mapping
Bayesian networks (BNs)	• Graphical structures • Indicates the probabilistic relationship, if any • Known as Bayes Nets or Belief Networks • Known as directed acyclic graphs (DAGs) as it cannot return through the path • Can deal with variables of multivalued nature simultaneously • Useful for prediction from knowledge gathered from prior observation of the relationship, if any
Training and learning	• Adaptive process of neural networks to learn • Data from examples are used • The difference (known as error) of target and processed outputs are observed and connecting weights are updated/adjusted accordingly
Components of ANNs	• Artificial neurons conceptually look like biological neurons • Connections and weights • Activation function: processing (normally non-linear) of the weighted sum to produce the output
Deep learning	• Neurons organized in multiple layers • Input layer: layer that receives external data • Output layer: layer that gives final output or result • Hidden layer: layers that exist between input and output layers • There are zero or more hidden layers.
DAG	• Connections of neurons in feedforward networks
Recurrent networks	• Networks having connections with the same or previous layers • Normally seen in Feedback Network
Learning rate	• Number of corrective per observation steps required to adjust errors • A high learning rate needs less training time, • A high learning rate may result in lower accuracy
Backpropagation	• One method of adjusting connection weights effectively by dividing using error obtained from learning among network connections
Learning paradigms	• Three main paradigms of learning in ANN are

(Continues)

Table 10.14 (Continued)

Item	Remarks
	(i) Supervised learning (ii) Unsupervised learning (iii) Reinforcement learning
Supervised learning	• Uses a set of paired inputs and outputs • Try to get desired output different cost functions like mean-squared error between the desired output and present output is minimized • Suitable for pattern recognition and classification, regression or approximation or function, recognition of gesture, analysis of sequential data like handwriting, speech, etc.
Unsupervised learning	• Input data, with a cost function, data function, and output data are used in learning. • Prior assumptions are made in the model • Function minimization or maximization is done • Useful in estimation compression, filtering, etc.
Reinforcement learning	• In learning, weight is adjusted in the network to minimize long-term cost • External reinforcement occurs in learning • Useful useful in games, control problems, sequential decisions, etc.
Self-learning	• No input of external advice • No input of external reinforcement from • E.g.: Crossbar Adaptive Array (CAA) that uses crossbar memory processing • Involves with behavior, emotional feelings, etc.
Convolutional neural networks (CNNs)	• Used for ○ visual data processing ○ two-dimensional data processing ○ long short-term memory (LSTM) ○ mixed low and high-frequency signals ○ speech recognition, ○ text-to-speech conversion ○ gaming: competitive networks like generative adversarial networks where multiple networks compete
Theoretical properties	• Computational power • Capacity: the ability to model a function • Convergence: reaching desired output by the error • Generalization and statistics • Training method • Hardware features

(Continues)

Table 10.14 (Continued)

Item	Remarks
	• Practical examples for training • Hybrid approach: neural network-integrated with another model
Adaptive Linear Neuron or Adaptive Linear Element (ADALINE)	• Single-layer ANN • Multiple nodes—each node has multiple input • Uses memistors having weight, bias, and summation functions • Difference between Adaline and standard perceptron (McCulloch–Pitts): • for Adaline, weights are adjusted by a weighted sum at the learning process • for standard perceptron, the sum is passed through the activation function and the output of the activation function is used to adjust weights
MADALINE (many ADALINE)	• Three-layer, fully connected, feed-forward ANN • Layers: input, output, and hidden) • Activation function is sign function. • Uses memistors • Three types of training are used • (MADALINE rules I, II, and III)
Some Feedforward Network	• Group method of data handling (GMDH) • Autoencoder • Probabilistic • Time delay • Convolutional • Deep stacking network (DSN)
Recurrent neural network (RNN)	• Fully recurrent • Simple recurrent • Reservoir computing • LSTM • Bi-directional • Hierarchical • Stochastic • Genetic scale
Other common types of ANN	• Instantaneously trained • Spiking • Regulatory feedback • Compound hierarchical-deep models • Deep predictive coding networks (DPCNs) • Multilayer kernel machine (MKM)
Dynamic network type	• Cascading

(Continues)

Table 10.14 (*Continued*)

Item	Remarks
	• Neuro-fuzzy • Compositional pattern-producing
Memory network type	• Memory networks • One-shot associative memory • Hierarchical temporal memory • Holographic associative memory • LSTM-related differentiable memory structures • Neural Turing machines • Semantic hashing • Pointer networks
GMDH	• Fully automatic structure of parameter optimization. • Activation functions (Kolmogorov–Gabor polynomials) allow multiplication and addition • Uses deep multilayer perceptron (MLP) • One type of supervised learning network • Training is done by regression analysis
Autoencoder, autoassociator or diabolo network	• Similar to an MLP • A single input layer, single-output layer • One or more hidden layers • Unsupervised learning to reconstruct its inputs • Useful for efficient coding
Probabilistic neural network (PNN)	• Four-layer feedforward neural network • Layers are input, output, hidden, and pattern/summation layer • Uses probability distribution function (PDF) useful for recognition and classification of patterns
Time-delay neural network (TDNN)	• Feedforward network • Handle sequential data • Can use perceptron • Weights may be trained by supervised learning • Used in the pattern recognition system
CNN, or ConvNet	• consists of one or more connected convolutional layers • Use MLPs • Allows input data having a 2D structure • Capsule Neural Networks (CapsNet) add additional networks (known as capsules) to CNN. The output of capsules are reused to provide a more stable final output

(*Continues*)

Table 10.14 (Continued)

Item	Remarks
DSN	• Useful in robot navigation, image recognition, video recognition, recommender, natural language processing, etc. • Block hierarchy consisting of modules of a simplified neural network • Each block is a simplified MLP having a single hidden layer
Tensor DSNs	• Extended DSN architecture • Uses covariance statistics in a bilinear mapping • Useful for regression, classification, etc.
Regulatory feedback networks	• Used to perform optimization in recognition process using feedback
Radial basis function (RBF) network	• RBFs are used in hidden function mapped with the input layer
General regression neural network (GRNN)	• Use memory neural network like PNN • Useful regression, approximation, etc.
Deep belief network (DBN)	• A probabilistic model with multiple hidden layers • Useful to train a deep neural network (DNN) • Offers faster ultimate convergence
RNN	• Data flow both in forward and backward • Useful in sequence processors
Reservoir computing	• Extended neural networks mapped with the input of higher dimension
The echo state network (ESN)	• Consists of sparse connection in the hidden layer • Useful at reproducing time series
LSTM	• Has vanishing problem • Can work well with long delays
Instantaneously trained neural networks (ITNN)	• Training takes a short duration
Spiking neural networks (SNN)	• Consider input values as well as input timing
DPCN	• A predictive model for coding • Extracts sparse features observations
MKMs	• Learning approach with the function of high nonlinearity by iterative use nonlinear kernels
Neuro-fuzzy network	• Fuzzy inference embedded with ANN
One-shot associative memory	• Can store new patterns without following re-training • Useful for a network of wireless sensors, GC, etc.
Pointer networks	• One type of DNN with reduced parameter
Encoder–decoder networks	• Deal with mapping of highly structured input to and output

References

[1] B. Nordman and K. Christensen, "Local power distribution with nanogrids", in *2013 International Green Computing Conference Proceedings*, Arlington, VA, 2013, pp. 1–8, DOI: 10.1109/IGCC.2013.6604464

[2] H. Samet, T. Ghanbari, E. Kazemzadeh, and E. Moshksar, "Full-selective contribution of sources in DC nanogrids using a technique based on buck/boost converter", *IET Energy Systems Integration*, 1, no. 1, 2019, pp. 14–22.

[3] H. Samet, E. Kazemzadeh, and T. Ghanbari, "Realising full-selective sources contribution in AC nanogrids", *IET Energy Systems Integration*, 2, no. 1, 2020, pp. 40–49.

[4] M. Heidari, T. Niknam, M. Zare, and S. Niknam, "Integrated battery model in cost-effective operation and load management of grid-connected smart nanogrid", *IET Renewable Power Generation*, 13, no. 7, May 2019, pp. 1123–1131.

[5] L. V. Bellinaso, C. D. Schwertner, and L. Michels, "Price-based power management of off-grid photovoltaic systems with centralised dc bus", *IET Renewable Power Generation*, 10, no. 8, 2016, pp. 1132–1139.

[6] R. Uhlig, S. Harnisch, M. Stötzel, M. Zdrallek, and T. Armoneit, "Profitability analysis of grid supporting EV charging management", *CIRED – Open Access Proceedings Journal*, 2017, no. 1, 2017, pp. 1945–1948.

[7] M. Sechilariu, N. Molines, G. Richard, H. Martell-Flores, F. Locment, and J. Baert, "Electromobility framework study: infrastructure and urban planning for EV charging station empowered by PV-based microgrid", *IET Electrical Systems in Transportation*, 9, no. 4, 2019, pp. 176–185.

[8] K. Khalkhali, S. Abapour, S. M. Moghaddas-Tafreshi, and M. Abapour, "Application of data envelopment analysis theorem in plug-in hybrid electric vehicle charging station planning", *IET Generation, Transmission & Distribution*, 9, no. 7, 2015, pp. 666–676.

[9] M. Gjelaj, S. Hashemi, P. B. Andersen, and C. Traeholt, "Optimal infrastructure planning for EV fast-charging stations based on prediction of user behaviour", *IET Electrical Systems in Transportation*, 10, no. 1, 2020, pp. 1–12.

[10] M. Gjelaj, N. B. Arias, C. Traeholt, and S. Hashemi, "Multifunctional applications of batteries within fast-charging stations based on EV demand-prediction of the users' behaviour", *The Journal of Engineering*, 2019, no. 18, 2019, pp. 4869–4873.

[11] M. Fazio, R. Ranjan, M. Girolami, J. Taheri, S. Dustdar, and M. Villari, "A note on the convergence of IoT, Edge, and Cloud Computing in smart cities", *IEEE Cloud Computing*, 5, no. 5, 2018, pp. 22–24, DOI: 10.1109/MCC.2018.053711663.

[12] B. Vaidya and H. T. Mouftah, "Smart electric vehicle charging management for smart cities", *IET Smart Cities*, 2, no. 1, 2020, pp. 4–13, DOI: 10.1049/iet-smc.2019.0076, Online ISSN 2631-7680.

[13] M. Rahmani-Andebili, H. Shen, and M. Fotuhi-Firuzabad, “Planning and operation of parking lots considering system, traffic, and drivers behavioural model”, *IEEE Transactions on Systems, Man, and Cybernetics: Systems*, 49, no. 9, 2019, pp. 1879–1892.

[14] E. S. Rigas, S. D. Ramchurn, N. Bassiliades, *et al.*, “Congestion management for urban EV charging systems”, in Proceedings of the SmartGridComm, Vancouver, BC, Canada, 2013, pp. 121–126.

[15] Y. Wang, Z. Su, Q. Xu, *et al.*, “A novel charging scheme for electric vehicles with smart communities in vehicular networks”, *IEEE Transactions on Vehicular Technology*, 68, no. 9, 2019, pp. 8487–8501

[16] R. Moghaddass, O. A. Mohammed, E. Skordilis, *et al.*, ‘Smart control of fleets of electric vehicles in smart and connected communities’, *IEEE Transactions on Smart Grid*, 10, no. 6, 2019, pp. 6883–6897.

[17] J. Timpner and L. Wolf, “Design and evaluation of charging station scheduling strategies for electric vehicles”, *IEEE Transactions on Intelligent Transportation Systems*, 15, no. 2, 2014, pp. 3122–3127

[18] X. Zhu, H. Han, S. Gao, *et al.*, “A multi-stage optimization approach for active distribution network scheduling considering coordinated electrical vehicle charging strategy”, *IEEE Access*, 6, 2018, pp. 50117–50130.

[19] H. M. Chung, W. T. Li, C. Yuen, *et al.*, “Electric vehicle charge scheduling mechanism to maximize cost efficiency and user convenience”, *IEEE Transactions on Smart Grid*, 10, no. 3, 2019, pp. 3020–3030.

[20] H. Qin and W. Zhang, “Charging scheduling with minimal waiting in a network of electric vehicles and charging stations”, in Proceedings of the VANET, Las Vegas, NV, USA, 2011, pp. 51–60.

[21] Y. Cao, T. Wang, X. Zhang, *et al.*, “Toward any casting-driven reservation system for electric vehicle battery switch service”, *IEEE Systems Journal*, 13, no. 1, 2019, pp. 2806–2816.

[22] B. Alinia, M. H. Hajiesmaili, and N. Crespi, “Online EV charging scheduling with on-arrival commitment”, *IEEE Transactions on Intelligent Transportation Systems*, 20, no. 12, 2019, pp. 4524–4537.

[23] N. Tucker and M. Alizadeh, “An online admission control mechanism for electric vehicles at public parking infrastructures”, *IEEE Transactions of Smart Grid*, 11, no. 1, 2020, pp. 161–170.

[24] Y. Cao, S. Liu, Z. He, *et al.*, “Electric vehicle charging reservation under pre-emptive service”, in Proceedings of the 2019 First International Conference on Industrial Artificial Intelligence (IAI), Shenyang, China, 2019, pp. 1–6.

[25] Z. Moghaddam, I. Ahmad, D. Habibi, *et al.*, “Smart charging strategy for electric vehicle charging stations”, *IEEE Transactions on Transportation Electrification*, 4, no. 1, 2018, pp. 76–88.

[26] M. S. Kuran, A. C. Viana, L. Iannone, *et al.*, “A smart parking lot management system for scheduling the recharging of electric vehicles”, *IEEE Transactions on Smart Grid*, 6, no. 6, 2015, pp. 2942–2953.

[27] Q. Ha and M. D. Phung, "IoT-enabled dependable control for solar energy harvesting in smart buildings", *IET Smart Cities*, 1, no. 2, pp. 61–70, DOI: 10.1049/iet-smc.2019.0052.
[28] S. Singh, A. Namboodiri, and M. P. Selvan, "Agent-based system to control the air-conditioner and EV charging for residents in smart cities", *IET Smart Cities*, 1, no. 2, pp. 71–80, DOI: 10.1049/iet-smc.2019.0060.
[29] V. Monteiro, J. C. Sousa, J. A. Afonso, *et al.*, "Innovative off-board EV home charging station as a smart home enabler: present and proposed perspectives", in Proceedings of the 2018 IEEE 16th International Conference on Industrial Informatics (INDIN), Porto, Portugal, 2018, pp. 966–971.
[30] K. Chaudhari, A. Ukil, K. N. Kumar, *et al.*, "Hybrid optimization for economic deployment of ESS in PV-integrated EV charging stations", *IEEE Transactions on Industrial Informatics*, 14, no. 1, 2018, pp. 106–116.
[31] A. S. Masoum, S. Deilami, P. S. Moses, *et al.*, "Smart load management of plug-in electric vehicles in distribution and residential networks with charging stations for peak shaving and loss minimisation considering voltage regulation", *IET Generation, Transmission & Distribution*, 5, no. 8, 2011, pp. 877–888.
[32] R. Wang, P. Wang, and G. Xiao, "Two-stage mechanism for massive electric vehicle charging involving renewable energy", *IEEE Transactions on Vehicular Technology*, 65, no. 6, 2016, pp. 4159–4171.
[33] A. Y. Saber and G. K. Venayagamoorthy, "Plug-in vehicles and renewable energy sources for cost and emission reductions", *IEEE Transactions on Industrial Electronics*, 58, no. 4, 2011, pp. 1229–1238.
[34] B. L. R. Stojkoska and K. V. Trivodaliev, "A review of internet of things for smart home: challenges and solutions", *Journal of Cleaner Production*, 140, 2017, pp. 1454–1464.
[35] M. D. Phung, M. De La Villefromoy, and Q. Ha, "Management of solar energy in microgrids using IoT-based dependable control", in 2017 20th International Conference on Electrical Machines and Systems (ICEMS), 2017, pp. 1–6.

Further reading

[1] "IEEE Smart Grid Vision for Vehicular Technology: 2030 and Beyond Roadmap," in *IEEE Smart Grid Vision for Vehicular Technology: 2030 and Beyond Roadmap*, pp. 1–17, 30 June 2015.
[2] "IEEE Vision for Smart Grid Controls: 2030 and Beyond Reference Model," in *IEEE Vision for Smart Grid Control: 2030 and Beyond Reference Model*, pp. 1–10, 12 September 2013, DOI: 10.1109/IEEESTD.2013.6598993.
[3] "IEEE Smart Grid Vision for Computing: 2030 and Beyond Roadmap," in *IEEE Smart Grid Vision for Computing: 2030 and Beyond Roadmap*, pp. 1–14, 11 January 2016.

[4] S. F. Bush; S. Goel, and G. Simard, "IEEE Vision for Smart Grid Communications: 2030 and Beyond Roadmap," in *IEEE Vision for Smart Grid Communications: 2030 and Beyond Roadmap*, pp. 1–19, 3 September 2013, DOI: 10.1109/IEEESTD.2013.6690098.

[5] "IEEE Grid Vision 2050 Reference Model," in *IEEE Grid Vision 2050 Reference Model*, pp. 1–15, 30 April 2013, DOI: 10.1109/IEEESTD.2013.6577591.

[6] G. Simard, "IEEE Grid Vision 2050," in *IEEE Grid Vision 2050*, pp. 1–93, 15 April 2013, DOI: 10.1109/IEEESTD.2013.6577603.

[7] K. Bhat; V. Sundarraj; S. Sinha, and A. Kaul, "IEEE Cyber Security for the Smart Grid," in *IEEE Cyber Security for the Smart Grid*, pp. 1–122, 16 September 2013, DOI: 10.1109/IEEESTD.2013.6613505.

[8] "IEEE Standard for Utility Industry Metering Communication Protocol Application Layer (End Device Data Tables)," in *IEEE Std 1377-2012 (Revision of IEEE Std 1377-1997)*, pp. 1–576, 10 August 2012, DOI: 10.1109/IEEESTD.2012.6264065.

[9] "IEEE Standard for an Architectural Framework for the Internet of Things (IoT)," in *IEEE Std 2413-2019*, pp. 1–269, 10 March 2020, DOI: 10.1109/IEEESTD.2020.9032420.

[10] "IEEE Approved Draft Standard for Test Methods for Surge Protectors and Protective Circuits Used in Information and Communications Technology (ICT) circuits, including Smart Grid Data Circuits," in *IEEE PC62.36/D38*, February 2016, pp. 1–146, 1 January 2016.

[11] "IEEE Standard Glossary of Stationary Battery Terminology," in *IEEE Std 1881-2016*, pp. 1–42, 30 August 2016, DOI: 10.1109/IEEESTD.2016.7552407.

[12] "IEEE Guide for Cloud Portability and Interoperability Profiles (CPIP)," in *IEEE Std 2301-2020*, pp. 1–62, 13 August 2020, DOI: 10.1109/IEEESTD.2020.9169938.

[13] "IEEE 802 Nendica Report: The Lossless Network for Data Centers," in *IEEE 802 Nendica Report: The Lossless Network for Data Centers*, pp. 1–29, 17 August 2018.

[14] "IEEE Standard for Adoption of Open Fog Reference Architecture for Fog Computing," in *IEEE Std 1934-2018*, pp. 1–176, 2 August 2018, DOI: 10.1109/IEEESTD.2018.8423800.

[15] "Cloud Computing Innovation in India: A Framework and Roadmap – 2. 0," in *Cloud Computing Innovation in India: A Framework and Roadmap – White Paper 2.0*, pp. 1–206, 31 December 2014, DOI: 10.1109/IEEESTD.2014.7039186.

[16] "IEEE Approved Draft Guide for the Use of Artificial Intelligence Exchange and Service Tie to All Test Environments (AI-ESTATE)," in *IEEE P1232.3/D3.2*, June 2014, pp. 1–96, 10 October 2014.

[17] "IEC 62243 Ed. 1 (IEEE Std 1232(TM)-2002): Standard for Artificial Intelligence Exchange and Service Tie to All Test Environments

(AI- ESTATE)," in *IEC 62243 First edition 2005-07 IEEE 1232*, pp. 1–128, 20 November 2002, DOI: 10.1109/IEEESTD.2002.8684709.

[18] "IEEE Standard for Artificial Intelligence and Expert System Tie to Automatic Test Equipment (AI-ESTATE): Overview and Architecture," in *IEEE Std 1232-1995*, pp. 1–32, 15 March 1996, DOI: 10.1109/IEEESTD.1996.80816.

[19] "IEEE Standard for Software Interface for Maintenance Information Collection and Analysis (SIMICA)," in *IEEE STD 1636-2009 (New Trial-Use Standard)*, pp. 1–38, 14 August 2009, DOI: 10.1109/IEEESTD.2009.5210055.

[20] "IEEE Standard for Software Interface for Maintenance Information Collection and Analysis (SIMICA)," in *IEEE Std 1636-2009 (Upgraded to Full Use)*, pp. 1–37, 14 August 2009, DOI: 10.1109/IEEESTD.2009.8654231.

[21] "IEEE Draft Standard for the Classification of Adaptive Instructional Systems," in *IEEE P2247.1/D20*, December 2020, pp. 1–28, 25 January 2021.

[22] Q. Yang, L. Fan, R. Tong, and A. Lv, "IEEE Federated Machine Learning," in *IEEE Federated Machine Learning – White Paper*, pp. 1–18, 15 June 2021.

Index

www.ingramcontent.com/pod-product-compliance
Lightning Source LLC
LaVergne TN
LVHW051049210926
848486LV00001B/52
* 9 7 8 1 8 3 9 5 3 4 8 2 9 *